T0140627

Lecture Notes in Geoinformation and Cartography

The Lecture Notes in Geoinformation and Cartography series provides a contemporary view of current research and development in Geoinformation and Cartography, including GIS and Geographic Information Science. Publications with associated electronic media examine areas of development and current technology. Editors from multiple continents, in association with national and international organizations and societies bring together the most comprehensive forum for Geoinformation and Cartography.

The scope of Lecture Notes in Geoinformation and Cartography spans the range of interdisciplinary topics in a variety of research and application fields. The type of material published traditionally includes:

- proceedings that are peer-reviewed and published in association with a conference;
- post-proceedings consisting of thoroughly revised final papers; and
- research monographs that may be based on individual research projects.

The Lecture Notes in Geoinformation and Cartography series also includes various other publications, including:

- tutorials or collections of lectures for advanced courses;
- contemporary surveys that offer an objective summary of a current topic of interest; and
- emerging areas of research directed at a broad community of practitioners.

More information about this series at http://www.springer.com/series/7418

Igor Ivan · Alex Singleton · Jiří Horák
Tomáš Inspektor
Editors

The Rise of Big Spatial Data

Springer

Editors
Igor Ivan
Institute of Geoinformatics
VŠB-Technical University of Ostrava
Ostrava, Moravskoslezsky
Czech Republic

Jiří Horák
Institute of Geoinformatics
VŠB-Technical University of Ostrava
Ostrava, Moravskoslezsky
Czech Republic

Alex Singleton
Department of Geography and Planning
University of Liverpool
Liverpool, Merseyside
UK

Tomáš Inspektor
Institute of Geoinformatics
VŠB-Technical University of Ostrava
Ostrava, Moravskoslezsky
Czech Republic

ISSN 1863-2246 ISSN 1863-2351 (electronic)
Lecture Notes in Geoinformation and Cartography
ISBN 978-3-319-83216-6 ISBN 978-3-319-45123-7 (eBook)
DOI 10.1007/978-3-319-45123-7

Softcover reprint of the hardcover 1st edition 2016

Technical editing, typesetting: Tomáš Inspektor

Printed on acid-free paper

This Springer imprint is published by Springer Nature
The registered company is Springer International Publishing AG
The registered company address is: Gewerbestrasse 11, 6330 Cham, Switzerland

GIS Ostrava 2016—The Rise of Big Spatial Data
March 16th–18th 2016, Ostrava, Czech Republic

Preface

Welcome to the dawn of the Era of Big Data. Big spatial data are characterised by three main features: volume beyond the limit of usual geo-processing, velocity higher than available by usual processes, and variety, combining more diverse geodata sources than usual. The popular term denotes a situation when one or more of these key properties reach a state when traditional methods of geodata collection, storing, processing, controlling, analysing, modelling, validating, and visualising fail to provide effective solutions. Entering the Era of Big Spatial Data requires finding solutions addressing all "small data" issues that will soon induce "big data" troubles. Resilience for Big Spatial Data means solving heterogeneity of spatial data sources (in topics, purpose, completeness, guarantee, licensing, coverage, etc.), large volumes (from gigabytes to terabytes and more), undue complexity of geo-applications and systems (i.e. combination of standa-lone applications with Web services, mobile platforms, and sensor networks), neglected automation of geodata preparation (i.e. harmonisation, fusion), insufficient control of geodata collection and distribution processes (i.e. scarcity and poor quality of metadata and metadata systems), limited capacity of analytical tools (i.e. domination of traditional causal-driven analysis), low performance of visual systems, inefficient knowledge discovery techniques (for transformation of vast amounts of information into tiny and essential outputs), and many more. These trends will even accelerate as sensors in the world become more ubiquitous. These proceedings should help to regulate the flood of spatial data.

Ostrava, Czech Republic

Igor Ivan

Programme Committee

Igor Ivan (VŠB-Technical University of Ostrava, CZE)—Chair
Hassan Karimi (University of Pittsburgh, USA)
Nick Bearman (Liverpool University, GBR)
Lex Comber (University of Leicester, GBR)
Mei-Po Kwan (University of Illinois, USA)
Atsuyuki Okabe (University of Tokyo, JPN)
Dani Aribas Bel (University of Liverpool, GBR)
Martin Raubal (ETH Zurich, CHE)
Alex Singleton (Liverpool University, GBR)
Jaroslav Hofierka (Pavol Jozef Safarik University in Kosice, SVK)
Bin Jiang (University of Gävle, SWE)
Juan Carlos García Palomares (Universidad Complutense de Madrid, ESP)
Jiří Horák (VŠB-Technical University of Ostrava, CZE)
Stewart Fotheringham (Arizona State University, USA)
Ákos Jakobi (Eötvös Loránd University, Budapest, HUN)
Cidália Costa Fonte (Universidade de Coimbra, PT)
James Haworth (University College London, GBR)
Marcin Stępniak (Polish Academy of Sciences, PL)
Hrvoje Gold (University of Zagreb, HR)
Gennady Andrienko (Fraunhofer Institute for Intelligent Analysis and Information Systems, GE)
Natalia Andrienko (Fraunhofer Institute for Intelligent Analysis and Information Systems, GE)
David W. Wong (George Mason University, USA)
Tao Cheng (University College London, GBR)
Vít Voženílek (Palacký Univerzity, CZE)
Harry J.P. Timmermans (Eindhoven University of Technology, NLD)
Soora Rasouli (Eindhoven University of Technology, NLD)
David O'Sullivan (Berkeley, University of California, USA)

Acknowledgments

The editors of this book would like to thank all the people who have been directly or indirectly involved in the development of this proceedings. First of all, we would like to thank all the authors who submitted papers. Special thanks go to all the authors whose work were finally included into this book. Next, we would like to thank the scientific committee members and all the reviewers who worked hard to read all the submitted papers, provided suggestions, and helped with the selection process. Each chapter was reviewed by at least two scientific committee members or other reviewers. As a final note, we would like to thank European Spatial Data Research Organisation (EuroSDR), International Society for Photogrammetry and Remote Sensing (ISPRS), European Association of Geographers (EUROGEO), Czech Association for Geoinformation (CAGI), Slovak Association for Geoinformatics (SAGI), Czech Geographical Society (ČGS), Miroslav Novák the President of the Moravian-Silesian Region, Tomáš Macura the Mayor of the City of Ostrava and Prof. Ivo Vondrák the Rector of VŠB—Technical University of Ostrava for they support.

Contents

Contributors

Emmanuel Alby Photogrammetry and Geomatics Group, ICube Laboratory UMR 7357, INSA, Strasbourg, France

Lucie Almášiová Department of Military Geography and Meteorology, Faculty of Military Technologies, University of Defence, Brno, Czech Republic

Lucie Augustinková Faculty of Civil Engineering, VSB-Technical University of Ostrava, Ostrava-Poruba, Czech Republic

Dalibor Bartoněk Institute of Geodesy, Faculty of Civil Engineering, Brno University of Technology, Brno, Czech Republic; European Polytechnic Institute, Kunovice, Czech Republic

Daniel Beran Section of Geomatics, Department of Mathematics, Faculty of Applied Sciences, University of West Bohemia, Plzeň, Czech Republic

Miroslav Blaženec Institute of Forest Ecology, Slovak Academy of Science, Zvolen, Slovak Republic

Peter Blišťan Institute of Geodesy, Cartography and Geographical Information Systems, Faculty of Mining, Ecology, Process Control and Geotechnology, Technical University of Košice, Košice, Slovak Republic

Fatma Canaslan Comut Turkey Disaster and Emergency Directorate (AFAD) of Denizli, Denizli, Turkey

Karel Dejmal Department of Military Geography and Meteorology, Faculty of Military Technologies, University of Defence, Brno, Czech Republic

Conrado Domínguez University of Las Palmas de Gran Canaria, Las Palmas de Gran Canaria, Spain

Süleyman Eken Department of Computer Engineering, Engineering Faculty, Kocaeli University, Izmit, Kocaeli, Turkey

Kyle Eyvindson Department of Forest Sciences, University of Helsinki, Helsinki, Finland

Jana Faixová Chalachanová Department of Theoretical Geodesy, Faculty of Civil Engineering, Slovak University of Technology in Bratislava, Bratislava, Slovak Republic

Pablo Fernández Division of Mathematics, Graphics and Computation (MAGiC), IUMA, Information and Communication Systems, University of Las Palmas de Gran Canaria, Las Palmas de Gran Canaria, Spain

Vladimír Fárek Czech Hydrometorological Institute Ústí Nad Labem, Kočkovská 18/2699, Ústí nad Labem, Czech Republic

Michal Gallay Institute of Geography, Faculty of Science, Pavol Jozef Šafárik University in Košice, Košice, Slovak Republic

Sławomir Goliszek Institute of Geography and Spatial Organization, Polish Academy of Sciences, Warsaw, Poland

Pierre Grussenmeyer Photogrammetry and Geomatics Group, ICube Laboratory UMR 7357, INSA, Strasbourg, France

Agung Budi Harto Geodesy and Geomatics, Faculty of Earth Science and Technology, Bandung Institute of Technology, Bandung, West Java, Indonesia

Jaroslav Hofierka Institute of Geography, Faculty of Science, Pavol Jozef Šafárik University in Košice, Košice, Slovak Republic

Markus Holopainen Department of Forest Sciences, University of Helsinki, Helsinki, Finland; Centre of Excellence in Laser Scanning Research, Finnish Geospatial Research Institute, Masala, Finland

Eija Honkavaara National Land Survey, Finnish Geospatial Research Institute, Masala, Finland

Jiří Horák Faculty of Mining and Geology, Institute of Geoinformatics, VŠB-Technical University of Ostrava, Ostrava-Poruba, Czech Republic

Martin Hubáček Department of Military Geography and Meteorology, Faculty of Military Technologies, University of Defence, Brno, Czech Republic

Juha Hyyppä National Land Survey, Finnish Geospatial Research Institute, Masala, Finland; Centre of Excellence in Laser Scanning Research, Finnish Geospatial Research Institute, Masala, Finland

Tomáš Inspektor Faculty of Mining and Geology, Institute of Geoinformatics, VŠB-Technical University of Ostrava, Ostrava-Poruba, Czech Republic

Igor Ivan Faculty of Mining and Geology, Institute of Geoinformatics, VŠB-Technical University of Ostrava, Ostrava-Poruba, Czech Republic

Ákos Jakobi Department of Regional Science, Faculty of Sciences, Eötvös Loránd University, Budapest, Hungary

Rastislav Jakuš Institute of Forest Ecology, Slovak Academy of Science, Zvolen, Slovak Republic

Karel Janečka Department of Mathematics, Faculty of Applied Sciences, University of West Bohemia, Pilsen, Czech Republic

Peter Jankovič Department of Mathematical Methods and Operations Research, Faculty of Management Science and Informatics, University of Žilina, Žilina, Slovakia

Karel Jedlička Section of Geomatics, Department of Mathematics, Faculty of Applied Sciences, University of West Bohemia, Plzeň, Czech Republic

Jan Ježek Section of Geomatics, Department of Mathematics, Faculty of Applied Sciences, University of West Bohemia, Plzeň, Czech Republic

Ľudmila Jánošíková Department of Mathematical Methods and Operations Research, Faculty of Management Science and Informatics, University of Žilina, Žilina, Slovakia

Harri Kaartinen National Land Survey, Finnish Geospatial Research Institute, Masala, Finland; Centre of Excellence in Laser Scanning Research, Finnish Geospatial Research Institute, Masala, Finland

Ville Kankare Department of Forest Sciences, University of Helsinki, Helsinki, Finland; Centre of Excellence in Laser Scanning Research, Finnish Geospatial Research Institute, Masala, Finland

Lukáš Karell Department of Theoretical Geodesy, Faculty of Civil Engineering, Slovak University of Technology, Bratislava, Slovakia

Mika Karjalainen National Land Survey, Finnish Geospatial Research Institute, Masala, Finland; Centre of Excellence in Laser Scanning Research, Finnish Geospatial Research Institute, Masala, Finland

Jan Kazak Department of Spatial Economy, Faculty of Environmental Engineering and Geodesy, Wrocław University of Environmental and Life Sciences, Wrocław, Poland

Ján Kaňuk Institute of Geography, Faculty of Science, Pavol Jozef Šafárik University in Košice, Košice, Slovak Republic

Jáchym Kellar Section of Geomatics, Department of Mathematics, Faculty of Applied Sciences, University of West Bohemia, Plzeň, Czech Republic

Juraj Kisztner Institute of Geological Engineering, VSB-Technical University of Ostrava, Ostrava, Czech Republic

Umut Kizgindere Department of Computer Engineering, Engineering Faculty, Kocaeli University, Izmit, Kocaeli, Turkey

Jiří Klepek Faculty of Mining and Geology, VSB-Technical University of Ostrava, Ostrava-Poruba, Czech Republic

David Kocich Institute of Geoinformatics, Faculty of Mining and Geology, VŠB-Technical University of Ostrava, Ostrava-Poruba, Czech Republic

Ľudovít Kovanič Institute of Geodesy, Cartography and Geographical Information Systems, Faculty of Mining, Ecology, Process Control and Geotechnology, Technical University of Košice, Košice, Slovak Republic

Aneta Krakovská Faculty of Mining and Geology, VSB-Technical University of Ostrava, Ostrava-Poruba, Czech Republic

Antero Kukko National Land Survey, Finnish Geospatial Research Institute, Masala, Finland; Centre of Excellence in Laser Scanning Research, Finnish Geospatial Research Institute, Masala, Finland

Marcin Kulawiak Department of Geoinformatics, Faculty of Electronics, Telecommunications and Informatics, Gdansk University of Technology, Gdansk, Poland

Marek Kulawiak Department of Geoinformatics, Faculty of Electronics, Telecommunications and Informatics, Gdansk University of Technology, Gdansk, Poland

Milan Lazecky IT4Innovations, Ostrava-Poruba, Czech Republic

Xinlian Liang National Land Survey, Finnish Geospatial Research Institute, Masala, Finland; Centre of Excellence in Laser Scanning Research, Finnish Geospatial Research Institute, Masala, Finland

Ville Luoma Department of Forest Sciences, University of Helsinki, Helsinki, Finland; Centre of Excellence in Laser Scanning Research, Finnish Geospatial Research Institute, Masala, Finland

Lukáš Marek Department of Geography, University of Canterbury, Christchurch, New Zealand

Jindra Marvalová Geomatics Section of Department of Mathematics, Faculty of Applied Sciences, University of West Bohemia, Pilsen, Czech Republic

Saptomo Handoro Mertotaroeno Geodesy and Geomatics, Faculty of Earth Science and Technology, Bandung Institute of Technology, Bandung, West Java, Indonesia

Eva Mertová Department of Military Geography and Meteorology, Faculty of Military Technologies, University of Defence, Brno, Czech Republic

Jean-Christophe Michelin SNCF Réseau, Saint-Denis, France

Tomáš Mildorf Section of Geomatics, Department of Mathematics, Faculty of Applied Sciences, University of West Bohemia, Plzeň, Czech Republic

Arnadi Murtiyoso Photogrammetry and Geomatics Group, ICube Laboratory UMR 7357, INSA Strasbourg, Strasbourg, France

Barbora Musilová Geomatics Section of Department of Mathematics, Faculty of Applied Sciences, University of West Bohemia, Pilsen, Czech Republic

Milan Muňko Department of Theoretical Geodesy, Faculty of Civil Engineering, Slovak University of Technology, Bratislava, Slovakia

Peter Márton Department of Mathematical Methods and Operations Research, Faculty of Management Science and Informatics, University of Žilina, Žilina, Slovakia

Martin Neruda Faculty of Environment, Jan Evangelista Purkyně University, Ústí nad Labem, Czech Republic

Kimmo Nurminen National Land Survey, Finnish Geospatial Research Institute, Masala, Finland

Sebastián Ortega Division of Mathematics, Graphics and Computation (MAGiC), IUMA, Information and Communication Systems, University of Las Palmas de Gran Canaria, Las Palmas de Gran Canaria, Spain

Lukáš Orčík Department of Telecommunication, Faculty of Electrical Engineering and Computer Science, VSB-Technical University of Ostrava, Ostrava, Czech Republic

Arnaud Palha Photogrammetry and Geomatics Group, ICube Laboratory UMR 7357, INSA, Strasbourg, France

Jana Palková Institute of Geodesy, Cartography and Geographical Information Systems, Faculty of Mining, Ecology, Process Control and Geotechnology, Technical University of Košice, Košice, Slovak Republic

Juraj Papčo Department of Theoretical Geodesy, Faculty of Civil Engineering, Slovak University of Technology in Bratislava, Bratislava, Slovak Republic

Daniele Perissin Lyles School of Civil Engineering, Purdue University, Lafayette, USA

Iva Ponížilová Faculty of Civil Engineering, VSB-Technical University of Ostrava, Ostrava-Poruba, Czech Republic

Handoko Pramulyo Geodesy and Geomatics, Faculty of Earth Science and Technology, Bandung Institute of Technology, Bandung, West Java, Indonesia

Gernot Pucher TraffiCon – Traffic Consultants GmbH, Salzburg, Austria

Jiří Pánek Department of Development Studies, Faculty of Science, Palacký University Olomouc, Olomouc, Czech Republic

Vít Pászto Department of Geoinformatics, Faculty of Science, Palacký University Olomouc, Olomouc, Czech Republic

Yuxiao Qin Lyles School of Civil Engineering, Purdue University, Lafayette, USA

Jan Růžička Institute of Geoinformatics, Faculty of Mining and Geology, VŠB-Technical University of Ostrava, Ostrava-Poruba, Czech Republic

Kateřina Růžičková Institute of Geoinformatics, Faculty of Mining and Geology, VŠB-Technical University of Ostrava, Ostrava-Poruba, Czech Republic

Ninni Saarinen Department of Forest Sciences, University of Helsinki, Helsinki, Finland; Centre of Excellence in Laser Scanning Research, Finnish Geospatial Research Institute, Masala, Finland

Jaisiel A. Santana Division of Mathematics, Graphics and Computation (MAGiC), IUMA, Information and Communication Systems, University of Las Palmas de Gran Canaria, Las Palmas de Gran Canaria, Spain

Jose M. Santana Imaging Technology Center (CTIM), University of Las Palmas de Gran Canaria, Las Palmas de Gran Canaria, Spain

Ahmet Sayar Department of Computer Engineering, Engineering Faculty, Kocaeli University, Izmit, Kocaeli, Turkey

Marek Strachota Czech Hydrometorological Institute Ostrava, K Myslivně 3/2182, Ostrava-Poruba, Czech Republic

Marcin Stępniak Institute of Geography and Spatial Organization, Polish Academy of Sciences, Warsaw, Poland

Jose P. Suárez Division of Mathematics, Graphics and Computation (MAGiC), IUMA, Information and Communication Systems, University of Las Palmas de Gran Canaria, Las Palmas de Gran Canaria, Spain

Vladislav Svozilík Institute of Geoinformatics, Faculty of Mining and Geology, VŠB-Technical University of Ostrava, Ostrava-Poruba, Czech Republic

Marta Sylla Department of Spatial Economy, Faculty of Environmental Engineering and Geodesy, Wrocław University of Environmental and Life Sciences, Wrocław, Poland

Szymon Szewrański Department of Spatial Economy, Faculty of Environmental Engineering and Geodesy, Wrocław University of Environmental and Life Sciences, Wrocław, Poland

Alejandro Sánchez Division of Mathematics, Graphics and Computation (MAGiC), IUMA, Information and Communication Systems, University of Las Palmas de Gran Canaria, Las Palmas de Gran Canaria, Spain

Topi Tanhuanpää Department of Forest Sciences, University of Helsinki, Helsinki, Finland; Centre of Excellence in Laser Scanning Research, Finnish Geospatial Research Institute, Masala, Finland

Jan Tesla Faculty of Mining and Geology, Institute of Geoinformatics, VŠB-Technical University of Ostrava, Ostrava-Poruba, Czech Republic

Agustín Trujillo Imaging Technology Center (CTIM), University of Las Palmas de Gran Canaria, Las Palmas de Gran Canaria, Spain

Jan Unucka Faculty of Mining and Geology, VSB-Technical University of Ostrava, Ostrava-Poruba, Czech Republic; Czech Hydrometorological Institute Ostrava, K Myslivně 3/2182, Ostrava-Poruba, Czech Republic

Mikko Vastaranta Department of Forest Sciences, University of Helsinki, Helsinki, Finland; Centre of Excellence in Laser Scanning Research, Finnish Geospatial Research Institute, Masala, Finland

Vít Voženílek Faculty of Science, Palacký University Olomouc, Olomouc, Czech Republic

Libor Váša Department of Computer Science and Engineering, Faculty of Applied Sciences, University of West Bohemia, Pilsen, Czech Republic

Ivo Winkler Czech Hydrometorological Institute Ostrava, K Myslivně 3/2182, Ostrava-Poruba, Czech Republic

Xiaowei Yu National Land Survey, Finnish Geospatial Research Institute, Masala, Finland; Centre of Excellence in Laser Scanning Research, Finnish Geospatial Research Institute, Masala, Finland

Vladislava Zelizňaková Institute of Geodesy, Cartography and Geographical Information Systems, Faculty of Mining, Ecology, Process Control and Geotechnology, Technical University of Košice, Košice, Slovak Republic

Václav Čada Geomatics Section of Department of Mathematics, Faculty of Applied Sciences, University of West Bohemia, Pilsen, Czech Republic

Renata Ďuračiová Department of Theoretical Geodesy, Faculty of Civil Engineering, Slovak University of Technology in Bratislava, Bratislava, Slovak Republic

Małgorzata Świąder Department of Spatial Economy, Faculty of Environmental Engineering and Geodesy, Wrocław University of Environmental and Life Sciences, Wrocław, Poland

Ján Šašak Institute of Geography, Faculty of Science, Pavol Jozef Šafárik University in Košice, Košice, Slovak Republic

Jan Šrejber Czech Hydrometorological Institute Ústí Nad Labem, Kočkovská 18/2699, Ústí nad Labem, Czech Republic

Dušan Židek Czech Hydrometorological Institute Ostrava, K Myslivně 3/2182, Ostrava-Poruba, Czech Republic

Abbreviations

ABA	Area-based approach
ADCP	Acoustic Doppler current profiler
ALS	Airborne laser scanning
API	Application programming interface
ARD	Average relative distance
BI	Business intelligence
BIM	Building information model/modelling
BJVSIS	Information system for Beijing virescence separator
CDH	Cloudera distribution for Hadoop
CLC	Corine Land Cover
CMEM	Comprehensive modal emission model
CODGIK	Polish Centre of Geodesic and Cartographic Documentation
CPU	Central processing unit
CR	Czech Republic
ČSN	Czech technical standard
ČUZK	Czech Office for Surveying, Mapping and Cadastre
DBH	Diameter at breast height
DEM	Digital elevation model
DFS	Distributed file system
DHTML	Dynamic Hypertext Markup Language
DSI	Digital stereo imagery
DSM	Digital soil map
DTE	Data terminal equipment
DTM	Digital terrain model
DTW	Dynamic time warping
EMS	Emergency medical service
ERS	European remote sensing
ETL	Extract, transform, load
ETR	Evidence of criminal proceedings

EU	European Union
FBD	Forest Big Data
FCD	Floating car data
FIWARE	Future Internet platform of the European Community
FLOPS	Floating-point operations per second
FME	Feature Manipulation Engine
FOT	Field operational test
FPMS	Forest protection management system
GCPs	Ground control points
GDAL	Geospatial Data Abstraction Library
GIS	Geographic information system
GNSS	Global navigation satellite system
GPS	Global Positioning System
GPU	Graphical processing unit
GSD	Ground sample distance
GTFS	General Transit Feed Specification
GUI	Graphical user interface
GWR	Geographically weighted regression
HDFS	Hadoop distributed file system
HPC	High-performance computing
HTML	Hypertext Mark-up Language
CHAID	Chi-squared automatic interaction detection
CHMs	Canopy height models
IBM	Image-based modelling
IMU	Inertial measurement unit
InSAR	Interferometric synthetic aperture radar
INSPIRE	Infrastructure for Spatial Information in Europe
IoT	Internet of Things
ITD	Individual tree detection
ITS	Intelligent transport systems
KCB	Konya closed basin
KVPs	Key-value pairs
LBSNS	Locationbased social networks
LCC	Land cover changes
LCF	Land cover flows
LDCM	Landsat Data Continuity Mission
LIDAR	Light detection and ranging
MAF	Mass airflow
MALP	Maximum availability location problem
MEXCLP	Maximum expected coverage problem
NIR	Near-infrared
NSDI	Dutch national spatial data infrastructure
NUTS	Nomenclature of Units for Territorial Statistics
OBD	Onboard diagnostic system
OBJ	Object files

OGC	Open Geospatial Consortium
OSM	OpenStreetMap
OSN	Online social networks
PCR	Police of the Czech Republic
PRF	Pulse repetition frequency
RAID	Redundant array of independent disks
RAM	Random-access memory
RANSAC	Random sample consensus
RMSE	Root-mean-square error
ROC	Receiver operating characteristic
RS	Remote sensing
RSO	Register of Statistical Units and Buildings
RTIARE	Registry of Territorial Identification, Addresses and Real Estate
RTK	Real-time kinematic
RUIAN	Registry of Territorial Identification, Addresses and Real Estate
SAR	Synthetic aperture radar
SD	Standard deviation
SfM	Structure from motion
SIFT	Scale-invariant feature transformation
SLC	Single look complex
SMDoS	Specialised military database of soils
SNCF	French National Railway Company
SP	Spherical photogrammetry
SQL	Structured Query Language
SVM	Support vector machines
SW	Software
TIN	Triangular irregular network
TLS	Terrestrial laser scanning
TS	Total station
UAV	Unmanned aircraft vehicle
UK	United Kingdom
UML	Unified Modelling Language
UPRNs	Unique property reference numbers
VBO	Vertex buffer objects
VDE	Virtual distributed ethernet
VOCs	Volatile organic compounds
VRAM	Video random-access memory
VRML	Virtual Reality Modelling Language
WFS	Web Feature Service
WMS	Web Map Service
XML	Extensible Mark-up Language
ZABAGED	Fundamental Base of Geographic Data of the Czech Republic

Application of Web-GIS for Dissemination and 3D Visualization of Large-Volume LiDAR Data

Marcin Kulawiak and Marek Kulawiak

Abstract The increasing number of digital data sources, which allow for semi-automatic collection and storage of information regarding various aspects of life has recently granted a considerable rise in popularity to the term "Big data". As far as geospatial data is concerned, one of the major sources of Big data are Light Detection And Ranging (LiDAR) scanners, which produce high resolution three-dimensional data on a local scale. The recent introduction of relatively low-cost LiDAR scanners has sparked a revolution in photogrammetry, as the technology offers data quality and cost-effectiveness which greatly outmatch traditional techniques. As a result, the volume and density of collected three-dimensional terrain data is growing rapidly, which in turn increases the pressure on development of new solutions dedicated to processing this data. This paper presents a concept system for web-based storage, processing and dissemination of LiDAR data in a geographic context. Processing and dissemination of LiDAR data in a web environment requires the data to be converted into a format which is fit for processing via DHTML. Because of this the presented system uses the emerging 3D Tiles open standard for serving LiDAR point clouds alongside reconstructed 3D models of buildings to remote users. The system provides remote presentation of LiDAR data in a web environment by means of Cesium, an open source GIS library for 3D visualization of geospatial data. The concept system has been tested using large volumes of LiDAR data collected in the Pomeranian region in Poland, with particular focus on the area of the Tricity, which is a large agglomeration consisting of Gdansk, Sopot and Gdynia. Presented results indicate that the combination of Cesium and 3D Tiles constitute a promising set of open standards for dissemination and visualization of LiDAR data in a web environment.

Keywords GIS · Web · LiDAR · 3D · Processing

M. Kulawiak (✉) · M. Kulawiak
Department of Geoinformatics, Faculty of Electronics, Telecommunications and Informatics, Gdansk University of Technology, Narutowicza Street 11/12, 80-233 Gdansk, Poland
e-mail: Marcin.Kulawiak@eti.pg.gda.pl

M. Kulawiak
e-mail: Marek.Kulawiak@eti.pg.gda.pl

I. Ivan et al. (eds.), *The Rise of Big Spatial Data*, Lecture Notes in Geoinformation and Cartography, DOI 10.1007/978-3-319-45123-7_1

1 Introduction

For quite some time, the term "Big data" has been gaining popularity due to the increasing number of digital data sources, which allow for semi-automatic collection and storage of information regarding various aspects of life. From the perspective of GIS, one of the major sources of Big data are Light Detection And Ranging (LiDAR) scanners, which collect geospatial information by illuminating their surroundings with a laser beam and measuring distance by analysing the reflected light. The application of LiDAR has seen a rapid increase in recent years, as the technology produces higher-quality results than traditional photogrammetric techniques at a lower cost. Because the laser scanning may be performed in a very dense manner, the resulting datasets tend to have a very high resolution. In consequence, the LiDAR revolution caused a rapid increase in the volume of collected data over the same area, as well as in the hardware and software requirements associated with processing this data.

This issue is becoming more pressing with the recent increase of available LiDAR data resulting from ongoing data collection initiatives. In many member states of the European Union the process of LiDAR data collection has seen a major boost with the adoption of Directive 2007/60/EC of the European Commission on the assessment and management of flood risks. This directive, issued on 26 November 2007, requires all Member States to identify possible sources of floods on their territories and perform their flood risk assessments, taking into account factors such as potential climate changes (European Commission 2007). Because successful flood modelling requires construction of an up-to-date Digital Elevation Model, the directive has sparked a remarkable increase in the volume of collected LiDAR data. At the same time, the adoption of the INSPIRE directive has dramatically increased the availability of these measurements to the scientific community.

In Poland, the data collected by LiDAR scanning over large areas is available in the form of high-resolution point clouds (average of 19 points/m^2) which often amount to terabytes in size. Although available freely for scientific use, these large volumes of data are very problematic in management and processing as well as storage. LiDAR point clouds are commonly stored in LAS format, which is an open standard of 3D data storage. Although LAS offers many advantages, such as storage of colour and classification information for every point, it is not a particularly convenient format for data processing and dissemination. This is because a single LAS file can only provide information regarding a relatively small subset of the Earth's surface, and the standard does not involve cross-referencing of LAS files. As a result, the users are forced to invent their own methods of organizing LiDAR measurements performed over a given area.

This situation is particularly problematic for researchers, who are often interested in particular elements of the 3D point clouds. Using binary LAS files for data exchange is inconvenient because the user has no means of remotely inspecting the data beforehand, and thus needs to ask the data provider for all information within a particular area of interest (which is then delivered in LAS format) before being able to have a look at the sought objects. Thus, in order to be successfully processed and disseminated in a web environment, the LAS files need to be converted into a format which is better fit for processing via DHTML. Recently, there have been several attempts at devising a method of presenting LiDAR data in the internet. Kuder and Zalik (2011) have presented a method of serving subsets of LiDAR data in their original format via a REST service, Krishnan et al. (2011) presented a web architecture for community access to LiDAR data, while Lewis et al. (2012) proposed a framework for LiDAR data storage, segmentation and web-based streaming. More recently, Mao and Cao (2013) proposed a method of remote 3D visualization of LiDAR data using HTML5 technologies, Nale (2014) designed a framework for processing, querying and web-based streaming of LiDAR data, while von Schwerin et al. (2016) presented a system for 3D visualization of pre-processed LiDAR data using WebGL. A common problem with these approaches is the lack of standardisation, which makes it difficult to implement the proposed solutions in a different environment.

This paper presents a concept system for web-based storage, processing and dissemination of big volumes of LiDAR data in a geographic context using emerging open standards.

2 Materials and Methods

The Department of Geoinformatics of the Gdansk University of Technology Faculty of Electronics, Telecommunication and Informatics conducts research in various areas concerning the geography of northern Poland, with particular focus on the Pomeranian region. The LiDAR data used in this work covers a large part of the Pomeranian region surrounding the Tricity area, which is a large agglomeration consisting of Gdansk, Sopot and Gdynia. A map of the investigated area, which measures approximately 1422 km^2, is depicted in Fig. 1.

The original LiDAR datasets have been obtained from the Polish Centre of Geodesic and Cartographic Documentation (CODGIK) in the form of LAS files (compliant with version 1.2 of the file format) containing point clouds. As mentioned in the previous section, the average scanning resolution of the utilized datasets is around 19 points/m^2, which translates to around 500 GB for the entire point cloud dataset. Because such a large volume of data would be unfit for direct dissemination and analysis, it was necessary to devise a methodology of processing it for the purposes of remote dissemination. Thus, the original LAS files needed to be parsed, and their contents analysed and extracted.

The structure of a LAS file consists of (ASPRS 2011):

- Public Header Block
- Variable Length Records

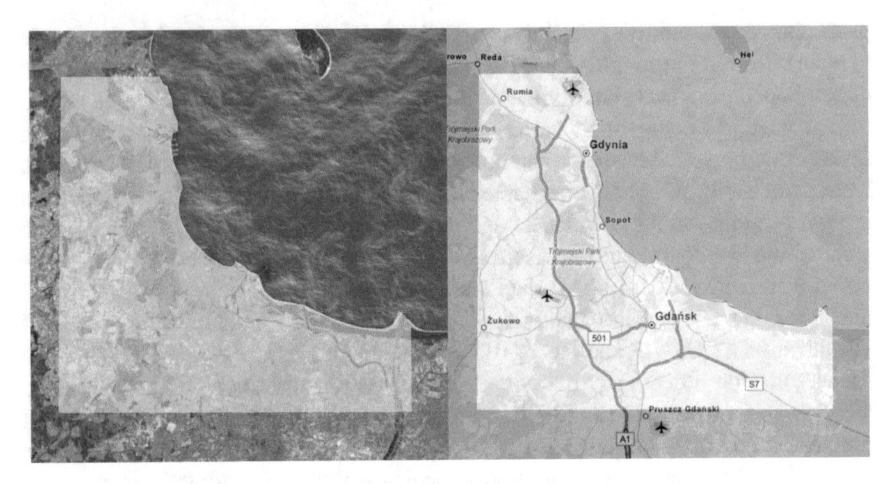

Fig. 1 The LiDAR data used in this work has been collected for the area of the Tricity in the Pomeranian region of Poland

- Point Data Records
- Extended Variable Length Records (optional, available in version 1.3 and later of the file format)

The Public Header Block contains information regarding file version, creation date, data bounding box, information regarding the projection of the point data, number of records as well as header size and offset to point data.

The Variable Length Records contain information regarding the source of data, such as the identification number of the author, as registered with the LAS specification managing body.

The actual results of LiDAR scanning are located in the Point Data Records, where they are stored in an array alongside additional information recorded during the process. There are several possible formats of Point Data Records. The basic information regarding a scanned point (such as location) may be stored in Format 0, while later format versions provide additional fields which may contain GPS Time (Format 1), three colour channels (Format 2), both of these (Format 3) and others.

The analysed files were provided with Point Data Record Format 3, which stores colour and classification data as well as time of capture for every recorded point. The contents of Point Data Record Format 3 are shown in Table 1.

In more recent file format revisions (1.3 and later), the file may also contain waveform data, which is stored either in the Extended Variable Length Records or an external file.

As it was already mentioned, successful processing and dissemination of LiDAR data in a web environment requires the LAS files to be converted into a format which is better fit for processing via DHTML. Thus far a common problem of proposed solutions was the lack of standardisation, which meant that the exchange of LiDAR data between two different web-based applications required one of them to be adapted to the other's method of data distribution.

Table 1 Contents of Point Data Record Format 3

Point Data Record field name	Description
X coordinate of the point	These need to be used in conjunction with the scale values and the offset values from the Public Header Block to determine the geographic coordinates of the point
Y coordinate of the point	
Z coordinate of the point	
Intensity	Optional value representing pulse return magnitude
Return number	Pulse return number for the given output pulse
Number of returns	Total number of returns for the given pulse
Scan direction flag	Represents the direction at which the scanner mirror was travelling at the time of the output pulse
Edge of flight line	Denotes whether the current point is the last one on a given scan line
Classification	Represents the "class" attribute of the point (only available if the point has been classified)
Scan angle rank	Represents the angle at which the laser point was output from the laser system, including the roll of the aircraft
User data	May contain custom information
Point source ID	Represents the file from which this point originated (if any)
GPS time	Represents the time at which the point was acquired
Red	The red image channel value associated with this point (if available)
Green	The green image channel value associated with this point (if available)
Blue	The blue image channel value associated with this point (if available)

In the case of two-dimensional raster and vector data, the issue of compatibility between different software systems has been solved with the advent of open standards in the form of Web Map Service (WMS) (De la Beaujardiere 2006) and Web Feature Service (WFS) (Vretanos 2005), published by the Open Geospatial Consortium (OGC). However, in the case of three-dimensional data, the Consortium's proposed CityGML standard (Gröger et al. 2012) for representing buildings, roads, vegetation, terrain topography and other elements of a three-dimensional city, although very complex in the case of data storage, is very limited in other areas. In particular, the standard does not define efficient methods of data dissemination and visualization, which means that these important issues must be solved by other means (Prandi et al. 2015). Fortunately, these concerns have been addressed in the emerging 3D Tiles standard. 3D Tiles is an open specification proposed by the authors of Cesium, an open source web-based GIS library for 3D visualization of geospatial data.

The 3D Tiles specification organizes three-dimensional geospatial data in a hierarchical tree-like structure, which consists of tiles which cover the entire surface of the Earth. Each tile in this structure has a bounding volume (defined by minimum and maximum longitude, latitude, and height relative to the WGS84 ellipsoid) which completely encloses its contents. Every tile may contain child tiles, which in turn may contain various types of 3D data as well as other child tiles. Consequently, the content of child tiles in the tree is completely enclosed in the parent's bounding volume. In the current version, the standard defines several supported types of tiles, which differ in their contents. Point Cloud tiles carry arrays of 3D points, Batched 3D Model tiles contain solid 3D shapes based on the GL Transmission Format (glTF), Instanced 3D Model tiles contain many instances of a single model, while Composite tiles may contain a combination of all the above. 3D tiles are organized in a tree hierarchy, where each tree may contain subtrees. A single tile may be divided into subtitles in several different ways, including the k-d tree subdivision in which each tile has two children separated by a splitting plane parallel to the x, y, or z axis, as well as the quadtree subdivision, which subdivides each tile into four uniformly subdivided children. In the current version of the standard, one may also apply subdivisions using octrees and grids. The latter, in particular, grant the highest flexibility by enabling definition of uniform, non-uniform as well as overlapping grids supporting an arbitrary number of child tiles. The difference between uniform and non-uniform subdivision methods is shown in Fig. 2.

The broad range of subdivision methods enables the user to choose the appropriate one depending on the task. For example, a non-uniform k-d tree subdivision can create a more balanced tree for sparse datasets, while more uniformly distributed data will be better represented using an octree (Cozzi 2015). Although the child tiles are completely enclosed in the parent's bounding volume, their individual bounding volumes may overlap. This enables one to more precisely adapt the tile subdivision to its contents. Whereas regular point clouds may use tight bounding volumes, large numbers of 3D models may benefit from child tile overlapping.

As mentioned previously, the point clouds used in this research are very dense, and thus it is impossible to visualize an area larger than 75,000 m^2 in a Web browser without engaging a level-of-detail mechanism. Because one of the areas of research conducted by the Department involves shape reconstruction, it was decided that by default the service would provide users with 3D models of buildings reconstructed from the original point clouds. In addition to improving the user's perception and recognition of geographical features, serving 3D models instead of point clouds considerably reduces the amount of transmitted data.

Once the data has been converted into the appropriate format, it must be placed in a data server which will make it available to any remote client compliant with the 3D Tiles standard. Figure 3 presents the architecture of the implemented Web-GIS for 3D visualization and dissemination of LiDAR data. The pre-processed 3D Tiles containing original LiDAR data (in point format) as well as reconstructed buildings (in glTF format) are stored on the Tile server. From there they are made available

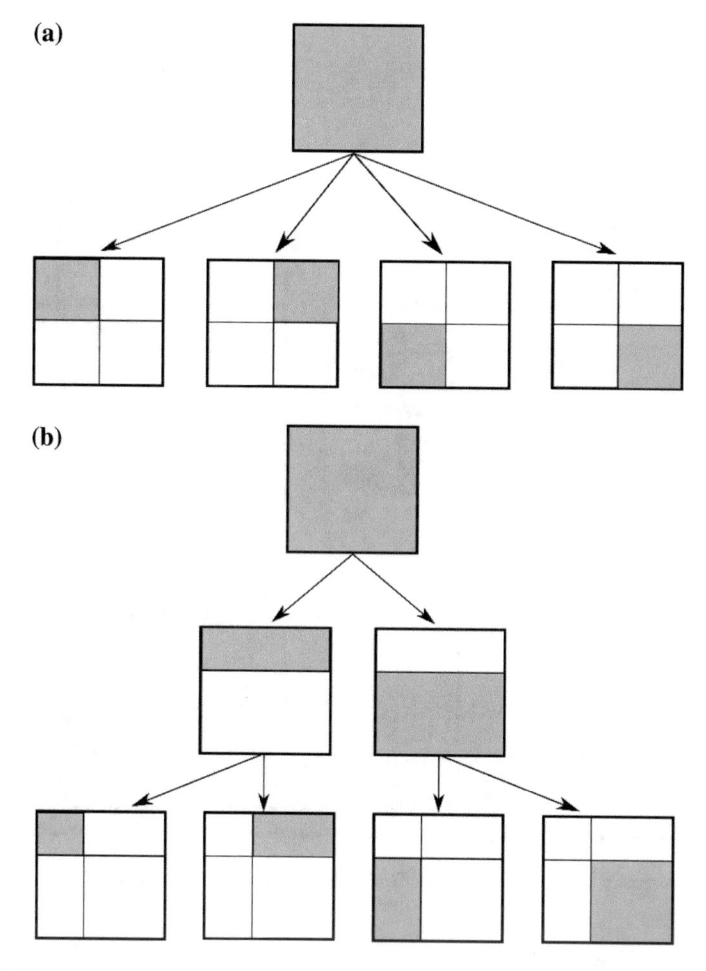

Fig. 2 Difference between a uniform quadtree (**a**) and an example non-uniform k-d tree (**b**) tile subdivision methodology

through web requests issued by remote users who launched the Cesium Client. The client is based on the standard Cesium Viewer application built from the experimental 3d-tiles branch (Cesium 3D-Tiles Code Branch 2015). The client provides the user with a 3D visualization of the Earth based on the WGS84 ellipsoid with overlaid 3D terrain height information and textured with satellite images. The application also provides the user with standard tools for zooming and panning the camera.

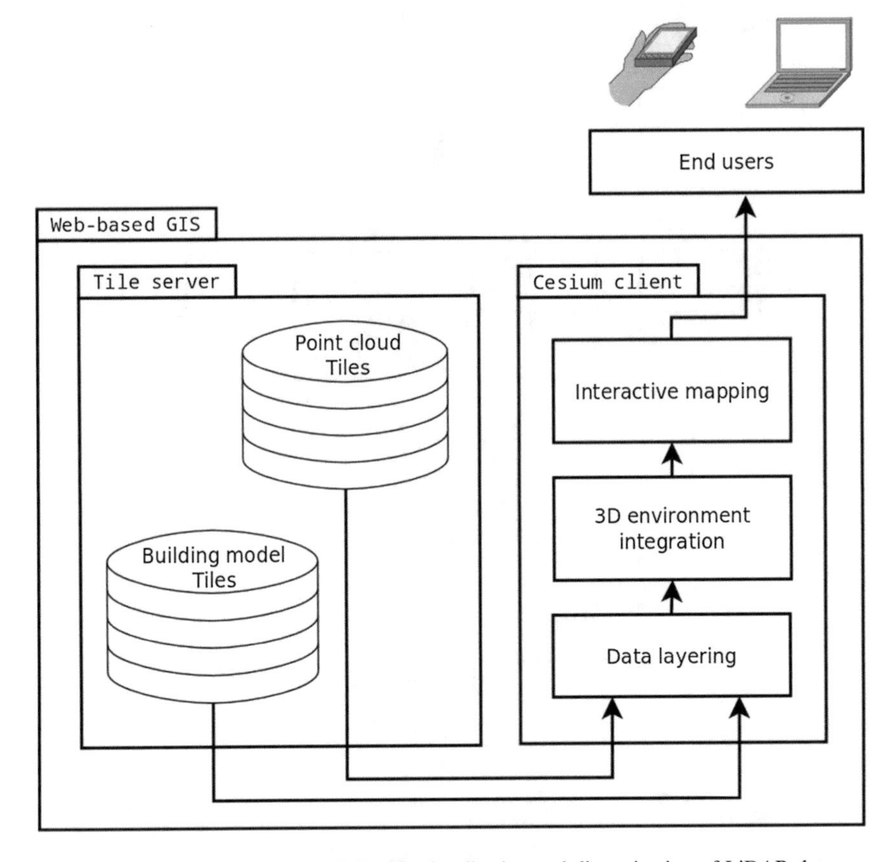

Fig. 3 Architecture of the web-GIS for 3D visualization and dissemination of LiDAR data

3 Results and Discussion

The original LiDAR data obtained from CODGIK contained a lot of information corresponding to the localization of individual scanning points in the geographic space. Because 3D Tiles use a different geographic coordinate system, many of the original attributes have been stripped during the conversion process. After conversion, the point data retains the following information:

- X, Y, and Z coordinates of the point
- Information about the "class" the point falls into
- Red: The Red image channel value associated with this point
- Green: The Green image channel value associated with this point
- Blue: The Blue image channel value associated with this point

Aside from recalculating the point coordinates, no other changes were necessary. Because the converted LiDAR data retains its original resolution, the dataset needed to be divided into chunks small enough to be easily transferred via the web. Thus the area of research was divided into a 1992 × 1826 tile grid, with each tile covering 400 m^2. Because a regular grid division was selected, around 43 % of the resulting tiles represent the area of the Gulf of Gdansk, and thus are empty. This results in over two million tiles carrying point cloud data, with an average tile size of around 250 KB.

For the purpose of shape reconstruction, the original LiDAR data was first converted into a regular point grid. Afterwards a classification algorithm was applied over this grid, using information on geographic location in addition to pre-classification data stored in the LAS file (if available). The adjacent grid cells classified as buildings were then clustered and simplified by removing non-significant elements of the cluster. The resulting point clusters were triangulated, and the Red, Green and Blue values of the original LiDAR points were used to interpolate a texture for the reconstructed buildings. Because the resulting datasets were significantly smaller in comparison to original point clouds, they were located within a 797 × 731 tile grid, with each tile covering 2500 m^2. Similarly to the point cloud case, only around 57 % of the grid space contains actual 3D building models. This results in over three hundred thousand tiles carrying 3D building information, with an average tile size of around 500 KB.

The resulting point cloud and 3D building datasets were placed on the server, ready to be streamed to remote users via the default client based on the Cesium library. The client provides a 3D model of the Earth presented in a basic top-down view, which can be manually modified using intuitive mouse controls. Initially the client presents the users with a view zoomed on the area of the Tricity. From this point users may continue exploring until they reach the area they are interested in. Because the current version of the system uses the basic level-of-detail algorithm provided by Cesium, the user needs to zoom in over an appropriately small area in order to be presented with data streamed via 3D Tiles. By default, once the camera view has been appropriately narrowed, the users are presented with 3D models of buildings which have been automatically reconstructed from original LiDAR data. Figure 4 presents a sample view of the reconstructed buildings in the area of Siedlicka street in Gdansk.

By design choice, the user may choose to change the type of 3D objects overlaid on the digital surface of the Earth. For this purpose the client application provides a simple menu which allows for selection of the currently visualized dataset. The menu, located in the upper left corner of the application window, contains two options:

- Point Cloud (representing original LiDAR data)
- Models (referring to the reconstructed 3D buildings)

The “Models” option is selected by default, but may be manually changed by the user at any time. When the “Point Cloud” option has been selected, the 3D building models shown in the application window are swapped for the LiDAR point clouds. Figure 5 depicts a view of coloured LiDAR point clouds representing the Faculty of

Fig. 4 A view of 3D building models in the area of Siedlicka street in the city of Gdansk as seen in the Cesium client

Fig. 5 A view of coloured LiDAR point clouds representing a fragment of the Gdansk University of Technology campus as seen in the Cesium client

Electronics, Telecommunication and Informatics alongside other fragments of the Gdansk University of Technology campus.

Because the system does not use simplified datasets (both buildings as well as point clouds are served in their native resolution), the 3D Tiles downloaded via the client may be readily processed in any other application compatible with the standard. Although currently Cesium is the only major application implementing the 3D Tiles specification, this is likely to change when the specification matures. A predicted steady growth in the standard's popularity is founded on the fact that Cesium, the biggest supporter of the technology, is also arguably the most advanced open source 3D GIS library currently available, and it is steadily becoming the platform of choice for Web-GIS developers (Kulawiak 2014).

As concerns possible future improvements to the presented system, optimization for deployment in a production environment is one of key aspects that should be considered. In the case of Cesium's 3D Tiles, system performance is a product of server throughput and client processing power. As concerns server throughput, the system provided adequate performance when streaming data from a RAID 5 disc array over a 100 Mbps Ethernet connection to a small userbase. This being said, implementing the system for a wider audience should involve storing 3D Tiles in a cloud environment, using server parallelization to provide increased throughput and ensure stable operation of the service with many simultaneous users. As concerns client-side processing, even though the current version of the system uses a development version of the Cesium library, it provided good performance on a 3.5 GHZ quad-core Intel CPU paired with a GeForce GTX 770 and 8 GB RAM when displaying roughly 1.6 million vertices overlaid on Cesium's Digital Terrain Model. The latest optimizations to Cesium source code indicate that the stable version of the library will provide similar levels of performance on significantly weaker hardware (Cesium Change Log 2016). As concerns other future developments, the applied process of reconstructing 3D building models could involve shape matching with existing 2D databases in order to provide each building with attributes such as name, identification number and function. This way they could be used e.g. in a more precise, three-dimensional implementation of established threat simulation algorithms (Kulawiak and Lubniewski 2014; Moszynski et al. 2015). In addition, currently the proposed solution only allows for one-way communication model, in which the client can only download, but not modify the 3D data which resides on the server. The updated version of the 3D Tiles specification promises to enable client-side data modification and upload in the second quarter of 2016.

Summing up, the results produced by the presented system indicate that the combination of Cesium and 3D Tiles constitute a promising combination of open standards for dissemination and visualization of LiDAR data in a web environment.

References

ASPRS (2011) LAS specification version 1.4. The American Society for Photogrammetry and Remote Sensing, Maryland, USA. http://www.asprs.org/a/society/committees/standards/LAS_1_4_r13.pdf. Accessed 12 Dec 2015

Cesium 3D-Tiles Code Branch (2015) https://github.com/AnalyticalGraphicsInc/cesium/tree/3d-tiles. Accessed 19 Dec 2015

Cesium Change Log (2016) https://github.com/AnalyticalGraphicsInc/cesium/blob/master/CHANGES.md. Accessed 10 Feb 2016

Cozzi P (2015) Cesium 3D-Tiles specification. https://github.com/AnalyticalGraphicsInc/3d-tiles#tiles.json. Accessed 19 Dec 2015

De la Beaujardiere J (ed) (2006) OpenGIS web map service (WMS) implementation specification ver.1.3.0. Open Geospatial Consortium, Inc. http://www.opengeospatial.org/standards/wms. Accessed 05 Feb 2016

European Commission (2007) Directive 2007/60/EC of the European parliament and of the council of 23 October 2007 on the assessment and management of flood risks. http://eur-lex.europa.eu/LexUriServ/LexUriServ.do?uri=OJ:L:2007:288:0027:0034:EN:PDF. Accessed 10 Oct 2015

Gröger G, Kolbe TH, Nagel C, Häfele K-H (eds) (2012) OGC city geography markup language (CityGML) encoding standard. Open Geospatial Consortium. http://www.opengeospatial.org/standards/citygml. Accessed 05 Feb 2016

Krishnan S, Crosby C, Nandigam V, Phan M, Cowart C, Baru C, Arrowsmith R (2011) OpenTopography: a services oriented architecture for community access to LiDAR topography. In: Proceedings of the 2nd international conference on computing for geospatial research and applications. ACM, p 7

Kuder M, Zalik B (2011) Web-based LiDAR visualization with point-based rendering. In: 2011 seventh international conference on signal-image technology and internet-based systems (SITIS). IEEE, pp 38–45

Kulawiak M (2014) Three-dimensional web geographic information system using web-based technologies. ICT Young Edycja 2014, 26–28 September, Gdansk University of Technology, Gdansk, Poland, pp 131–134, ISBN 978-83-60779-10-1

Kulawiak M, Lubniewski Z (2014) SafeCity—a GIS-based tool profiled for supporting decision making in urban development and infrastructure protection. Technol Forecast Soc Change 89:174–187

Lewis P, Mc Elhinney CP, McCarthy T (2012) LiDAR data management pipeline: from spatial database population to web-application visualization. In: Proceedings of the 3rd international conference on computing for geospatial research and applications. ACM, p 16

Mao B, Cao J (2013) HTML5 based 3D visualization of high density LiDAR data and color information for agriculture applications. Social media retrieval and mining. Springer, Berlin, pp 143–151

Moszynski M, Kulawiak M, Chybicki A, Bruniecki K, Bieliński T, Łubniewski Z, Stepnowski A (2015) Innovative web-based geographic information system for municipal areas and coastal zone security and threat monitoring using EO satellite data. Mar Geodesy 38(3):203–224

Nale A (2014) Design and development of a generalized LiDAR point cloud streaming framework over the web. Università degli Studi di Padova, Italy

Prandi F, Devigili F, Soave M, Di Staso U, De Amicis R (2015) 3D web visualization of huge CityGML models. Int Arch Photogramm Remote Sens Spat Inf Sci 40(3):601–605

von Schwerin J, Richards-Rissetto H, Remondino F, Spera MG, Auer M, Billen N, Loos L, Stelson L, Reindel M (2016) Airborne LiDAR acquisition, post-processing and accuracy-checking for a 3D WebGIS of Copan, Honduras. J Archaeol Sci Rep 5:85–104

Vretanos PA (ed) (2005) OpenGIS web feature service (WFS) implementation specification ver. 1.1

Design and Evaluation of WebGL-Based Heat Map Visualization for Big Point Data

Jan Ježek, Karel Jedlička, Tomáš Mildorf, Jáchym Kellar and Daniel Beran

Abstract Depicting a large number of points on a map may lead to overplotting and to a visual clutter. One of the widely accepted visualization methods that provides a good overview of a spatial distribution of a large number of points is a heat map. Interactions for efficient data exploration, such as zooming, filtering or parameters' adjustments, are highly demanding on the heat map construction. This is true especially in the case of big data. In this paper, we focus on a novel approach of estimating the kernel density and heat map visualization by utilizing a graphical processing unit. We designed a web-based JavaScript library dedicated to heat map rendering and user interactions through WebGL. The designed library enables to render a heat map as an overlay over a background map provided by a third party API (e.g. Open Layers) in the scope of milliseconds, even for data size exceeding one million points. In order to validate our approach, we designed a demo application visualizing a car accident dataset in the Great Britain. The described solution proves fast rendering times (below 100 ms) even for dataset up to 1.5 million points and outperforms mainstream systems such as the Google Maps API, Leaflet heat map plugin or ESRI's ArcGIS online. Such performance enables interactive adjustments of the heat map parameters required by various domain experts. The described implementation is a part of the WebGLayer open source information visualization library.

J. Ježek (✉) · K. Jedlička · T. Mildorf · J. Kellar · D. Beran
Section of Geomatics, Department of Mathematics, Faculty of Applied Sciences, University of West Bohemia, Univerzitní 8, 306 14 Plzeň, Czech Republic
e-mail: jezekjan@kma.zcu.cz

K. Jedlička
e-mail: smrcek@kma.zcu.cz

T. Mildorf
e-mail: mildorf@kma.zcu.cz

J. Kellar
e-mail: kellar@students.zcu.cz

D. Beran
e-mail: dberan@students.zcu.cz

I. Ivan et al. (eds.), *The Rise of Big Spatial Data*, Lecture Notes in Geoinformation and Cartography, DOI 10.1007/978-3-319-45123-7_2

Keywords Kernel density map · Heat map · WebGL · GPU · Visualization

1 Introduction

Depicting a point symbol on a map is the most common approach to display a location of a point feature. However, when considering a big amount of point features, such technique leads to over-plotting that may overwhelm users' perception and cognitive capacities. A relevant visual approach that helps to interpret spatial distribution of point data is the kernel density map (sometimes called heat map). This colored isarithmic map can be used to depict the density of spatial features. Efficient exploration of point data through the density map often requires interactions such as zooming or heat map parameters adjustment and filtering. These interactions demand low latency responsiveness even in the case of large dataset. However, calculation of such a map might be too complex to achieve this without advanced techniques.

In this paper we propose a system for an efficient kernel density estimation and heat map generation for the web by utilizing hardware acceleration through WebGL. We designed a system that enables visualization of a kernel density map for a dataset of up to 1.5 million points in a time that enables interactive frame rate. This Web-based system enables to display a heat map on top of a background map managed by a third party mapping library (e.g. Open Layers, Leaflet or Google Maps). Such speed enables fast interactions for zooming or calculation parameters adjustments even for the specified size of data.

In this paper, a detailed description of the kernel density map creation through hardware acceleration is given and several implementation considerations are discussed. For the demonstration purposes, a visualization application for up to 1.5 million points (traffic accidents) has been designed. We compared the solution with relevant systems such as the GoogleMap heat map API, Leaflet heat map plugin or ArcGIS Online heat map function. Furthermore we collected feedback for the effectiveness and usefulness of provided heat map interactions by domain experts (traffic engineers).

The outline of this paper is as follows. Section 2 sheds light on theoretical background related to density estimation as well as related mapping systems. Section 3 defines a basic methodology and terms for kernel density estimation used in our approach. Section 4 describes the details about the proposed GPU-based implementation. Section 5 shows the performance results and compares our approach with existing systems. This chapter also summarizes the user feedback. Then the conclusion is given.

2 Related Work

Visualization of big amount of geospatial point data is a topic tightly connected to various research areas. The number of point features that can be visualized on the screen is limited by the display size and resolution as well as by human perception abilities. In order to reduce the data size, there are two main approaches: (1) aggregate and sample the data, where data are grouped based on their spatial or semantic proximity to reasonable number of clusters (Fig. 1c), or (2) partition the space to disjoint areas and aggregate the data located in each area. These areas can be defined by spatial semantics (e.g. by province boundaries as depicted in Fig. 1a), or artificially by regularly dividing the space into cells (Fig. 1b).

The most relevant spatial clustering methods widely accepted in GIS were summarized by Fotheringham and Rogerson (1994). The general problem of spatial point patter analysis is described by O'Sullivan and Unwin (2014), where the methods are divided to distance-based and density-based. The cartographic aspects and relevant visualization techniques are given by Slocum (2008). Clusters are most often visualized as a point symbol map where the amount of data is encoded to a visual variable (e.g. to the size of the symbol). Particular applications of these methods alongside with their advantages and disadvantages for big point data are further discussed by Kuo et al. (2006) and Skupin (2004).

In addition to clustering techniques that represent data as discrete values, there is a density map approach that represents the point distribution as a smooth continuous surface (Fig. 1d). The simple point density estimation method works on a predefined grid. Each cell of the grid has a specified circular neighbourhood used to calculate the cell density value as the ratio of the number of features that are located within the search area to the size of the area (Silverman 1986). The other approach

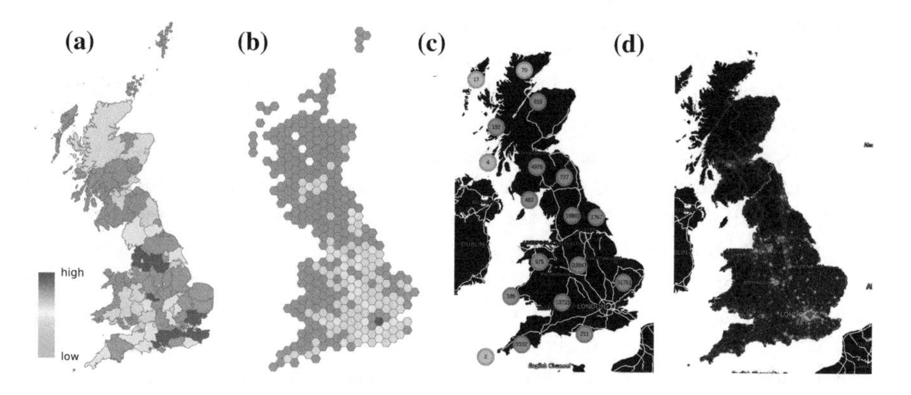

Fig. 1 Example of point density visualization through data reduction based on space partitioning (**a**, **b**), clustering (**c**) and the heat mapping (**d**). The maps visualize the same dataset containing 100,000 car accidents in Great Britain

is the kernel density estimation. Instead of considering a circular neighbourhood around each cell, as used in the simple method, a circular neighbourhood around each spatial feature is drawn (the kernel). Then a mathematical equation is applied that goes from 1 at the position of the feature point to 0 at the neighbourhood boundary (Fotheringham et al. 2000). The final surface is given by summing the values of the kernels.

In order to visualize such a density surface, the isarithmic map approach can be used. Specifically a contour map, hypsometric tints (isopleth map) or a continuous-tone map (heat map) (Slocum 2008). The application examples of density estimation and its visualization can be found in various domains, e.g. road safety (see Wongsuphasawat et al. 2009; Anderson 2009; Dillingham et al. 2011), or a criminality mapping (Chainey et al. 2008). Considering the above mentioned applications, the most widely used technique for visualization of a large number of points in these domains is the kernel density estimation visualized as a continuous-tone map, which is commonly called a heat map. Dillingham et al. (2011) validates the heat map as an effective visual technique for road accident data. Furthermore, a strong advantage of heat maps over the clustering approach is its ability to preserve outliers (Liu et al. 2013). A cartographic heat map has its background in the information visualization (Infovis) domain, where the heat map is used in slightly different manner. The history of heat maps, as considered in Infovis, is summarized by Wilkinson and Friendly (2009).

The kernel density estimation and its visualization using a heat map might be an expensive process for big data. Therefore, an efficient system architecture and algorithm design is desired. One of the relevant speed-up techniques is the algorithm parallelism, enabled by a utility of the graphical processing unit (GPU) [as demonstrated by Srinivasan et al. (2010)]. The GPU-based approach proofs its benefits in various similar scenarios. For example Fekete and Plaisant (2002) use the GPU approach to produce a treemap of 1 million records. Liu et al. (2013), Lins et al. (2013) present a visualization of billions of records through linked views by using the GPU approach and Hennebohl et al. (2011) demonstrate the GPU-based kriging. However, a GPU-based heat map designed for the purpose of web mapping has not been investigated so far.

In order to achieve an immediate response of visualization interactions, a value of 100 ms response time is considered (Card et al. 1983). Even though the most recent research shows the benefits of using a heat map for various scenarios, widely accepted web GIS tools enable only limited scalability and the interactivity tends to be too slow for a heat map visualization. Mainstream GIS system as well as the geovisualization systems [e.g. the GeoViz toolkit (Hardisty and Robinson 2011), the CDV (Dykes 1998) or similar] usually do not allow to effectively visualize more than 100,000 points. Our work follows various successful systems focused on big data processing by using a graphical hardware.

3 Geological Characteristic

The proposed approach is based on the kernel density estimation (KDE). The KDE is a basic method for estimation of a density surface from irregularly distributed point data. The KDE involves placing a symmetrical surface over each point. Afterwards, an evaluation of the distance from the point to the reference location is done and then a sum of the value for all the surfaces for that reference location is calculated. This allows to place a kernel over each observation, and summing these individual kernels provides the density estimate for the distribution of the points. An estimation of the relative density around a location s can be obtained from the following:

$$\lambda_s = \sum_{i=1}^{n} \frac{1}{\pi r^2} k(d_{is}, r, m) \tag{1}$$

where λ_s is the density estimate at the location s; n is the number of observations, r is the search radius (bandwidth) of the KDE (only points in this radius are used) and k is the weight function. k is a function of the point i at the distance d_{is}, the radius r and the weighting parameter m. An example of k is $k(d_{is}, r, m) : 1 - \frac{d_{is}^m}{r^m}$.

4 GPU-Based Approach

A kernel density estimation surface is usually represented in the form of a raster. In order to produce such a surface and visualize it through a heat map, these main steps must be performed: (1) project the data from an input coordinate system to the actual screen viewport, (2) render the kernel around each observation, (3) summarize the values for each pixel of the output raster and (4) use color gradient to calculate the final color for the visualization.

The complexity of the algorithm depends on the number of observations, the raster size and also the radius value. For interactive visualization that supports zooming, it might be important to enable the user to change the resolution as well as to adopt the radius or the weighting parameter (m) of the kernel function in the runtime. Furthermore, application of color gradient can be customized to be used just within the user defined interval.

In our approach we focus on the GPU-based technique. A GPU enables to process the input data in parallel by using thousands of cores of the GPU hardware. Modern browsers supports WebGL as an API for GPU-based programming. On the lower level, WebGL is designed to render geometry primitives (points, lines, triangles). WebGL uses shader programs that run on the GPU as parallel processes: vertex shader assign a screen position to the input data and fragment shader that computes a color of every pixel of a geometry primitive and writes that in the frame buffer as a pixel. During this process, the blending function evaluates how to

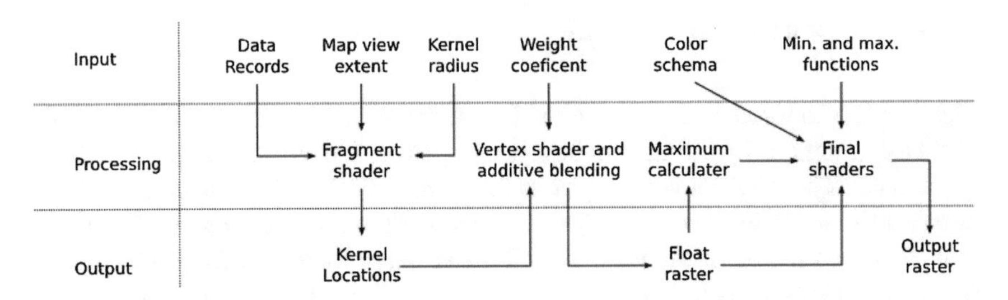

Fig. 2 Processing workflow schema

combine the new pixel color with an old color contained in the framebuffer. Due to the GPU architecture, the shaders run in parallel and allow efficient data processing on the one hand. On the other hand, the transmission from CPU to GPU is an expensive process that might slow down the performance.

In order to produce a heat map by using WebGL, we designed a workflow depicted in Fig. 2. The resulting heat map is considered as an overlay over user defined background map that supports zooming and panning. Such a background map is supposed to have a projected coordinate system and an actual geographic extent as well as dedicated viewport on the screen (map view extent) expressed by the height and width in pixels. The input data records are considered to be expressed in the same projected coordinate system. In order to display them into the viewport, only scaling and translation need to be applied. The resolution of the heat map is equal to the resolution of the current screen viewport. In our approach, all data are uploaded to the GPU during the initialization phase and the translation and scaling are performed according to the actual zoom level or panning operation within the vertex shader. This allows to transmit just information about the map extent from CPU to GPU during panning and zooming interactions.

In order to render the kernel around each observation, the circle area for every point is allocated according to the kernel radius. For each pixel within this kernel the value of the kernel function is calculated by the fragment shader and this value is assigned to the pixel (a float raster is used). Our implementation allows to set a weight coefficient (n) of the kernel function.

Afterwards, the WebGL enables to set the blending function that influences how newly calculated pixels should be combined with the pixels already stored in the framebuffer. In the case of a heat map, an additive blending is set to calculate the sum of the values for a particular pixel.

Afterwards, the resulting float raster is processed and the maximum value from all the pixels is withdrawn. For this purpose, a GPU technique based on generating series of mipmap of original texture is used (depicted as the maximum calculator in Fig. 2). This maximum is an input to specify the final range to be displayed by the heat map.

Finally, the produced density surface is processed once again by the WebGL shaders and a particular color schema is applied to produce the final heat map. For such a purpose, a user defined color scheme is set (e.g. green, yellow, red) and the

Fig. 3 GPU-based visualization of a heat map of 1,494,275 car accidents in Great Britain. Radius size 4 px

final value of the pixel is interpolated by using color gradient. One of the well-known and often used color schemas is the color gradient from green across yellow to red (e.g. a default schema in Google Maps Heat map API). It is useful to use different transparency for locations with low density and locations with high density so that the low density places do not overplot the background map while the high density places are less transparent. The final visualization of the car accident dataset by applying the mentioned color schema and transparency is depicted in Fig. 3.

The implementation of the described technique has been done as a part of the WebGLayer javascript library (see http://jezekjan.github.io/webglayer). The WebGLayer is JavaScript, WebGL based library for coordinated multiple views visualization. The implementation provides input variables including the radius, the kernel function coefficient, the range of minimum and maximum values of the kernel density estimation to be visualized through color gradients.

5 Comparison of the Developed GPU-Based Approach to Current Solutions

In order to validate our approach we implemented a demonstration application, a heat map of car accidents of the Great Britain. For this purpose, the road accident data from the data.gov.uk portal were used (see https://data.gov.uk/dataset/road-accidents-safety-data/resource/626a5e18-be29-4fcf-b814-9c4edb4f3cc6). The developed GPU-based solution was compared with three widely used techniques. The demo application is available online at: http://home.zcu.cz/~jezekjan/wgl_git/examples/heatmap/.

5.1 Methodology

In order to evaluate the performance and scalability of the solution, subsets of various sizes of the road accident dataset were prepared:

- Datasets of a size of 10,000, 15,000, 25,000 and 27,000 road accidents in the Birmingham metropolitan area. The datasets from 10,000 to 25,000 were diluted for the purposes of this test, the dataset of 28,000 points shows all road accidents in Birmingham between the years 2005 and 2014.
- Datasets of ranging from 100,000 to 1,494,275 points (the maximum is the size of the dataset) that represents road accidents in Great Britain in the same time period. The datasets were diluted for the purpose of this test and present always top-k elements sorted by a timestamp.

Afterwards, four applications displaying heat maps of these datasets were implemented. In addition to the GPU-based application, other three solutions were used: ArcGIS Online, Google Maps API and the Leaflet heat map plugin. For each technology, the rendering time needed to redraw the complete heat map was measured. The measurements were done with respect to the radius value that affects the rendering time in most cases. Other available libraries such as OpenLayers has not been evaluated in details as their performance seems to be similar with the mentioned solutions.

5.2 ArcGIS Online

ArcGIS Online (a web GIS platform that allows the user to use, create, and share maps) was used (see http://doc.arcgis.com/en/arcgis-online/reference/what-is-agol.htm) for two reasons. Firstly, ArcGIS Online is a widely spread GIS platform. Secondly, it offers a heat map visualization technique as an example of data driven visualization (see http://blogs.esri.com/esri/arcgis/2015/03/03/whats-new-arcgis-online-March-2015/ or heat map section at the page http://doc.arcgis.com/en/arcgis-online/create-maps/change-style.htm), ~heat map as an overlay on another data (e.g. background map). The following text shortly describes the technical challenges that were faced and the results that were achieved.

During uploading and setting up feature services from source layers, a limitation of ArcGIS Online was identified In ArcGIS Online, by default, the maximum number of records returned by the server is set to one thousand (see http://support.esri.com/en/knowledgebase/techarticles/detail/42901). This can be changed following this procedure (see http://support.esri.com/fr/knowledgebase/techarticles/detail/44060), however the technical article says, that: "*Increasing the value impacts performance if there is a large amount of data added to a web map. Hence, querying and displaying data from the web map may take a significant amount of time*". This was experienced during the testing and it can be clearly seen from Table 1. To redraw the heat map the radius change was performed. e.g. when the radius was changed from 5 to 6 pixels, the redraw delay was measured and written into under the radius 6 px value (see Table 1). The value was measured for all the datasets.

When data were successfully loaded for visualization, experiments with changing the radius started, starting with the smallest dataset and keeping the zoom to the Birmingham area. The achieved results are mentioned in Table 1. Values for the datasets 10–27 k were acquired. The values of the 100,000 dataset measurement could not be taken into account because the rendering ended with this warning: "*The layer did not draw completely as there are too many features to display in this area of the map*". The ArcGIS Online platform doesn't provide an exact time measuring tools, therefore, a stopwatch was used. The web application used for tests is accessible online at: http://arcg.is/1lmRQ3t.

Table 1 Comparison of ArcGIS Online, Google Maps and GPU solution

Dataset	ArcGIS Online		Google Maps		Leaflet	GPU	
Number of points	Radius change		Radius change		Radius	Radius	
	5–10 px (s)	10–20 px (s)	5–10 px (s)	10–20 px (s)	2–20 px	5 px	20 px
10,000	0.6	0.9	<1	<1	0.066	<0.032	
15,000	0.7	1.3	<1	<1	0.084		
20,000	0.8	1.6	<1	<1	0.110		
28,000	0.9	2.1	<1	<1	0.139	0.032	0.037
100,000	Layer did not load properly		<1	<1	0.180	0.046	0.080
250,000			1.5	2.5	0.341	0.046	0.138
500,000			3	4	1.382	0.052	0.253
750,000			5	6	1.520	0.067	0.349
1,000,000			6.5	8	1.914	0.068	0.420
1,494,275			–	–	–	0.100	0.603

5.3 Google Maps

Another way to create a heat map is provided by the Google Maps Javascript API v3. It allows the user to use the heat map layer as an overlay on the top of the map. Data can be rendered either according to their coordinates or weight or a combination of both of them. The appearance of the heat map layer can be customized by changing such options as the radius of data points, the maximum intensity of the heat map, the color schema or the opacity (see https://developers.google.com/maps/documentation/javascript/heatmaplayer). Although the heat map layer can render a large number of point data, it may result in reduced performance. Again, a stopwatch was used, showing that the performance is better than in ArcGIS Online, but still losing interactivity and dynamics when a dataset of 100 k and more points is used. See Table 1 for all measured times. The map application can be found online at: http://home.zcu.cz/~kellar/webglayer/heatmap.php.

In addition to the heat map layer, which provides the client-side rendering of heat maps, the Google Maps Javascript API allows to use the server-side rendering via Fusion Tables. In this case, data are rendered on the server and sent to the user in the form of a raster tile overlay. This solution provides better performance for larger datasets, but obviously limits the interactions in runtime (e.g. to change the radius or maximum parameters is not possible). The Fusion Tables Layer allows to display first 100,000 rows of data in a table (see https://developers.google.com/maps/documentation/javascript/fusiontableslayer#limits). Due to these differences, the Fusion Tables Layer was not included in the test.

5.4 Leaflet Heat Map Plugin

The last considered solution is the Leaflet heat map plugin v0.2.0 provided by Vladimir Agafonkin (see https://github.com/Leaflet/Leaflet.heat). Leaflet as well as the heat map plugin are open source products used within the MapBox platform (see https://www.mapbox.com). The plugin is based on HTML and canvas 2D rendering technique and is inspired by more general tool heatmap.js (see http://www.patrick-wied.at/static/heatmapjs). Due to the open source nature of the project, it was possible to precisely measure the rendering time by modifying the source code. During the testing, no dependency between radius value and rendering time was observed. The measure time is an average time for various radius possibilities (from 2 to 22), while rendering the map 10 times for each radius. The map application can be found online at: http://jachymkellar.github.io/heatmap/.

5.5 *Test Result and Comparison*

The performance of the GPU-based solution was evaluated by measuring the time needed for the whole rendering workflow. As the testing methodology suggests, during the tests, rendering times for datasets of different size and of different radius of the kernel were measured. For the test the zoom level that displays the whole dataset (zoom level 9) was used and a resolution of the map was 1631 × 965. All the tests were performed on a mahine with the following configuration: Google Chrome 46.0.2490.80 (64-bit), Ubuntu Linux on Lenovo T430s, Intel Core i7-3520M CPU @ 2.90 GHz × 4, RAM 3.6, graphics: NVIDIA, GF117M GeForce 610M/710M/820M/GT 620M/625M/630M/720M. The rendering workflow was repeated 50 times and the average time was calculated. The Google Chrome web browser was setup to enable all available graphics hardware accelerations (under chrome://gpu).

The results are available in Table 1. The results demonstrate that the ArcGIS Online solution is not suitable for big data. It slows down to more than a second when using larger radius on datasets around 15,000 points. The 100,000 dataset didn't even load properly.

Google Maps API can handle higher amount of data than ArcGIS Online, but it starts to have significant lags around 250,000 points (2.5 s for 20 px radius). Even though Google Maps API can work with 1,000,000 dataset, the rendering time is 26 s.

Considering all three concurrent solutions, the Leaflet heat map plugin was the fastest one. The time below 100 ms that is considered as suitable for interactive visualization was reached for dataset below 20,000 points. In contrast to all the other solutions, no dependency on the value of the radius was observed.

Comparing all four solutions, the proposed GPU-based approach gives the best results: the interactive response time was reached even for the case of visualizing the whole dataset (1,492,475 points) for the case of the radius below 6 pixels. The dependency on the radius size is significant, however, the use of big radius does not usually provide valuable visualization and leads to the loss of details as the peaks of the surface are smoothed in such a case.

The detailed results of the performance measurements of the GPU-based solution are depicted in Fig. 4. The GPU solution reacts immediately [t < 100 ms as defined in (Card et al. 1983)] for 100,000 dataset with 20 px radius or even 1,400,000 dataset with 6 px.

5.6 *User Feedback*

In addition to the performance and scalability testing, we asked for feedback from domain experts using the tested tools on a daily basis. We discussed the car accidents visualization with experienced traffic engineers from EDIP s.r.o to validate

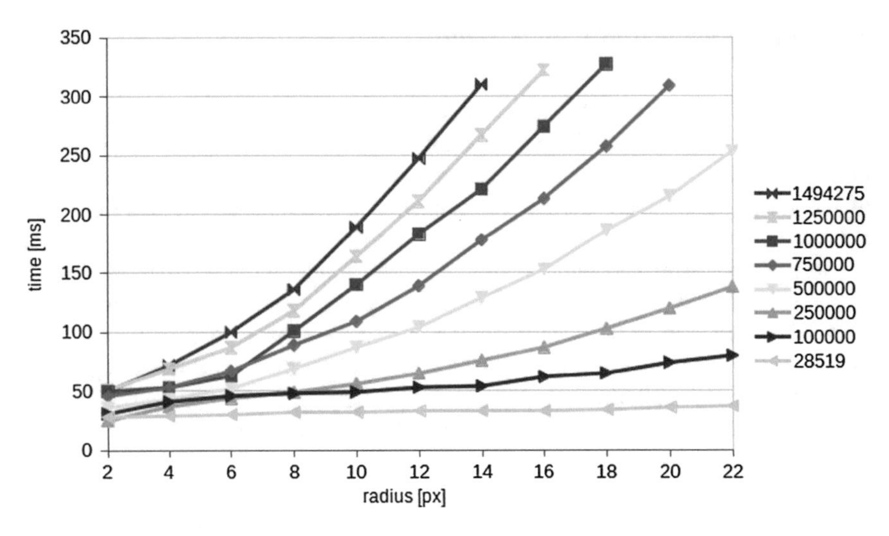

Fig. 4 Rendering times for GPU-based visualization

the usability of the interactive heat map using the GPU-based visualization. We received this comment: "*An interactive heat map enables a user to browse large dataset and might reveal hidden patterns. Such insights serve a valuable input to better formulate precise traffic engineering statistical analysis and safety management. State-of-the art tools that we used so far were not able to provide such interactions for the larger datasets*".

Another feedback was gathered from traffic specialists from the Birmingham City Council, fully aware about the traffic accidents hotspots by daily experience. The feedback was positive and especially highlights the interaction responsiveness that increases the usability of the tool and encourages curiosity about the visualized problem and the data. The people with local knowledge usually immediately tend the explore the known places and validate their knowledge about dangerous crossroads and junctions.

6 Conclusions

This paper demonstrates a novel approach of creating kernel density estimation and heat map rendering suitable even for data over 1 million points by using an accelerated graphics hardware—GPU. Benchmark results prove the interactive responsiveness of our system (below 100 ms) even for large data by using a commodity hardware. The solution outperforms the mainstream state-of-the-art systems such as the Leaflet heat map plugin, Google Maps API or ESRI's ArcGIS Online. Feedback from domain experts proves the benefits and usability of user interactions that are cannot be achieved by concurrent solutions. The proposed algorithm and workflow was implemented as a part of the WebGLayer javascript open source visualization library.

Acknowledgments The authors of this paper are supported by the European Union's Competitiveness and Innovation Framework Programme under Grant Agreement No. 620533, the OpenTransportNet project.

References

Anderson TK (2009) Kernel density estimation and k-means clustering to profile road accident hotspots. Accid Anal Prev 41(3):359–364. doi:10.1016/j.aap.2008.12.014

Card SK, Newell A, Moran TP (1983) The psychology of human-computer interaction. L. Erlbaum Associates Inc., Hillsdale. ISBN:0898592437

Chainey S, Tompson L, Uhlig S (2008) The utility of hotspot mapping for predicting spatial patterns of crime. Secur J 21(1):4–28. doi:10.1057/palgrave.sj.8350066

Dillingham I, Mills B, Dykes J (2011) Exploring road incident data with heat maps. In: Geographic information science research UK 19th annual conference (GISRUK 2011), 27–29 April 2011, University of Portsmouth, Portsmouth

Dykes J (1998) Cartographic visualization. J R Stat Soc Ser D 47(3):485–497. doi:10.1111/1467-9884.00149

Fekete J-D, Plaisant C (2002) Interactive information visualization of a million items. In Information Visualization, 2002. INFOVIS 2002. IEEE Symposium on, pp 117–124. IEEE. doi:10.1109/INFVIS.2002.1173156

Fotheringham AS, Brunsdon C, Charlton M (2000) Quantitative geography: perspectives on spatial data analysis. Sage, Thousand Oaks

Fotheringham S, Rogerson P (1994) Spatial analysis and GIS. CRC Press, Boca Raton. ISBN:780748401048

Hardisty F, Robinson AC (2011) The geoviz toolkit: using component-oriented coordination methods for geographic visualization and analysis. Int J Geogr Inf Sci 25(2):191–210. doi:10.1080/13658810903214203

Hennebohl K, Appel M, Pebesma E (2011) Spatial interpolation in massively parallel computing environments. In: Proceedings of the 14th AGILE international conference on geographic information science (AGILE 2011)

Kuo R, An Y, Wang H, Chung W (2006) Integration of self-organizing feature maps neural network and genetic k-means algorithm for market segmentation. Expert Syst Appl 30(2):313–324. doi:10.1016/j.eswa.2005.07.036

Lins L, Klosowski JT, Scheidegger C (2013) Nanocubes for real-time exploration of spatiotemporal datasets. Visual Comput Graph IEEE Trans 19(12):2456–2465. doi:10.1109/TVCG.2013.179

Liu Z, Jiang B, Heer J (2013) Immens: real-time visual querying of big data. In: Computer graphics forum, vol 32. Wiley, Hoboken, pp 421–430

Silverman BW (1986) Density estimation for statistics and data analysis. Chapman and Hall, Boca Raton. ISBN:978-0412246203

Skupin A (2004) The world of geography: visualizing a knowledge domain with cartographic means. Proc Natl Acad Sci 101(Suppl. 1):5274–5278. doi:10.1073/pnas.0307654100

Slocum TA (2008) Thematic cartography and geovisualization, 3rd edn. Prentice Hall, Upper Saddle River. ISBN:978-0132298346

Srinivasan BV, Qi H, Duraiswami R (2010) Gpuml: graphical processors for speeding up kernel machines. In: Siam conference on data mining. Workshop on high performance analytics-algorithms, implementations, and applications
O'Sullivan D, Unwin D (2014) Geographic information analysis, 2nd edn. Wiley, Hoboken. ISBN:978-0-470-28857-3
Wilkinson L, Friendly M (2009) The history of the cluster heat map. Am Stat 63(2):179–184
Wongsuphasawat K, Pack M, Filippova D, Van Daniker M, Olea A (2009) Visual analytics for transportation incident data sets. Transp Res Record: J Transp Res Board 2138:135–145

Open Source First Person View 3D Point Cloud Visualizer for Large Data Sets

Arnaud Palha, Arnadi Murtiyoso, Jean-Christophe Michelin, Emmanuel Alby and Pierre Grussenmeyer

Abstract The use of laser scanning techniques has become a common way to measure the real world. Millions of points could be generated by this system, which brings forth the problem of visualizing and eventually performing analyses on them. In the open source domain, various software packages exist to visualize 3D point clouds. However, most of these software packages do not allow the real-time rendering of large point clouds. Few of them also allow visualization in an immersive manner. The research aims to create an open source viewer for large 3D point clouds, which enables a dynamic and immersive visualization. In order to do so, rendering and point cloud management strategies must be implemented to avoid overloading the computer's memory. The rendering is done using OpenGL engine by utilizing the graphic card's memory in order to perform faster visualization. The program consists of a pre-processing stage in which the point cloud files are divided into quadtrees and then subsampled using the random tree sampling method. The visualization itself will calculate the distance between the point of view and the center of nodes generated by the pre-processing stage. The amount of points rendered within each node will depend on this distance; the farther away the node is from the point of view, the fewer points are rendered. A loading and unloading function enables the point cloud to be rendered dynamically. With the point cloud

A. Palha · A. Murtiyoso · E. Alby · P. Grussenmeyer (✉)
Photogrammetry and Geomatics Group, ICube Laboratory
UMR 7357, INSA, Strasbourg, France
e-mail: pierre.grussenmeyer@insa-strasbourg.fr

A. Palha
e-mail: arnaud.palha@insa-strasbourg.fr

A. Murtiyoso
e-mail: arnadi.murtiyoso@insa-strasbourg.fr

E. Alby
e-mail: emmanuel.alby@insa-strasbourg.fr

J.-C. Michelin
SNCF Réseau, 6 Avenue François Mitterrand, 93574
Saint-Denis, France
e-mail: jean-christophe.michelin@reseau.sncf.fr

I. Ivan et al. (eds.), *The Rise of Big Spatial Data*, Lecture Notes
in Geoinformation and Cartography, DOI 10.1007/978-3-319-45123-7_3

management algorithm implemented, the resulting program is able to load large point clouds generated by a mobile laser scanner using an ordinary computer. The resulting program as well as the source code will be available for the public due to its open source nature.

Keywords Visualization · Large point cloud · Immersive · Quadtree · Random tree sampling

1 Introduction

Massive 3D point clouds such as those generated by LiDAR are usually difficult to visualize using computers with low configurations. Several open source point cloud processing software such as CloudCompare (http://www.danielgm.net/cc/) and MeshLab (http://meshlab.sourceforge.net/) have improved communication and exchange within the 3D community (laser scanning, photogrammetry, and computer vision). Unfortunately these software packages do not allow the real-time rendering of point clouds and are limited by the size of the point cloud.

This project was initiated from the needs of the French National Railway Company (SNCF) to visualize large data sets of point clouds measured using mobile laser scanners. The objective of this project is therefore to create an open source viewer that loads 3D pre-divided node files. This loading will be done on-the-fly and concerns only the visible points. In order not to overload the RAM, real-time sub-sampling algorithms were implemented which allows the computer to load the point cloud in a much reduced size while still maintaining its topographic information for measurement purposes.

The design of the viewer will largely take inspiration from video games, mainly the first person genre as to give a more immersive environment. The open source nature of the work is essential since today only commercial solutions of the same type are available on the market. An example of such commercial solution is the Australian software EuclideonGeoverse (http://www.euclideon.com/products/geoverse/), which offers similar features.

Immersive visualization is a type of visualization where the camera shows a head-track and stereoscopic view of the surrounding (Kreylos et al. 2008). It has several advantages, such as a direct interaction between the data and the user, increased quality control capability and accuracy due to the ease of use as well as a potential for a better analysis on the 3D data.

The viewer uses the principle of the free-fly camera, commonly used in first person video games. The principle of this camera is to emulate the user's eyes' field of view, hence its name. The user controls the movement of the camera through the keyboard and mouse. The first person perspective is often used in shooter video games, as it gives the player an experience of directly being in the scene. This is also exactly the reason why the research draws inspiration from the first person video game genre. The visualization of the point cloud in this research is meant to be immersive. The first person system is therefore very well suited to give the user a closer and more interactive view of the point cloud as well as to render the viewer more immersive.

As regards to point cloud management, the approach to be taken in this research will be based on a distance-based sub-sampling of the raw data. The idea is to divide the massive point cloud into smaller quadtree nodes (rectangles with a part of the point cloud inside them). The program will then measure the distance between the user (the camera) and the nodes' centers. The closest nodes will be visualized in full resolution, while farther ones will be sub-sampled. This way, the time to visualize the point cloud will be greatly reduced, as the computer will no longer need to load the whole point cloud at the same time.

2 State of the Art

Various algorithms have been used in several published works. The majority of these publications concern the Level of Detail (LoD) data structure (division of the large point cloud into smaller entities) and the memory management problem with respect to the rendering issue. Both Richter and Döllner (2014) and van Oosterom et al. (2015) noted that in today's existing visualization systems, the most referred data structures used are the quadtree and the octree.

Kovac and Zalik (2010) and Zhang and Ma (2012) prefer to use the quadtree division to visualize their terrestrial data sets. Goswami et al. (2013) and de la Calle et al. (2012) used the kd-tree approach, while Kreylos et al. (2008) and more recently Elseberg et al. (2013) proposed the use of the octree subdivision method which is a derivative of the kd-tree.

Another approach proposed by Meng and Zha (2004) and Klein et al. (2004) involves the random tree sampling method (Fig. 1). In this method, the point structure within the point cloud file is randomized, and then divided into files of equal number of points. The appropriate reduced files can be called whenever necessary, for example only one reduced file is called for rendering during camera

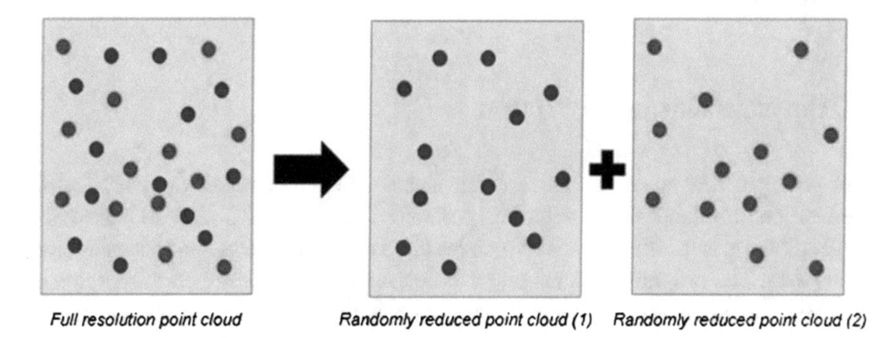

Fig. 1 Illustration of the random sampling subdivision of a full resolution point cloud

movement. A rendering of all the reduced files at the same time gives a full resolution of the point cloud.

Richter and Döllner (2014) concluded that the octree approach is more suitable for 3D point clouds with a varying resolution, distribution and density. Quadtrees, however, are more commonly used for datasets with a 2.5 dimensional characteristic. Therefore in this research the octree approach will not be too suitable since the dataset is mainly in the form of elevation models. The quadtree will be used to further divide the data in the first place into smaller nodes, and then followed by an implementation of the random tree sampling on the quadtree nodes to address the multi LoD facet of the problem.

The principal rendering engine used in off-line visualization of 3D data is OpenGL. Johansson et al. (2015) used this engine to render an interactive BIM (Building Information Model) in real time. The authors also noted the requirement to migrate from older versions of OpenGL to newer ones in order to use the VBO (Vertex Buffer Objects) to render the objects faster by using the graphic card's memory. The older OpenGL enables the loading of data directly from the CPU's memory (RAM). This is a very straight-forward method which is easy to implement, however the loading takes much more time. The VBO approach is therefore more recommended (Wand et al. 2008; Kovac and Zalik 2010; Johansson et al. 2015). Despite this, de la Calle et al. (2012) warn about the limitations of OpenGL. In their tests, it is concluded that depending on the hardware used, OpenGL can handle only up to a few million points at the same time.

Another possibility is to use the WebGL environment which is a JavaScript written engine based on the OpenGL ES 2.0. It also has the capabilities to directly access the graphic card's memory (Callieri et al. 2015). One notable example of the implementation of WebGL is the online visualizer Potree (Liu and Boehm 2015).

The web-based approach is an interesting choice which enables the data to be diffused for a large community. However, in this research the main data to be tested are point clouds measured for the internal purposes of the SNCF. The off-line approach by using OpenGL will therefore be primarily used in order to create an executable application which may be opened by computers without the aid of the internet. This however, does not change the open-source nature of the resulting program and its codes.

3 Development of the Viewer

The principal tool used in this research is the OpenGL engine. OpenGL is the premier environment for developing portable, interactive 2D and 3D graphics applications. It is an environment which is suited for the rendering and visualization of 3D data. The programming language to be used in interaction with this engine is the C++. The choice of programming language is based on its open source nature, but also on its robustness, maneuverability and existing libraries and functions.

Other dependencies used include the SDL and liblas libraries for C++. SDL is a set of C functions which provides simple interaction between the user and the computer via the keyboard and mouse. It is widely used in the development of video games, but SDL is limited to 2D environments. By coupling SDL's capabilities for human-machine interaction and OpenGL's 3D rendering, we can create an interactive and immersive platform. This model has been used in many implementations of OpenGL in video games.

The library liblas is a set of C/C++ functions which translates the common LAS format of laser scanning products into the program. This will then enable us to analyze and perform manipulations on the data such as sub-sampling and quadtree division as well as visualization via OpenGL.

3.1 Rendering Methods

In rendering the scene of a 3D object in OpenGL, two approaches are available. Older versions of OpenGL use a traditional approach of loading the data into the computer's RAM, and then reading the data from this internal memory. It is one of the simplest ways to render in OpenGL. However, this method has a disadvantage in that the RAM can become easily overloaded with large amounts of data. This is because this approach actually takes the longer path to render an object in the computer screen. The data will have to be loaded first into the computer's memory and then passed onto the graphics card to be rendered.

Starting from OpenGL 3.0, a new protocol for scene rendering was implemented. The use of VBO is recommended. The VBO bypasses the CPU memory by loading the object directly into the graphic card's memory (Video RAM or VRAM). The VRAM is admittedly lower than the normal RAM; however it is very efficient since the loading time can be cut by dismissing the loading of files in the RAM first before rendering them in the screen (Fig. 2).

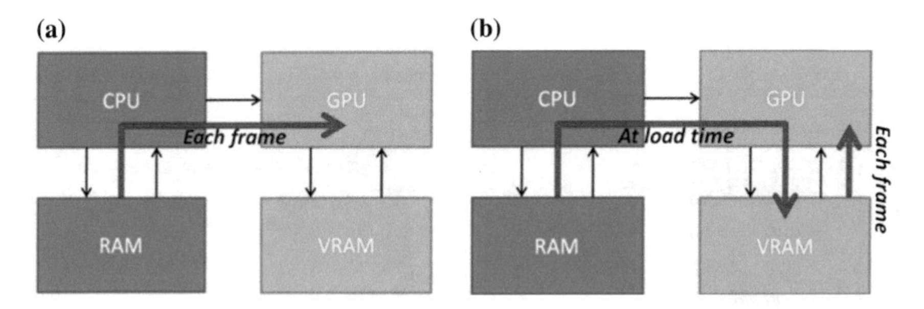

Fig. 2 The two different approaches in rendering a scene in OpenGL (modified from Yaldex 2015): **a** the regular and older RAM-based method loads the data from the RAM and then renders each frame in the same manner, **b** the VBO approach loads the data from the RAM only once and passed it to the VRAM

3.2 Data Pre-processing

The first part of the program is the pre-processing of the raw data. In this part, the point cloud will be subdivided into quadtree nodes. The resulting nodes will then be subsampled randomly to create several point clouds with limited number of points.

The initial input parameters are written by the user in a parametrisable text file. This file includes the path for the results of the pre-processing (quadtree nodes with their randomly sub-sampled files), the limit of points requested for each sub-sampled file, and the path for the raw LAS point cloud (Fig. 3).

This module will retrieve information on the quadtree node (level n) file's limits (X_{min}, X_{max}, Y_{min}, Y_{max}) using the liblas library from the files' header. Using this information, Algorithm (a) calculates new limits for the nodes by simply dividing the intervals of X and Y respectively by two. In this case of quadtree sampling, the Z axis is not taken into account, due to the data's characteristics which lean more towards a 2.5D data (i.e. a plane representation of the terrain with lack of 3D details) rather than a full 3D model. Taking into account the Z axis would have changed the quadtree into an octree. Each new quadtree node generated will be identified by an alphabet identifier.

Afterwards, another function is called to shuffle the order of points inside the quadtree files. From this shuffled file, smaller files are created by simply truncating the quadtree file into the desired number of points. In Algorithm (b), NumFilePoints is the number of points in the node, while NumTreePoints is the threshold number of points for each random-sampling. The subsampled files will be identified by their parent quadtree node and an Arabic number indicating their order of creation.

Metadata are stocked in an indexing ASCII file as to give quick access to these data. This indexing file contains a list of the quadtree nodes and their supporting information such as the 2D center, the number of files for each node as well as the path to the required file. The radius between the user's point of view and the nodes are computed with the data in this index file.

(a)

```
Data: Quadtree node (level n), X_min, X_max, Y_min, Y_max
Result: Quadtree nodes (level n + 1) (A.las, B.las, C.las,D.las)
initialization;
load Quadtree node (level n) (.las);
while ReadNextPoint() do
    if X < (( X_max-X_min)/2)+X_min then
        if Y < (( Y_max-Y_min)/2)+Y_min then
            write point into A.las;
        else
            write point into B.las;
        end
    else if X > (( X_max-X_min)/2)+X_min then
        if Y < (( Y_max-Y_min)/2)+Y_min then
            write point into C.las;
        else
            write point into D.las;
        end
end
```

(b)

```
Data: Quadtree node (A.las), NumFilePoints, NumTreePoints
Result: Random-sampled files (A_R_0.las, A_R_1.las, etc.)
initialization;
load A.las;
random_shuffle;
for int i=0;i<=NumFilePoints/NumTreePoints;i++ do
    int j=0;
    while j<NumTreePoints and ReadNextPoint() do
        write point into A_R_[i].las;
        j++;
    end
end
```

Fig. 3 General algorithms for **a** quadtree subdivision and **b** random tree sampling

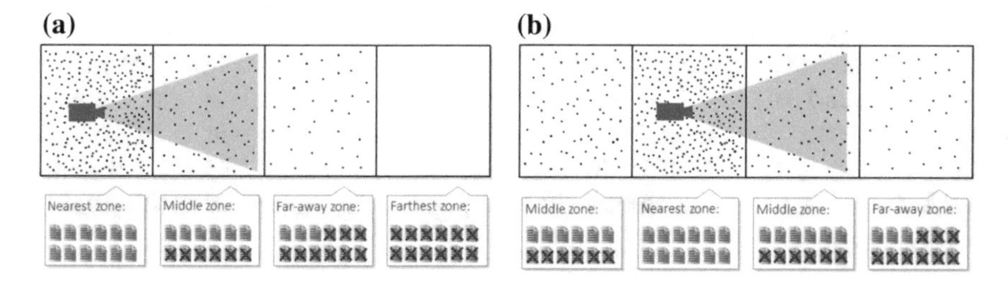

Fig. 4 The point cloud visualization zones based on the number of random-sampled files loaded. In **a** the camera is located at the *left*-most node, and the number of files loaded depends on other nodes' distance to the camera. In **b** as the camera moves to the *right*, the zoning changes and the distances between nodes are recalculated accordingly

3.3 Visualization Strategies

In order to be able to load more points, a point cloud visualization strategy has to be implemented. Loading all points at the same time would not be effective, even by using VRAM. Indeed, it must be noted that even though it is more effective than ordinary RAM, the VRAM usually has lower memory. Such strategy is often referred to as the out-of-core rendering method (Meng and Zha 2004; Kovac and Zalik 2010; Richter and Döllner 2014). A conceptual drawing of the distance-based visualization strategy is shown in Fig. 4.

The algorithm divides the space into four categories. The 2D distance between the camera and the center of each node is computed for all the nodes created by the pre-processing. The first distance category is called the "nearest" zone in which the highest density of points is loaded. This means that all the random-sampled files produced from the pre-processing stage for nodes within the set radius will be loaded. Farther away is called the "middle" zone. Already in this zone a reduced number of files are loaded. The "far-away" zone loads the least amount of files. The last zone is called the "farthest" zone in which no points are loaded. The radius as well as the number of files to be loaded for each zone can be modified by the users according to their needs. This algorithm is computed dynamically for each frame. This means that when the camera moves forward, the distance between the camera and each node is recalculated and the zones are re-attributed accordingly (see Fig. 4).

4 Results and Discussions

The first analyzable results are those from the pre-processing stage, which involves two simultaneous methods. These results are then used as input for the visualizing program with the help of a text file (called the project file), which describes the available point clouds to load. This descriptive file is generated automatically at the end of the pre-processing phase.

Tests were performed on a first data set of a sample of 16 million points (around 500 MB) and another second much larger main data set with over 300 million points (around 10 GB). Both data were obtained from a mobile laser scanning system. The computer used for the tests is an ordinary laptop with 2.7 GHz Inter Core i7 processor, 4 GB of RAM with 1333 MHz frequency, 384 MB integrated VRAM Intel HD Graphics 3000 (650 MHz), and a hard drive with 5400 RPM.

4.1 Pre-processing Results

The pre-processing generates quadtree nodes for all input files and then subsamples these nodes into several more sparse files. This is the longest stage of processing, since the visualization will only use the results of this pre-processing. It should be noted that the pre-processing takes less than a minute for the first data set (about 500 MB) with the computer specifications used, while the second data set (about 10 GB) takes around 40 min to finish.

Figure 5 illustrates the division of a point cloud file into 4 quadtree nodes. The process is straightforward and the results are then used for a further sampling using the random-tree method. Figure 6 shows the results of this sampling process. The quadtree node is divided into several smaller files having a limited number of points (the limit is determined by the user). The idea is to load only a certain number of files depending on the distance between the camera and the node concerned. In Fig. 6, the point cloud in (a) is the result of the loading of a majority of the smaller files at the same time, which is equal to nearly full-resolution. Point cloud (b) loads only half of the amount of files generated by the pre-processing, while (c) loads only one subsampled file. For these tests, each data was divided into quadtree nodes of 50 $\times$ 50 meters and then random sampled into files of 300,000

Fig. 5 Results of the quadtree sampling with each node in different *colors* to showcase the division

Fig. 6 Results of the random tree sampling. The point cloud in **a** with the highest density shall be loaded for the nearest node, while **b** is less dense and will be loaded for the middle-distanced node. The least dense **c** will be loaded for the far-away nodes. Farthest nodes will not show any points

points. This configuration was chosen because it offers a balance between point density and performance for this particular test computer. These values were empirically determined.

4.2 Visualization and Performance Analysis

The program manages to handle both data sets, albeit several issues are still present. Some screenshots of the loaded point cloud inside the program are shown by the Fig. 7. In this figure, two perspectives of the point cloud are shown. The first involves a bird's eye overall view, which is attained simply by moving the camera to the zenith of the point cloud. Note that this is not the default point of view of the viewer. The second one involves the first person view, in which the user can move around using the keyboard as well as rotate the camera using the mouse.

The visualizing module of the program was first used to test the first data set. No rendering problem was encountered in this case, and the algorithm manages to load and unload the defined subsampled files dynamically. A performance graph of this test can be seen in Fig. 8 where the abscissa denotes the observation of scenes or steps rendered and the ordinate the time required for the rendering. Despite some

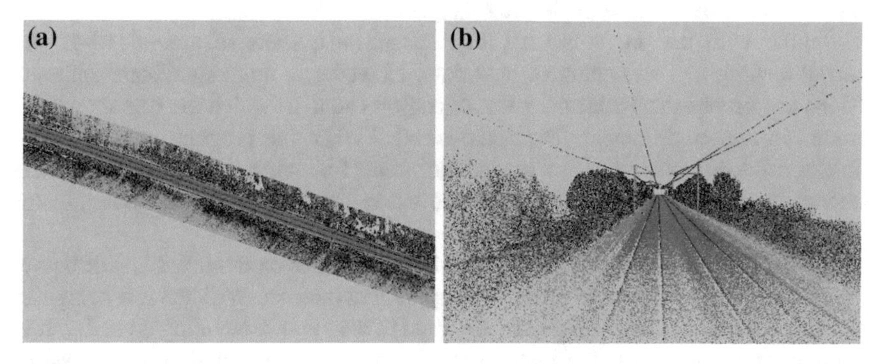

Fig. 7 Visualization results from the second data set in the program. In **a** a small part of the 300 million points is shown from a bird's eye view, while **b** shows a first person perspective of the data

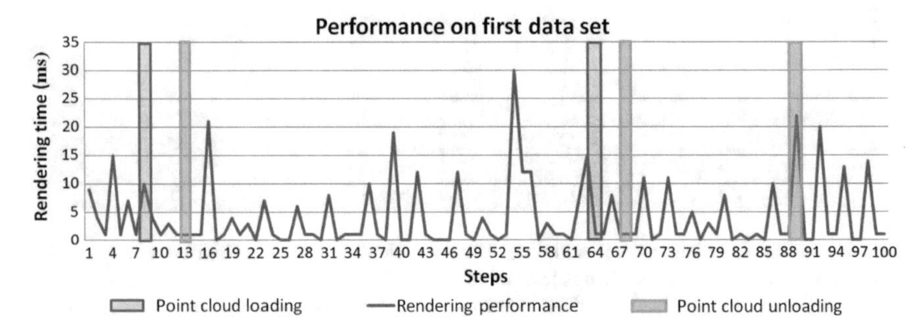

Fig. 8 Performance graph for the first data set with events marked by *colored bars*. Six subsampled files were loaded for the nearest, two for medium, and one for far-away zones

minor spikes in the graph, the rendering time is rather stable. No other systematic spikes can be observed except during loading and unloading of files, in which the rendering time takes in average about 10–20 ms longer than normal rendering. However, one larger spike amounting to 30 ms can be observed around the 54th step. At this state we are unable to detect the cause of this solitary spike, whether it is accidental or systematic.

The second (main) data set, tested with the same configuration as the previous one, shows large anomalies with the rendering time. This significantly slows down the performance of the program. The low specifications of the computer are presumed to be unable to handle such amount of data, considering that the VRAM is very limited and even the RAM is low. However, this condition enables us to perform some performance analysis based on the amount of points loaded. Another test is therefore conducted with a reduced number of points to determine the correlation between the implementation of our algorithms and the large spikes in rendering time.

In this analysis, rather than using the same number of files to be loaded for each distance category as in the previous case, a reduced number is used. Different scenarios with this loading configuration were tested, in which the rendering time is taken into account. In Fig. 9 two scenarios are observed in which the rendering time is analyzed as in the case of the first data set.

Similar with the previous case, Fig. 9a shows a spike of around 10–20 ms during loading and unloading of new data. However, a more significant spike of 40 ms can be observed after the second loading event, in which the camera moves around in this environment. The spike receded when the program unloads some files and reloads other files to render the scene following the movement of the camera. A preliminary hypothesis is made where the loading and unloading is not related to the unsystematic spikes.

In Fig. 9b, the amount of files is greatly increased even though it is still below the number of files loaded in the case with the first data set. With the test computer specifications, some large spikes are observed in the graph amounting up to 5 s. It is quite evident from this graph that the loading/unloading events do not necessarily correspond with the spikes. It should be noted that each loading/unloading event

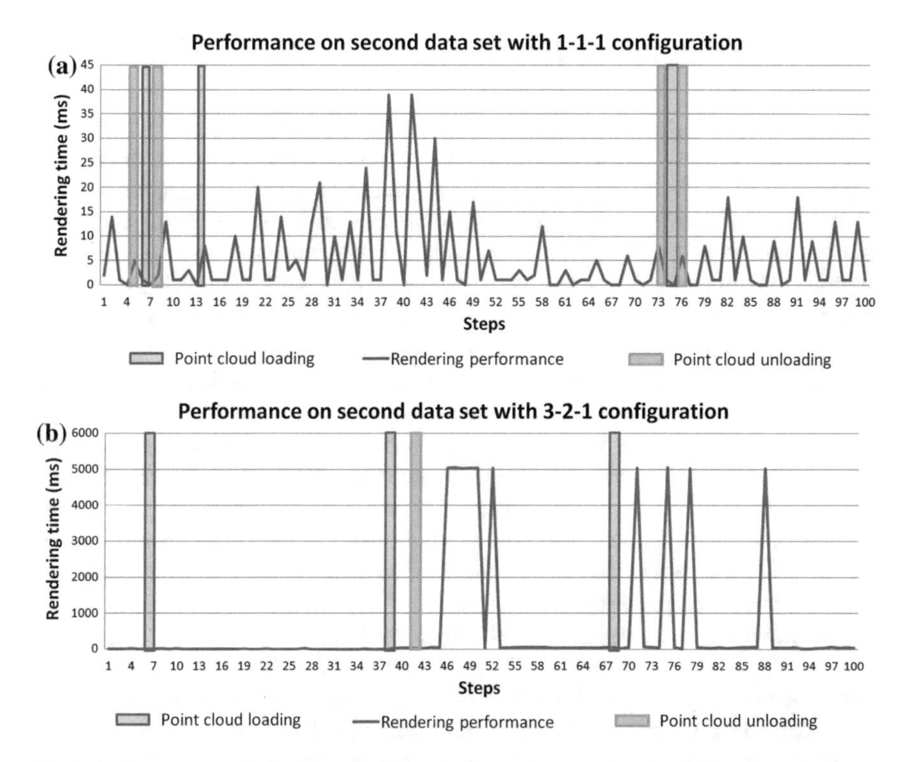

Fig. 9 Performance graph for the second data set. In **a** only one subsampled file is loaded for each zone, whereas in **b** three were loaded for the nearest, two for the middle, and one for the far-away zone

does not necessarily concern the same amount of points. There are no significant spikes observed after the first loading event, and the spikes occurring from steps 70 up to 90 do not reflect a systematic behavior in regards to the loading event which occurred just before them. The time used in the loading and unloading events themselves is still around 10–20 ms, even though the graph's scale cannot show this fact.

These anomalies are therefore not linked to the rendering time during the loading/unloading events themselves, which are more systematic in nature (of around 10–20 ms during each event). It is most probably due to the differing number of points rendered in each scene which depends on the camera's position. In this case the limitation of the test computer's VRAM is very much felt, since only 384 MB are allocated for the graphics card. It should also be noted that the VRAM is integrated with the RAM, which further reduces performance. Further benchmarking tests with other computer specifications should be performed to verify this analysis.

5 Conclusions and Future Works

The method developed to visualize large point clouds and large spatial data in this research has the advantage of being able to adapt the data to the computer's characteristics. With this program, a point cloud can be inspected using an ordinary computer. The density can be lowered if the device is not powerful enough to process massive 3D data. With this adaptive characteristic the viewer could improve 3D point cloud data sharing and collaborative work. The program has also turned out to be able to handle large point clouds, up to 300 million points by implementing the algorithm using an ordinary laptop.

The tests and the performance analysis show that the rendering time is largely independent from the loading and unloading algorithm. This is shown by the systematic rendering time of the loading/unloading events which is almost always around 10–20 ms. However, the anomalies are more likely due to the graphics card which is not designed to process massive 3D data. Other benchmarking tests using stronger computer specifications should be performed to assess this analysis. A beta testing of the program is also planned in order to obtain an objective perspective on the performance of this program.

The program currently only works with LAS format datasets, therefore the compatibility with other standard 3D formats should be added in the future to create a more accessible program for the 3D community. Also, the software was designed to process data which are mostly horizontal with few variations in elevation (2.5D structure). The pre-processing algorithm could be further improved by using octree rather than quadtree sampling to optimize the 3D visualization for point clouds with large variations of elevation.

Even though for now the program simply pre-processes the data and visualizes them, its open source nature makes it a good base to develop professional applications. Since the program keeps the original information from the raw data, tools to extract geometric information from the point cloud can be added eventually (distance measurements, coordinate inquiry, etc.).

Acknowledgments This research has been conducted during a Master research project at INSA Strasbourg, France (Graduate School of Science and Technology) in collaboration with SNCF Réseau (French National Railway Company). The LiDAR data used in the tests were furnished by SNCF Réseau.

References

Callieri M, Dellepiane M, Scopigno R (2015) Remote visualization and navigation of 3D models of archeological sites. ISPRS Int Arch Photogramm Remote Sens Spat Inf Sci XL-5/W4:147–154

de la Calle M, Gómez-Deck D, Koehler O, Pulido F (2012) Point cloud visualization in an open source 3D Globe. ISPRS Int Arch Photogramm Remote Sens Spat Inf Sci XXXVIII(5):135–140

Elseberg J, Borrmann D, Nüchter A (2013) One billion points in the cloud—an octree for efficient processing of 3D laser scans. ISPRS J Photogramm Remote Sens 76:76–88

Goswami P, Erol F, Mukhi R, Pajarola R, Gobbetti E (2013) An effcient multi-resolution framework for high quality interactive rendering of massive point clouds using multi-way kd-trees. User Model User Adapt Interact 29(1):69–83

Johansson M, Roupé M, Bosch-Sijtsema P (2015) Real-time visualization of building information models (BIM). Autom Constr 54:69–82

Klein J, Krokowski J, Wand M (2004) The randomized sample tree: a data structure for interactive walk-throughs in externally stored virtual environments. Presence Teleoper Virtual Environ 13 (6):617–637

Kovac B, Zalik B (2010) Visualization of LiDAR datasets using point-based rendering technique. Comput Geosci 36(11):1443–1450

Kreylos O, Bawden G, Kellogg L (2008) Immersive visualization and analysis of LiDAR data. Adv Vis Comput 846–55

Liu K, Boehm J (2015) Classification of big point cloud data using cloud computing. ISPRS Int Arch Photogramm Remote Sens Spat Inf Sci XL-3/W3:553–557

Meng F, Zha H (2004) An easy viewer for out-of-core visualization of huge point-sampled models. In: Proceedings 2nd international symposium on 3D data processing, visualization and transmission, Thessaloniki, 2004. IEEE

Richter R, Döllner J (2014) Concepts and techniques for integration, analysis and visualization of massive 3D point clouds. Comput Environ Urban Syst 45:114–124

van Oosterom P, Martinez-Rubi O, Ivanova M, Horhammer M, Geringer D, Ravada S, Tijssen T, Kodde M, Gonçalves R (2015) Massive point cloud data management: design, implementation and execution of a point cloud benchmark. Comput Graph 49:92–125

Wand M, Berner A, Bokeloh M, Jenke P, Fleck A, Hoffmann M, Maier B, Staneker D, Schilling A, Seidel HP (2008) Processing and interactive editing of huge point clouds from 3D scanners. Comput Graph 32:204–220

Yaldex (2015) Algorithms in game programming. http://www.yaldex.com/game-programming/0131020099_app02lev1sec10.html. Accessed 17 Dec 2015

Zhang H, Ma J (2012) The organization and visualization of point cloud data based on the octree. In: Zhang Y (ed) Future wireless networks and information systems. Springer, Berlin, pp 425–432

Web-Based GIS Through a Big Data Open Source Computer Architecture for Real Time Monitoring Sensors of a Seaport

Pablo Fernández, Jose M. Santana, Sebastián Ortega, Agustín Trujillo, Jose P. Suárez, Jaisiel A. Santana, Alejandro Sánchez and Conrado Domínguez

Abstract Numerous activities and processes of a wide nature occur in a modern seaport. To name a few, goods, travelers transportation, fishing, rescue and protection agents and increasing demanding human habits around tourism. In the paradigm of "The Internet of Things" we present "Puerto de la Luz" SmartPort a solution for real time monitoring of sensor data in a seaport infrastructure. We describe the computer architecture and the enriched Internet application that allows the user to visualize and manage real-time information produced in the seaport environment. The Big Data management is based on the FIWARE platform. The entire system is implemented in the "Puerto de La Luz" seaport, in Las Palmas de

P. Fernández (✉) · S. Ortega · J.P. Suárez · J.A. Santana · A. Sánchez
Division of Mathematics, Graphics and Computation (MAGiC), IUMA,
Information and Communication Systems, University of Las Palmas de Gran Canaria,
Campus de Tafira, 35017 Las Palmas de Gran Canaria, Spain
e-mail: pablo.fernandez@ulpgc.es

S. Ortega
e-mail: mxmeater@gmail.com

J.P. Suárez
e-mail: josepablo.suarez@ulpgc.es

J.A. Santana
e-mail: jaisiel@gmail.com

A. Sánchez
e-mail: alemagox@gmail.com

J.M. Santana · A. Trujillo
Imaging Technology Center (CTIM), University of Las Palmas de Gran Canaria,
Campus de Tafira, 35017 Las Palmas de Gran Canaria, Spain
e-mail: josemiguel.santana@ulpgc.es

A. Trujillo
e-mail: agustin.trujillo@ulpgc.es

C. Domínguez
University of Las Palmas de Gran Canaria, Juan de Quesada,
35001 Las Palmas de Gran Canaria, Spain
e-mail: conradodt@ulpgc.es

I. Ivan et al. (eds.), *The Rise of Big Spatial Data*, Lecture Notes
in Geoinformation and Cartography, DOI 10.1007/978-3-319-45123-7_4

Gran Canaria, Canary Islands, Spain. We remark the architecture for processing and visualizing the streaming data coming from two sensors physically deployed on the neighborhood of the seaport and underline new features to truly manage Big Data.

Keywords Open source · GIS · Big data · Seaport · Sensors · Web

1 Introduction

The fast growth of data is due to improvement in digital sensors, Internet facilities, software and hardware that have evolved in the last 20 years. This advances allow to collect and treat huge volume of data in a very efficient and fast way, using new techniques and algorithms. According to Halevi and Moed (2012), the term Big Data, first used by Roger Magoulas, includes all these concepts and processes. In Emani et al. (2015) many Big Data technologies and definitions are described. Sensors produce huge volumes of data continuously over time, and this leads to new computational challenges to overcome, challenges of sensor data analytics and the different areas of research in this context.

SmartPort is a visualization and data management system, supported by an Internet web, that receives and shows the data of "Puerto de la Luz" seaport. In Suárez et al. (2015a, b), a first introduction to the entire system can be found. "Puerto de la Luz", is one of the main seaports of Spain and the first of the geographical area of West Africa. With more than 16 km. of docks, this seaport serves as the crossroads between Europe, Africa and America.

Technologies and computer architectures, both software and hardware, are now available allowing to mount a smart system in a seaport where officers and users can exploit a huge amount of data. Recently, the Future Internet platform of the European Community (FIWARE) architecture (Ramparany et al. 2014) has been released as a platform that allows the validation of new concepts, technologies, business models, applications and services in big scale. The University of Las Palmas de Gran Canaria, as a partner of the FIWARE project, has the goal of developing innovative projects using this new technology. This platform belongs to the Program Future Internet Public Private Partnership (FI-PPP), which is a program of public-private cooperation in the field of FI technologies, funded by the European Commission involving more than 152 European companies and organizations (Villaseñor and Estrada 2014).

In a general way, there are numerous sources of available data within a seaport area. These sources can be grouped into two sets, depending on their nature. The first set is composed of human made resources and seaport activities, including all the human elements within the seaport. These elements are important for the seaport authority who has to manage them. They are also important for the seaport users, who could make use of them. A second set comprises the natural environment and includes both static and dynamic parameters of the natural surroundings. See Suárez et al. (2015a, b).

1.1 Computer and Sensor Architecture in "Puerto de la Luz" Smartport

The sources where the acquisition of data is mainly focused are sensors connected to the Internet. Currently the "Puerto de la Luz" seaport has deployed the Aandela Instruments series 3791–3798 of water level sensor and Geonica Datamar-2000C tide gauge. The seaport also has meteorological sensors such as Geonica PTH-4000 for air temperature and relative humidity, Geonica 05106 for wind speed and directions, and Geonica 52203 for tipping bucket rain gauge.

We conducted the project through the following three requirements:

- Collecting the generated data source sensors.
- Analyzing and processing the measurements.
- Visualizing all these data in a 3D geospatial environment.

Figure 1 illustrates the system architecture of the "Puerto de la Luz" SmartPort solution which integrates both backend and frontend components as server-side applications and databases, FIWARE module, virtual globe and web visualization.

1.2 Managing Context Data with the Orion FIWARE Component

Every time the sensors read newer values, they are sent to the Orion Context Broker. Orion Broker uses a Publish/Subscribe requests, providing the NGSI9 and NGSI10 interfaces. Using these interfaces, clients can perform several operations:

- Registering context producer applications, e.g. a temperature sensor within a room.
- Updating context information, e.g. send temperature updates.

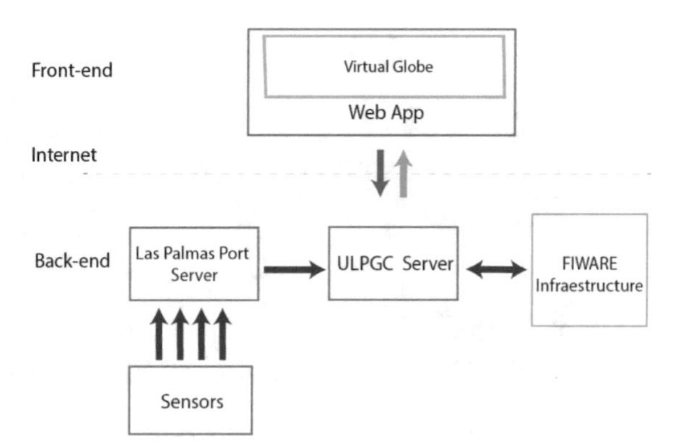

Fig. 1 "Puerto de la Luz" SmartPort computer and sensor architecture

- Being notified when changes on context information take place (e.g. the temperature has changed) or with a given frequency (e.g. get the temperature each minute).
- Querying context information. The Orion Context Broker stores context information updated from applications, so queries are resolved based on that information.

2 Managing Big Data with the Cosmos Fiware Component

Cosmos is an implementation of the Big Data GE, allowing the deployment of private computing clusters based on Hadoop ecosystem. Current version of Cosmos is capable of:

- I/O operations regarding Infinity, a persistent storage cluster based on Hadoop Distributed File System (HDFS).
- Creation, usage and deletion of private computing clusters based on MapReduce and SQL-like querying systems such as Hive or Pig.
- Manage the platform, in many aspects such as services, users, clusters, etc. from the Cosmos API or the Cosmos CLI.

Cygnus, as part of Cosmos, is a component devoted to receiving context data from Orion Context Broker GE and storing it in a HDFS, see Fig. 2.

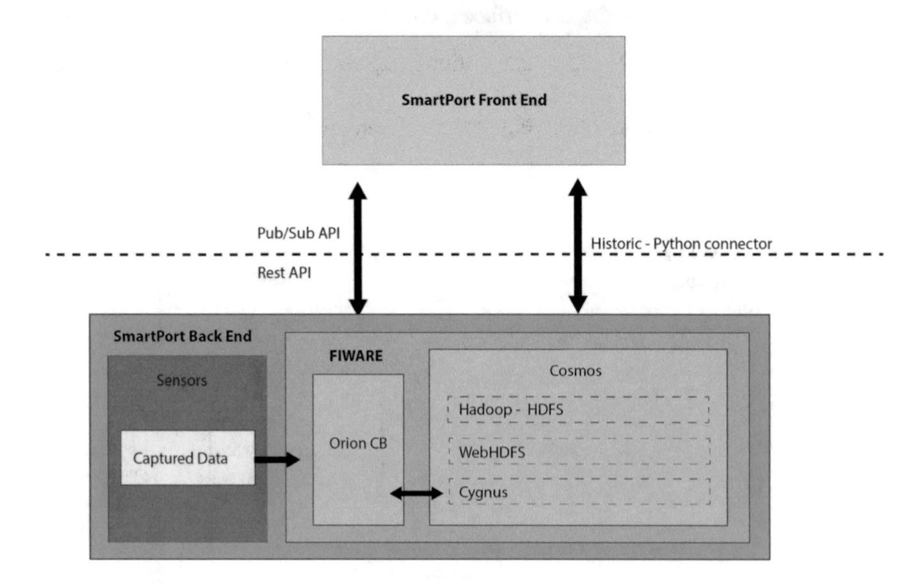

Fig. 2 Cosmos big data connection in SmartPort

We upload all the data in chronological order to Orion and relying on Cygnus for sending the data to Cosmos. The solution that is implemented consists in adding data to Cosmos using the WebHDFS API. To preserve system security in the front-end application, the Cosmos instance is only accessible from machines within the same network. Then, making requests directly from the client was dismissed and thus, another abstraction layer was necessary.

The technologies and techniques used to process and store the data allow to achieve high data availability and ensure that the whole system is scalable. However, some technologies like Hive or Hadoop imply a performance trade off, as showed in Suárez et al. (2015a, b). The proposed architecture minimizes the impact of these modules by integrating intermediate layers. This way the data transactions achieve the speed that the interactive user interface demands.

3 Real-Time Ships Location Tracking Through Automatic Identification System (AIS)

SmartPort offers the possibility to visualize ships in the surroundings of the seaport. The source information that allows displaying the location of ships is provided by the Automatic Identification System (AIS) data streaming which is recollected and provided by the seaport authority. In Fig. 3 two web interfaces of SmartPort are shown: (a) perspective view and (b) top view of Las Palmas Puerto de la Luz seaport.

The AIS system protocol defines a set of messages that are sent by the ships using a VHF transceiver and later displayed by information systems located in the own ships or shore stations.

AIS technology has proven to be helpful for different scenarios:

- Automated identification and discovery of neighboring ships position
- Complementing radar sensors in avoiding collisions between ships.
- Monitoring fleets, for example, by seaport authorities.
- Ships traffic services.
- Maritime rescue and security.

In the case of SmartPort, the seaport authority has given access to a web data stream of AIS messages. The service feeds the stream with AIVDM/AIVDO text messages known as sentences. The following message is an example:

!AIVDM,1,1,B,177KQJ5000G?tÒK>RA1wUbN0TKH,0*5C

In SmartPort the AIS stream is received by a custom Python script which opens the HTTP stream and starts decoding the sentences as they arrive. The decoding of each sentence is done by a Python script provided by the GPSD developers tools. Once the sentence is decoded, the main script saves it in a parsed stream in order to have it accessible for possible comparisons with the backed up raw stream. After

Fig. 3 Two web interfaces of SmartPort: **a** perspective view and **b** *top view* of Las Palmas Puerto de la Luz seaport

the packet is backed up, the script sends a push request to the Orion GE instance in order to publish it as context data, see for example:

```
with open(backup_file, "w") as ais_parsed:
  for (raw, parsed, bogon) in parse_ais_messages(f, scaled, skiperr, verbose):
   if not bogon:
     date_str = datetime.datetime.now().isoformat()
     json_dump = ("{" + ",".join(map(lambda x: '"' + x[0].name + '":"' + str(x[1]) +'"',
                   parsed)) + ', "read_date":\"'+ date_str +'"}').encode('string-escape')
     ais_parsed.write( json_dump + "\n" )
     message  = json.loads(json_dump)
     mtype = int(message["msgtype"])
     # Take out mmsi from dictionary as it will be the id of the entity
     id = message["mmsi"]
     try:
       del message['mmsi']
     except KeyError:
                    pass
     # Get the proper params for the message type
     params = getMessageParams(message)
     if( params["type"] == 'NOT_DEFINED' ):
       # We are not interested in base station messages
       if ( (mtype == 4) | (mtype == 11) ): pass
       # We are not interested in AIS network messages
       elif ( mtype == 20 ): pass
     else:
       if ( mtype == 8  ):
         message = decodeType8Data(message)
     sendToOrion(url, id, params["type"], params["translation"],
               params["units"], message,  False)
```

At this point, messages are accessible as context data in the Orion GE instance, allowing the front end to easily query the data of interest. Full architecture of SmartPort’s AIS processing is shown at Fig. 4.

The type of messages are dependent of the data being transmitted and the category of the ship that is transmitting, having large ships a distinct set of messages than the smaller ones.

From the AIS stream, we capture, among other variables, the following data:

- International Maritime Organization (IMO) ID
- Origin
- Destination
- Ship type
- Ship dimensions
- Last known location
- Estimated time of arrival

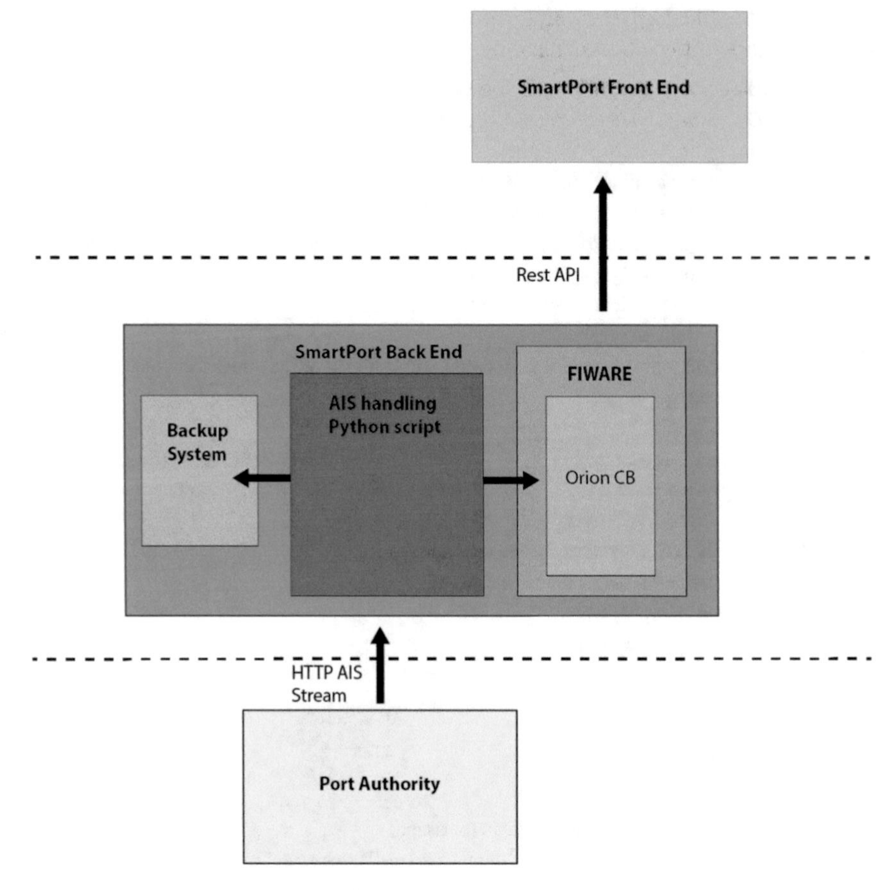

Fig. 4 AIS data stream handling architecture

- Data Terminal Equipment (DTE)
- Positioning information
- Message type
- Positioning information
- Last transmission

Given this information, the SmartPort frontend renders different 3D models based on the ship type and scales it based on its dimensions (Fig. 5). From the positioning, we can also detect the ship orientation and replicate it in the model representation.

Fig. 5 The draft of a 3D model of cargo ship

4 Interaction Between the Back-End and the Front-End of Smartport

The front-end of Smart Port is developed using HTML5 based technologies. HTML5 brings a richer experience to the user due to all its new possibilities and one of the most remarkable is the canvas element, which allows to draw graphics on a web page.

One of the most important interactions of the front-end application are the requests to the back-end of SmartPort. In order to create an interactive solution, the communication between the client and the server has been developed using AJAX (Asynchronous JavaScript and XML) (see sample code below), allowing us to exchange data without reloading the page. This substantially increases the speed and usability of the application.

```
var jqxhr =$.ajax({ url:proxy+url,
        dataType: 'json',
          accepts: {
             json: 'application/json'},
          headers: {
            "Content-type": 'application/json',
            "Accept": 'application/json'
          }})
          .done(function(data) {
            response=data;
            if("errorCode" in data)
            {alert("Cannot get ship data");
             setTimeout(function(){initBoatPanelData()},200000);
             return; }
            processBoatData(response);
          })
          .fail(function(data) {
            console.log("failure: "+JSON.stringify(data));
          });
```

Since some application components are located inside different iframes, they need to exchange information between them, we have developed a small JavaScript API called iframeCommunicationAPI that simplifies this process.

In order to use this API, each element that could receive messages must register an input with an action that will be performed when a message arrives. Once an input is registered, an element can send a message to it specifying the target iframe name, the target input name and the data to be transmitted.

4.1 GIS Based Visualization and Sensor Data Output via Web

One of the main requirements of Smart Port is the visualization of seaport-related geographic information. To accomplish some of those requirements we use the Glob3 Mobile library as a JavaScript component which is compatible with browsers that support HTML5 (including mobile devices). Glob3 Mobile provides an easy-to-use API that allows to interact with the virtual 3D globe, see Trujillo et al. (2013), Suárez et al. (2015a, b). For some components such as the alert manager, which do not need the 3D capabilities of the Glob3 Mobile library, we use Leaflet 2D mapping library due to its performance and simplicity (Fig. 6).

Another important feature of Smart Port is the graphical visualization of the historical data from sensors. For that purpose we used the Highstocks library, which allows to render interactive timeline charts easily.

Some interesting features provided by the library are:

- selecting time ranges
- selecting different zoom levels
- comparing different sensors
- exporting to different file formats

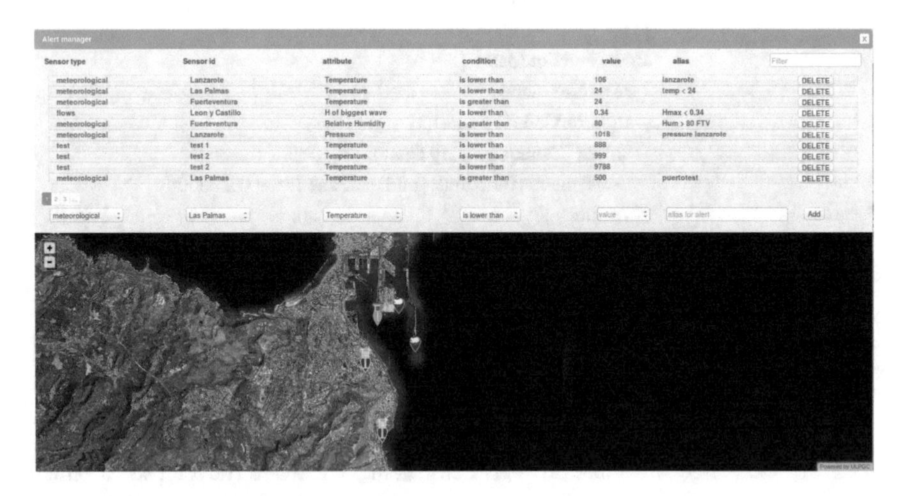

Fig. 6 Leaflet map in Alert manager

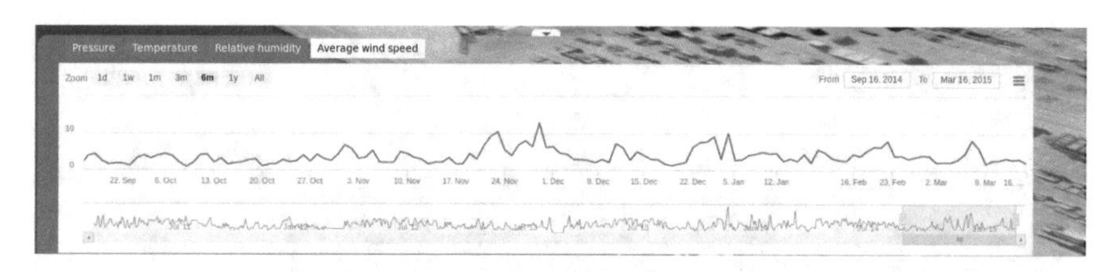

Fig. 7 Sample of historical wind speed data from a sensor

- calculating average values (Fig. 7)

We noticed that some modules of the application, like the alert notification, could have a large list of elements; therefore, in order to facilitate the interaction with them, we use the List.js library. List.js allows to add dynamic behavior to HTML lists. Besides, another main functionalities are searching and sorting items; applying filters and pagination.

5 Smartport in Action

Although SmartPort is still under development, a release candidate version is currently being used by the Puerto de la Luz Port Authority as a Decision Support System (DSS). All the information from AIS and sensors, as well as external services like WMS, is gathered in a unified service, a virtue highlighted by the ap-plication users. In the past, Smartport components either were not properly supported or had support in separated applications.

A new functionality is the representation of the vessels in the 3D map and the georeferentiation of their data. This provides an interactive and intuitive overview of the port status to technicians in their day-to-day work, showing the number of ingoing and outgoing vessels, as well as providing real-time information about the environmental conditions around them.

It is worth noting the utility of visualizing the sensors historic data. This is achieved via graphs and exporta-ble files, which afterwards are used by the technicians to make queries about the state of the port sensors based on temporal data series.

At the time of taking decisions, thanks to the alert system, technicians are capable of deciding on the need-ed actions to take upon. For example, in the case of a current direction change, or when an alert for very high waves persists on time, the port can easily identify the trend and notify to the vessels the need to change their location.

In Fig. 8, we can see the complete interface of the application that is accessible to the port technicians.

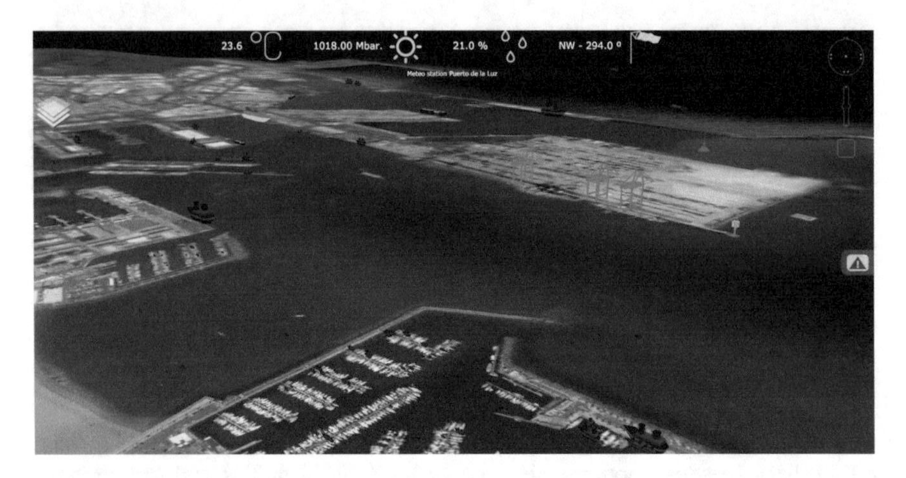

Fig. 8 Full display of SmartPort application

6 Conclusions

SmartPort is a system able to handle, manage and visualize geographic data. Thanks to the functionality of generating graphs, we have the capability to obtain a history of data from sensors of the Puerto de la Luz de las Palmas GC. This is extremely useful, since the quick access to all the information generated by sensors allows us to analyze, look for patterns, and in a close the future, foresee possible situations that require some action by the port. We would like to highlight the alert manager; a system that allows to monitor and display sensor information, and take live decisions when a sensor takes a certain value. We have done specific systems for static or dynamic data in the latter case with Big Data solutions such as Hadoop, integrated into the GE Cosmos and implementing a specific high efficient architecture. One of the next steps in the application is to improve the analytical features. We consider it essential to give intelligence to the implementation and to continue the steps towards a geographic decision-making application.

Acknowledgments This work has been supported by the Spanish Ministry of Economy and Competitiveness (MINECO) Project RTC-2014-2258-8, by the European Commission FP7 Project "FIWARE: Future Internet Core Platform" FP7-2011-ICT-FI 285248 and by FP7 Grant Agreement No. 632893 Project "Future Internet, Core. FI-Core". The second author wants to thank Agencia Canaria de Investigación, Innovación y Sociedad de la Información, for the grant "Formación del Personal Investigador-2012 of Gobierno de Canarias".

References

Emani CK, Cullot N, Nicolle C (2015) Understandable big data: a survey. Comput Sci Rev 17:70–81
Halevi G, Moed H (2012) The evolution of big data as a research and scientific topic: overview of the literature. Res Trends 30:3–6
Ramparany F, Galán Márquez F, Soriano J, Elsaleh T (2014) Handling smart environment devices, data and services at the semantic level with the FI-WARE core platform. In 2014 IEEE International Conference on Big Data (Big Data). Washington, DC, USA, pp 14–20
Suárez JP, Trujillo A, Domínguez C, Santana JM, Fernández P (2015a) Managing and 3d visualization of real-time big geo-referenced data from Las Palmas port through a flexible open source computer architecture. In: GISTAM 2015, 1st international conference on geographical information systems theory, applications and management, Barcelona Spain, April 28–30. Scitepress, Lisbon Portugal, pp 72–82
Suárez JP, Trujillo A, Santana JM, de la Calle M, Gómez-Deck D (2015b) An efficient terrain level of detail implementation for mobile devices and performance study. Comput Environ Urban Syst 52:21–33
Trujillo A, Suárez JP, de la Calle M, Gómez-Deck D, Pedriza A, Santana JM (2013) Glob3 mobile: an open source framework for designing virtual globes on ios and android mobile devices. In: Progress and new trends in 3D geoinformation sciences. Springer, Berlin, pp 211–229
Villaseñor E, Estrada H (2014) Informetric mapping of "big data" in FI-WARE. In: Proceedings of the 15th annual international conference on digital government research, Aguascalientes, Mexico, June 18–21. ACM, New York, USA, pp 348–349

Deriving Traffic-Related CO_2 Emission Factors with High Spatiotemporal Resolution from Extended Floating Car Data

Gernot Pucher

Abstract Despite the deployment of more efficient vehicle technologies, global CO_2 emissions related to transportation have increased by 250 % between 1970 and 2010 due to a rise of vehicle ownership, traffic volumes and congestion. CO_2 is the most common of the anthropogenic greenhouse gas emissions and is a main contributor to climate change. Fine-scaled information on the spatial and temporal distribution of traffic-related CO_2 emissions can support decision making processes with regard to emission mitigation measures. For the purpose of providing such information, commonly traffic emission models are applied. However, such models are often restricted in their spatiotemporal resolution due to a lack of adequate input data. A potential data source could be provided by the extended floating car data (xFCD) approach, where vehicle operation parameters like fuel consumption are read out on-trip via the vehicle's onboard diagnostic system and get correlated with vehicle positions and timestamps at short recording intervals. In this work, the potential of fuel consumption recordings from xFCD for quantifying traffic-related CO_2 emissions is evaluated. For this, an extensive database of GNSS-trajectories from vehicles (FCD) and xFCD fuel consumption measurements were recorded in the city of Salzburg, Austria. Using this input data, a set of averaged driving patterns for road segments, 15-min intervals and weekdays was derived. A similarity measurement algorithm was performed on these patterns, so that the most representative vehicle speed profile with fuel consumption recordings could be identified. The results indicate that the elaborated methodology can be applied for calculating representative, plausible and consistent CO_2 emission factors from xFCD fuel consumption recordings with high spatial and temporal resolution. This shows the potential of the systematic usage of xFCD for the purpose of estimating traffic related CO_2 emissions.

Keywords Traffic emission models · CO_2 emissions · Extended floating car data · Vehicle probe data

G. Pucher (✉)
TraffiCon – Traffic Consultants GmbH, Strubergasse 26, 5020 Salzburg, Austria
e-mail: pucher@trafficon.eu

I. Ivan et al. (eds.), *The Rise of Big Spatial Data*, Lecture Notes in Geoinformation and Cartography, DOI 10.1007/978-3-319-45123-7_5

1 Introduction and Objectives

In the year 2011, 600 million motorized passenger cars were traveling on the roads worldwide and a further significant rise of vehicle ownership is expected in the future, particularly in emerging countries such as China or India (Annema et al. 2011). Although this motorization has led to a significant rise of individual mobility over the century (Dargay et al. 2007), it negatively effects the natural and built environment as well as human health. The transportation sector contributes significantly to the anthropogenic emission of carbon monoxide (CO), nitrogen oxides (NO_x), volatile organic compounds (VOC) and carbon dioxide (CO_2). In most motorized vehicles, fossil fuels are used as energy for propulsion due to their high energy content. Emissions are formed as unwanted byproducts of the imperfect combustion of hydrocarbons of fossil fuels (Van Wee et al. 2013).

While CO_2 is not directly harmful to human health, anthropogenic CO_2 emissions are considered to be the major contributor to climate change (IPCC 2013). CO_2 is a common greenhouse gas, which absorbs thermal radiation from the earth's surface and thus causes the heating of the atmosphere. Although advancements in vehicle technology have led to the propagation of more fuel efficient and low-emission vehicles, transportation is the major sector with the strongest continuing growth in CO_2 emissions. The correlation of improvements in energy efficiency with increases of energy consumption is commonly described as 'rebound-effect', as the enhanced efficiency sets incentives for an intensified energy consumption (Brookes 1990). In the case of transportation, this results in a rise in vehicle ownership and a tendency to buy larger cars. Moreover, also continuing traffic growth and resulting traffic congestions contribute to an intensified emission of CO_2, as vehicles need more energy for propulsion during stop-and-go driving patterns (Capiello 2002). As a result, global CO_2 emissions caused by transport have increased by 44 % between 1990 and 2008, with motorized road traffic being responsible for 74 % CO_2 emissions from transport. In 2010, the transportation sector released 14 % of anthropogenic GHG emissions (Kok et al. 2011).

The availability of fine-scaled information on emissions related to traffic could support policy makers to set up effective emission mitigation measures. Knowing the spatial distribution and temporal shift of emissions within a road network, it is possible to efficiently target problematic areas and thus maximize impact and minimize costs of operational efforts (Gurney et al. 2012). However, most current emission models are applicable rather on a national or regional scale than on the microscopic level of single roads. The exact localization of traffic emissions is mainly limited due to the restricted availability of appropriate data at the local scale (Gately et al. 2013; Smit et al. 2008).

With the proliferation of communication and information technologies and their use in road transportation, new services came in use, which are commonly summarized under the term 'Intelligent Transport Systems' (ITS) (Ezell 2010).

Widespread examples of ITS applications are onboard navigation systems, which determine a vehicle's position in a road network based on GNSS coordinates. GNSS (Global Navigation Satellite Systems) has become the primary technology for the determination of a vehicle's position. Satellites are sending signals to the earth with information on the exact time the message was sent (every satellite is equipped with at least 4 atomic clocks, the time is annotated in the Universal Time Coordinated, UTC) and the satellite position. Through receiver devices, the geographic longitude and latitude as well as the altitude on a specific point on earth can be determined with an accuracy between around 12 m and 1 mm (Zogg 2011). GNSS devices are also used in the 'Floating Car Data' (FCD) approach, where vehicles which participate in regular road traffic, serve as mobile sensors for the collection of on-trip movement data. In this way, it is possible to determine a vehicle's speed and direction on the travelled road segment during a specific time. This information is for example used for dynamic traffic management or for the traffic-sensitive routing of navigation systems (Messelodi et al. 2009).

The concept of FCD, where usually only timestamps, coordinates and GNSS-specific quality parameters are transmitted in a specific recording interval, is further enhanced by the approach of 'extended Floating Car Data' (xFCD). Data collected on-trip are extended by additional parameters, including for instance fuel consumption or engine revolutions per minute. These parameters are sourced directly from the vehicle electronics through reading out data from a vehicle's onboard diagnostic system (OBD) using standardized message protocols like CAN-bus (Breitenberger et al. 2004). From fuel consumption, which is calculated from the mass air flow (MAF) of the engine, on-trip CO_2 emissions can be determined and related to the respective coordinates and timestamps of the recording interval. This enables a microscopic localization of vehicle CO_2 emissions both in space and time. The application of xFCD in traffic emission modeling could therefore potentially contribute to overcome current limitations due to restricted availability and resolution of input data. Thus, the main objective of this work is to develop, implement and evaluate a methodology to derive plausible and consistent emission factors from onboard fuel consumption recordings of passenger vehicles (xFCD) for the quantification of traffic induced CO_2 emissions on the level of road segments and short time intervals.

2 Current State of Research

Li et al. (2010) developed a road segment centered vehicle emission model for the estimation of greenhouse gases on highways in Beijing using FCD from more than 20,000 taxis as well as stationary detector data as primary inputs. Emission factors were calculated based on vehicle specific power (VSP) and engine stress (ES) of the vehicles. Through the use of FCD, spatiotemporal patterns of daily greenhouse gases could be estimated in the road network of Beijing.

Bert et al. (2007) conducted a study comparing results from CO_2 emission calculations based on FCD and traffic sensor data. The study area was the road network of the city center of Lausanne, Switzerland. CO_2 emission factors were calculated based on driving state (accelerating, decelerating, idling and cruising) and vehicle speed. The results showed a tendency that FCD-based emissions were more similar to simulation results the higher the penetration rate of probe vehicle was.

In a study by Yu et al. (2012), taxi emissions in the urban area of Shenzhen, China, were observed based on FCD. Detailed taxi operation information is derived from taxis equipped with GNSS receivers. The emission model comprehensive modal emission model (CMEM) was used to calculate emissions from the taxi FCD at each implemented speed and acceleration category. Through FCD, it was also possible to relate amount of emissions to specific driving pattern (idling, cruising) and to a specific time of day.

Several further studies focus on evaluating eco-driving and eco-routing recommendations through onboard fuel consumption data (Jakobsen et al. 2013; Marquette et al. 2012; Litzenberger et al. 2014). Crowd sourcing approaches towards collecting fuel consumption and corresponding emission data were pursued in the two projects 'Fueoogle' and 'EnviroCar'. Vehicles participating in the data collection process are equipped with Bluetooth adapters, which are connected to the OBD-2 interface, reading out vehicle sensor data. These data then get transmitted to a smartphone and to a central processing server. Further analyses are made using these data, including a fuel-efficient routing application (Pham et al. 2009).

Based on literature review, it can be concluded that data from mobile sources have already been used in several studies to support the estimation of traffic emissions or evaluate driving patterns with regard to emissions. However, a systematic approach to utilize xFCD-based fuel consumption data to derive valid emission factors for quantifying traffic emissions on the level of road segments and in their temporal variation has not been undertaken so far (Krampe et al. 2013).

3 Data Sources

The collection of on-trip fuel consumption data from vehicle sensors (xFCD) was conducted with a gasoline-driven BMW MINI Cooper R56. For reading out data from vehicle sensors, the device tinxi® Bluetooth EOBD OBDII was used. It is a low-cost vehicle diagnostics system, which is attached to the OBD-2 interface of a vehicle. It reads out 15 different kinds of sensor data, including engine speed (rpm), vehicle speed (km/h) and fuel consumption (l/100 km). Via Bluetooth, the device was connected to a smartphone or a computer and reads out fuel consumption, timestamps and coordinates on-trip at a transmission rate of 1 s. The data recording process was conducted between November 2013 and October 2014 in the city of

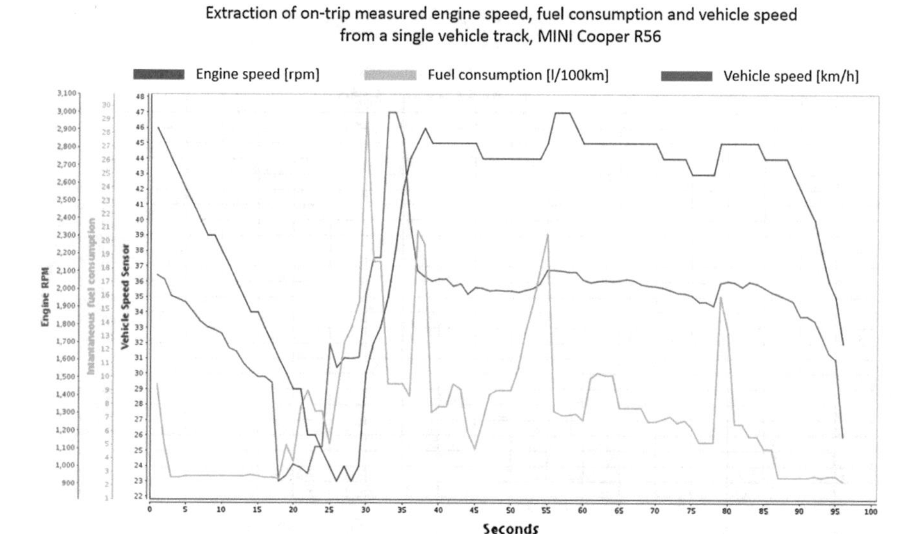

Fig. 1 Extraction of on-trip measured vehicle sensor parameters, MINI Cooper R56, gasoline fuelled

Salzburg and surroundings, resulting in 128 single vehicle trajectories recorded at 1 Hz, with overall 513 km of road covered. Figure 1 shows an exemplary extraction from a vehicle trajectory with recorded vehicle sensor parameters. In this case, vehicle speed (km/h), engine speed (rpm) and fuel consumption (l/100 km) are depicted. It can be observed that a plausible, correlated behaviour exists between these parameters: During a phase of deceleration, fuel consumption and engine speed are low or decline. While vehicle speed rises, also the other parameters incline. Also the effect of gear shifting is recognizable, with sharply falling engine speed and fuel consumption despite increasing speed.

The following Fig. 2 shows a plot of vehicle speed in km/h and engine speed in rpm. The colour gradient of the data points is based on fuel consumption in l/100 km, with blue colour for 0 or no fuel consumption to red colour for the highest measured values. It can be seen that data points with high fuel consumption have a tendency to cumulate especially during high engine speeds, as it would be expected. Also, the influence of gear shift behaviour and vehicle speed on engine speed and thus on fuel consumption can be observed.

For the further quantification of CO_2 emissions, also cross-section traffic volume counts from stationary road-side detectors, data on registered vehicles in Salzburg from 'Statistics Austria' representing traffic composition as well as GNSS trajectories from vehicles participating in regular traffic were used.

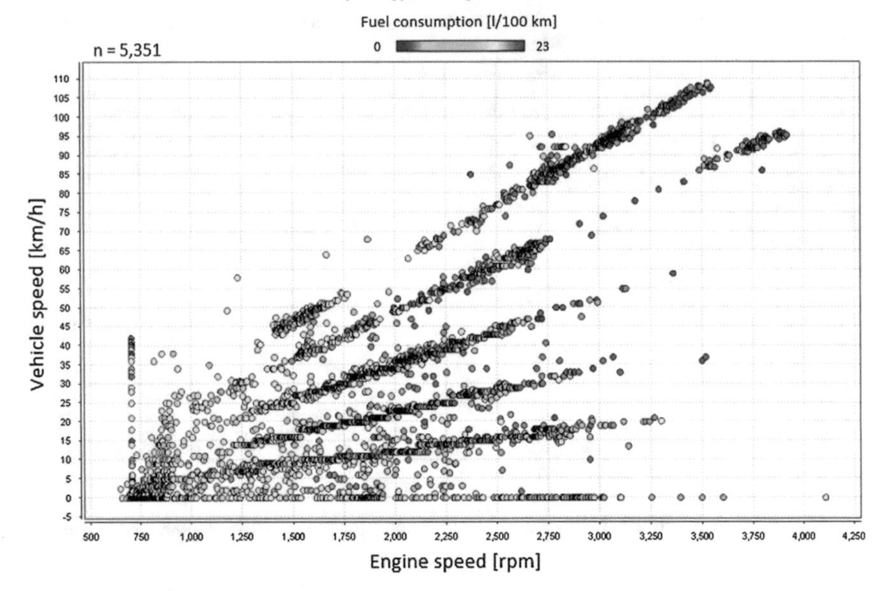

Fig. 2 Plot of vehicle speeds (km/h), engine speeds (rpm) and fuel consumption (l/100 km) of an extract of the recorded set of xFCD

4 Methodology

Various factors have impact on the amount of fuel combusted on-trip by a vehicle. Driving kinematics and the resulting driving patterns are among the most significant factors (Alessandrini et al. 2012). Especially patterns with repeating phases of acceleration and deceleration during low engine load cause higher fuel consumptions. Such patterns are for instance typical for urban driving or during limited traffic flow quality. Various studies elaborating real world or laboratory fuel consumption tests further show the influence of driving patterns on the amount of combusted fuel. Fuel consumption values tend to be distinctively higher during stop-and-go traffic, and the acceleration pattern has a much higher influence than average speed values. Fuel consumption can increase by about 80 % during stop-and-go traffic compared to free flow traffic conditions. This is due to the dependence of fuel consumption on the effective motor pressure and the revolution rate of a vehicle's crankshaft (Treiber et al. 2007).

Because of the obvious correlation between driving patterns, it is assumed that through the definition of a representative driving pattern with fuel measurements for a road segment and a time interval, also a representative emission factor can be obtained. In this way, the dynamics of driving are incorporated as main determinant for calculated emission factors. Here, the term 'driving pattern' describes the sequence of acceleration, deceleration as well as the resulting course of vehicle velocity over time. A driving pattern on a road segment can also be regarded as time series. A time series $T = t1, \ldots,$ is defined as an ordered set of p real-valued variables.

For determining the similarity of time series, usually statistical distance measures are applied. Given two time series $T1$ and $T2$, the distance $D(T1, T2)$ between them is calculated as a similarity function. Various distance measures haven been proposed in literature, with the Euclidean distance being the most frequently used one. The Euclidean distance aligns time series in point-to-point manner. In this way, point i in time series X is compared with point in time series Y (Lin et al. 2009). The Euclidean distance is easy to compute, is parameter-free and performs generally well compared to more complex methods, especially for shorter time series data. However, the accuracy of this approach diminishes in case local time shifting is observed between sequences, as the points of the compared sequences are regarded as fixed. For this reason, time series which might appear to have similar shapes but are slightly shifted in terms of time, might have a high measured distance (Ding et al. 2008). Such time shifts can be expected between driving patterns of different vehicles, as phases of acceleration and deceleration, for example at highway ramps, might set in at different points in time due to individual driving behaviors.

This limitation of the Euclidean distance approach for similarity measurement can be overcome by applying a dynamic time warping (DTW) algorithm, which is used especially in speech recognition (Müller 2007). In DTW, a potential time shift between two or more time series is considered by stretching or compressing locally until the minimum distance between them is obtained. The similarity function is then calculated by summing the heights of the aligned points of the compared time series, resulting in a real number which quantifies their similarity. Formally, DTW is computed as follows (Petitjean et al. 2012):

$$D(A_i, B_j) = \delta(a_i, b_j) + \min \left\{ \begin{array}{c} D(A_{i-1}, B_{j-1}) \\ D(A_i, B_{j-1}) \\ D(A_{i-1}, B_j) \end{array} \right\}$$

where:

A_i Sequence with ⟨a_1, …, a_i ⟩
B_j Sequence with ⟨b_1, …, b_i ⟩
δ Distance between elements of the sequence.

In the developed approach, the similarity search is conducted between all xFCD driving patterns on a road segment and a typical driving pattern, which is representative for the road segment at the observed time interval. In order to define this representative driving pattern, a single driving pattern out of all available driving patterns for the specific spatio-temporal context is calculated. For this, the global averaging strategy 'DTW Barycenter Averaging' (DBA) for multiple time series is applied. The sequences are averaged all together, hence obliterating the effects of order on the calculation outcomes. The DBA algorithm minimizes the sum of

DTW-calculated squared distances of an average times series to the set of all incorporated time series. It is computed as the sum of Euclidean distances between a point and points of the sequences aligned to it according to the DTW calculation. This partial sum is minimized for each point by calculating the barycenter of the associated set of points. The average sequence is the result of the computation of a barycenter for each data point in the sequence. The DBA is defined as follows (Petitjean et al. 2011):

$$C_T^{'} = barycenter(assoc(C_T))$$

$$barycenter\{X_l, \ldots, X_\propto\} = \frac{X_l + \cdots + X_\propto}{\propto}$$

C_T The average sequence with ⟨C_l, …, C_T ⟩ at iteration i
$C_T^{'}$ The average sequence with ⟨C_l^′, …, C_T^′⟩ at iteration i + 1

assoc Association of each point of the average sequence to one or more points of the set of sequences.

As stated before, the xFCD vehicle trajectories are constituted by sequences of points, which bear both spatial (latitude, longitude) and temporal (timestamp) information. However, recorded values from single vehicle tracks can't be simply adopted as representative for the identified spatio-temporal dimension due to the influence of individual driving behaviour. Thus, a broader set of vehicle trajectories have to be aggregated spatio-temporally in order to decrease the impacts of individual, non-representative patterns (Jackson et al. 2009). Another reason for the aggregation of xFCD is the resulting smoothing effect, which further contributes to reduce the influence of potentially erroneous data points (Lou et al. 2009). As basic spatial aggregation unit, road segments of the digital OpenStreetMap road network graph are used. To reference data points to the road segments, a 'map-matching' procedure is applied. Temporal similarity of xFCD is commonly determined by grouping the data into daily time intervals (Krampe 2006). Accordingly, data points of all incorporated vehicle trajectories, which were recorded at 1 Hz, are assigned to 15-min intervals based on the GNSS-timestamps, resulting in 96 daily intervals. All calculations for deriving representative emission values were conducted based on the described spatio-temporal reference units.

As emission factors from xFCD fuel consumption data are only representative for the specific vehicles which conduct the recordings, further adjustment factors have to be incorporated in order to make assumptions about the emission output of an entire traffic system per observed spatio-temporal unit. The vehicle specific emission factor from fuel consumption has to be transformed in such a way that it is representative for the expected overall traffic composition on a road network. As no thoroughly empirical on-site measurements of traffic composition were available, data from the annual report of the year 2012 on automobile stocks per county and make of car published by the national statistics authority is used. The calculated vehicle specific emission factor is correlated with the expected mix of passenger vehicles in the traffic composition on a road network by applying a rule of three. For this, standardized fuel consumption values from the measuring procedure determined in European Community regulation 715/2007 of the vehicle used for recording xFCD and the assumed statistical average of the traffic composition of passenger vehicles in Salzburg, Austria, is incorporated. Another utilized adjustment factor is the absolute traffic volume, which are derived from detector measurement at road cross sections.

5 Results

The developed methodology is applied and evaluated based on two case studies of road stretches in Salzburg, Austria. The first case study represents an inner-urban road stretch, while the second case study is conducted on an important artery road to the inner-city of Salzburg. In this paper, the results of the second case study are introduced.

The observed road stretch is a 1.68 km long part of the Alpenstraße between Anif and the administrative city boundary of Salzburg, Austria (see Fig. 3). The driving direction is northbound towards the city of Salzburg, with an allowed driving speed of 70 km/h. The road side detector, which is situated near the exit of the administrative limits of Anif, measures 5705 cars passing by on average on a weekday, with peaks during morning and especially evening traffic. Unlike the road in case study 1, the pattern of the average travel time shows a distinct peak during morning traffic, with the highest travel times of 157 s in average between 07:30 and 08:15 am. It can be seen that the number of vehicles passing the detector cross section declines during this morning peak, the traffic volume is less due to a restricted traffic flow. Another peak can be observed during evening traffic between 04:00 and 05:15 pm, with an average travel time of 120 s. Unlike during the morning peak, this does not lead to a significant decline of traffic volume. In the daily average, the time to pass this road stretch is 84 s. It can be seen that during night phases, low traffic volume and short travel times occur. Thus, the traffic quality on the observed road stretch of the Alpenstraße varies highly in the course of a weekday, which is assumed to have also impact on computed emissions. For the studied road stretch, 5803 FCD vehicle trajectories recorded at 1 Hz with information on acceleration and vehicle speed are available in the data basis, with an average number of 60 trajectories per aggregated 15-min interval for weekdays. The data basis of xFCD trajectories with fuel consumption recordings comprises 38 trips.

In Fig. 4, the daily course of predicted fuel consumption per calculation method for a MINI Cooper R56 on the road stretch of case study 2 per 15-min intervals on weekdays is depicted. A morning peak of fuel consumption can be observed, with increasing values around 6:30 am. The maximum fuel consumption for traveling over the road segment is predicted between 08:15–08:30 am, with 8.64 l/100 km.

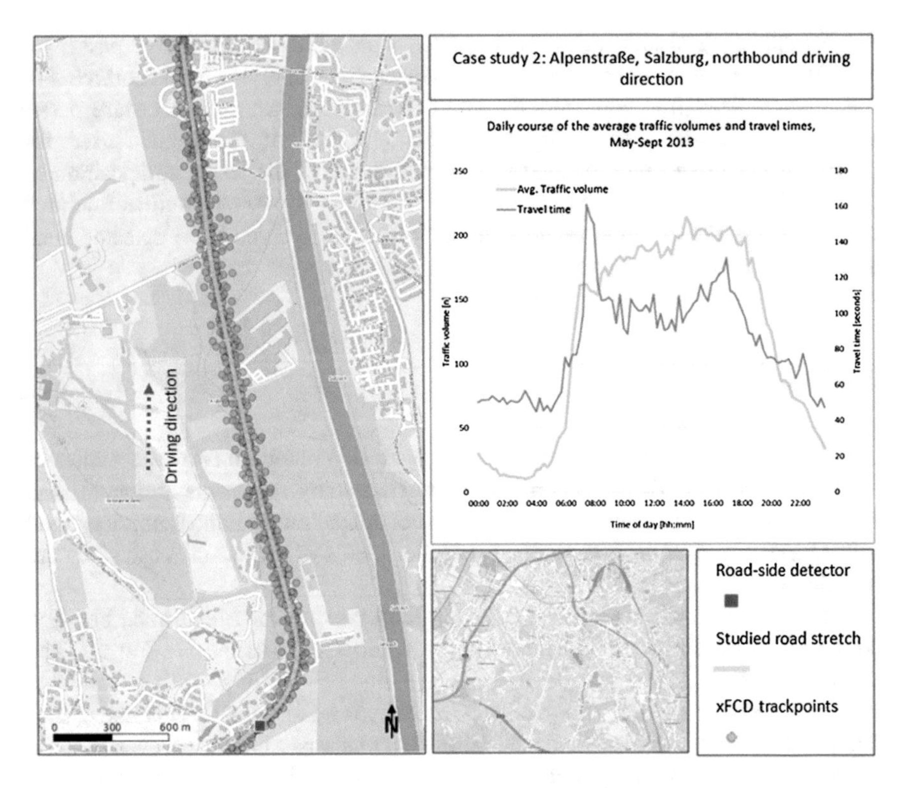

Fig. 3 Road segment of the inter-urban case study of Alpenstraße, Salzburg, Austria

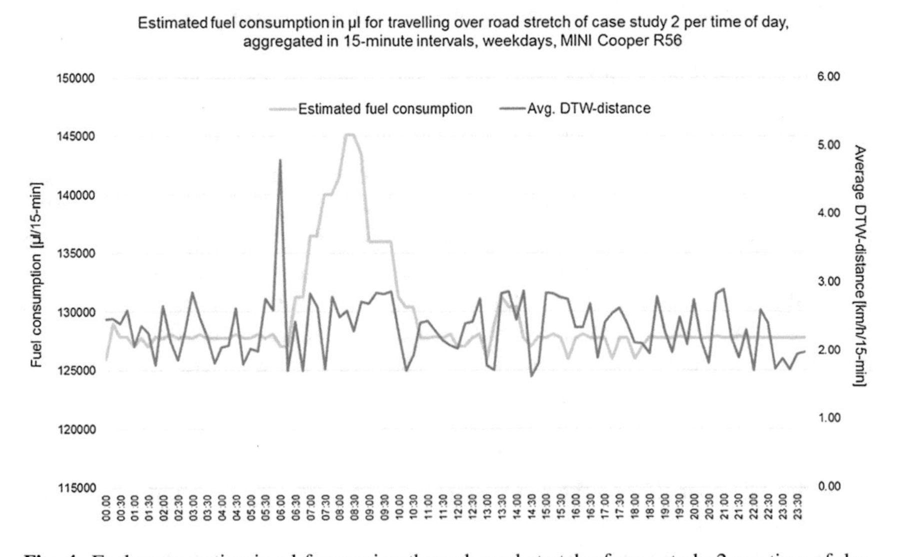

Fig. 4 Fuel consumption in µl for passing through road stretch of case study 2 per time of day, aggregated in 15-min intervals, weekdays, MINI Cooper R56

This peak in fuel consumption corresponds to the highest travel time derived from the FCD recordings. This indicates that the reduced traffic flow, which leads to more dynamic driving patterns, causes increasing calculated fuel consumptions. However, the smaller evening peak in travel time around 17:00 does not show any effects on the curve of predicted fuel consumption values. The average daily fuel consumption for traveling over the observed road stretch on weekdays is predicted highest with 7.70 l/100 km by the developed methodology.

For the evaluation of the driving pattern approach, the DTW-distances are utilized. It gives the difference in km/h per recording (conducted in 1 Hz frequency) of a matched driving pattern with fuel consumption to the representative driving pattern on a road segment and 15-min interval. For the case study, the minimum DTW-distance is derived at 14:30–14:45 (1.62 km/h per second), the maximum at 06:00–06:15 (4.80 km/h per second). For all 15-min intervals, the average distance between a reference driving pattern and the matched fuel consumption pattern is 2.34 km/h per second. The obtained distance is considered to be tolerable and it is assumed that for the observed road segment, the fuel consumption patterns do adequately match the representative driving patterns.

Based on the derived emission factors from fuel consumption, the absolute CO_2 emission values are calculated. This is done by introducing a traffic composition adjustment factor for fuel consumption, so that the values from the vehicle used for recording are adjusted according to the general traffic composition. Data on traffic composition is derived from vehicle ownership statistics in Salzburg. Moreover, the overall traffic volume is incorporated by using traffic volume data from a stationary road side detector. The resulting predicted daily course of CO_2 emissions is depicted in Fig. 5. As it can be expected, the extent of CO_2 emissions are highly

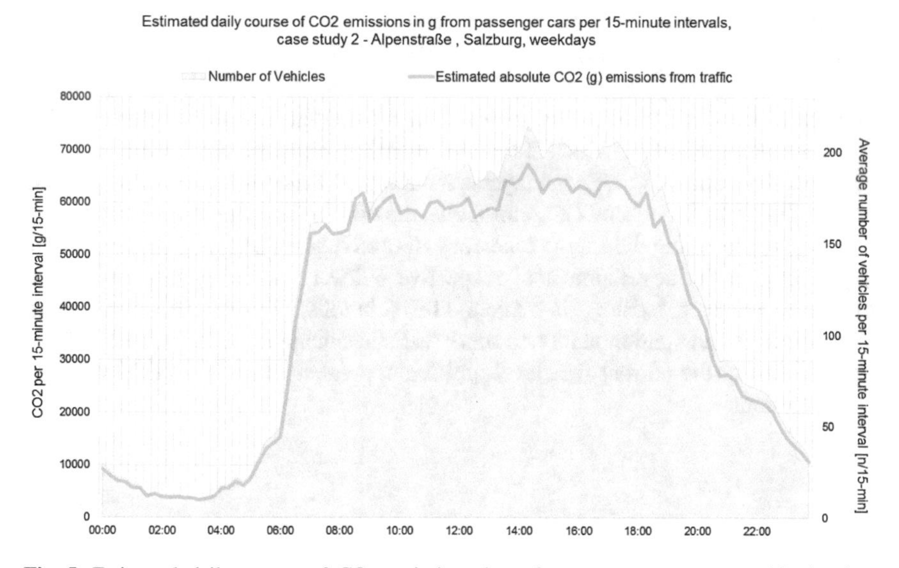

Fig. 5 Estimated daily course of CO_2 emissions in g from passenger cars per 15-min time intervals, case study 2—Alpenstraße from Anif to Salzburg, weekdays

coupled with the traffic volume, which usually starts to increase steeply from 05:45 on and declines again around 19:30. The level of CO_2 emissions is estimated to be between 50,100 and 67,300 g/15-min during daytime. The peak is predicted for 14:15–14:30 with 67,264.5 g of CO_2 emissions. Another peak can be observed between 08:45–09:00, with 63,613.8 g of CO_2 emission. The overall daily amount of CO_2 emitted on the observed 1.68 km long road stretch on regular weekdays is predicted to be 3,713,806 g. From a typical vehicle traveling over this road stretch, 191 g/km of CO_2 are emitted on daily average.

6 Conclusion

Based on the evaluation of the developed methodology in the case studies, the elaborated driving pattern approach is considered to have high potential for calculating representative, plausible and consistent CO_2 emission factors from xFCD fuel consumption recordings with a high spatial and temporal resolution. In general, the obtained CO_2 emissions showed values within plausible ranges and are closely correlated to the traffic flow quality in the respective interval, as it would be expected. Moreover, the daily profile of estimated CO_2 emissions and the fluctuations of values between subsequent 15-min intervals appeared to be reasonable. However, to further determine the validity and feasibility of the developed methodology, also additional case studies would have to be conducted, involving larger and diverse vehicle fleets for the recording of fuel consumption data, as well as a wider study area. A suitable framework for this could be a field operational test (FOT).

Nevertheless, the potential of the systematic usage of xFCD for the purpose of estimating traffic related CO_2 emissions could be shown. A better understanding of the spatio-temporal occurrences of CO_2 emissions through the provision of fine-scaled information based on mobile fuel consumption measurements can thus be a valuable building block for implementing dynamic and more efficient ecologically-orientated traffic management strategies. Besides the intended purpose of quantifying traffic-related CO_2 emissions, the developed methodology could also be used for evaluating the energy efficiency of road segments at specific times of the day. Provided the development of respective software modules, this could contribute to enhance existing in-vehicle ITS solutions, like eco-friendly routing alternatives for navigation units or a more realistic calculation of kilometers left for traveling using the current fuel level and the appointed route of a vehicle's navigation system.

References

Alessandrini A, Cattivera A, Filippi F, Ortenzi F (2012) Driving style influence on car CO_2 emissions. In: Conference proceedings: 20th international emission inventory conference —"emission Inventories—Meeting the challenges posed by emerging global, national, and regional and local air quality issues", Tampa, FL

Annema JA, Kok R, Van Wee B (2011) Cost-effectiveness of greenhouse gas mitigation in transport: a review of methodological approaches and their impact. Clean Cook Fuels Technol Dev Econ 39:7776–7793

Bert E, Chung E, Dumont AG (2007) Traffic emission using floating car and traffic sensor data. In: Conference proceedings: 6th European congress on intelligent transport systems

Breitenberger S, Grüber B, Neuheurz M (2004) Extended floating car data—Potenziale für die Verkehrsinformation und notwendige Durchdringungsraten. The Information Technology and Innovation Foundation. In: Straßenverkehrstechnik, 10, 522–531

Brookes L (1990) The greenhouse effect: the fallacies in the energy efficiency solution. Energy Policy 18(2):190–201

Capiello A (2002) Modeling traffic flow emissions. Massachusetts Institute of Technology, Boston

Dargay J, Gately D, Sommer M (2007) Vehicle ownership and income growth, worldwide: 1960–2030. Institute for Transport Studies, University of Leeds, Leeds

Ding H, Trajcevski G, Scheuermann P, Wang X, Keogh F (2008) Querying and mining of time series data: experimental comparison of representations and distance measures. In: Conference proceedings: VLDB endowment, pp 1542–1552

Ezell S (2010) Intelligent transportation systems. The Information Technology and Innovation Foundation, Washington DC

Gately CK, Hutyra IR, Brondfield MN (2013) A bottom-up approach to on-road CO_2-emissions estimates: improved spatial accuracy and applications for regional planning. Environ Sci Technol 47:2423–2430

Gurney KR, Razlivanov I, Wing IS, Yang S (2012) Quantification of fossil fuel CO_2-emissions on the building/street scale for a large U.S. city. Environ Sci Technol 46:12194–12202

IPCC—Intergovernmental Panel on Climate Change (2013) Climate change 2013: the physical science basis. Contribution of working group I to the fifth assessment report of the intergovernmental panel on climate change. Cambridge University Press, Cambridge/ New York

Jackson E, Aultman-Hall L (2009) Analysis of real-world lead vehicle operation for the integration of modal emissions and traffic simulation models. J Transp Res Board 2158:44–53

Jakobsen K, Mouritsen S, Torp K (2013) Evaluating eco-driving advice using GPS/CANBus data. In: Proceedings of the 21st ACM SIGSPATIAL international conference on advances in geographic information systems, pp 44–53

Kok R, Annema JA, van Wee B (2011) Cost-effectiveness of greenhouse gas mitigation in transport: a review of methodological approaches and their impact. Energy Policy 39 (12):7776–7793

Krampe S (2006) Nutzung von Floating Traveller Data (FTD) für mobile Lotsendienste im Verkehr. Schriftenreihe des Instituts für Verkehr, Universität Darmstadt

Krampe S, Leitinger S, Pucher G, Rehrl K (2013) FCD Modellregion Salzburg - Einsatz und Nutzen von Extended Floating Car Data im Bundesland Salzburg. In: Strobl J, Blaschke T, Griesebner G (eds) Angewandte Geoinformatik 2013. Wichmann Verlag, Berlin/Offenbach, pp 450–455

Li Q, Chang X, Cui X, Tang L, Li Z (2010) A road segment based vehicle emission model for real-time traffic GHG estimation. In: Conference proceedings: TRB 90th annual meeting 2011, No. 11-1174, pp 1–16

Lin J, Li Y (2009) Finding structural similarity in time series data using bag-of-patterns representation. In: Winslett M (ed) Scientific and statistical database management. Springer Verlag, Berlin, Heidelberg, pp 461–477

Litzenberger M, Dünnebeil G, Cagran B, Murg A, Öttl D, Van Poppel M, Orthofer R (2014) Correlation of vehicle acceleration and roadside black carbon concentration. In: 20th international transport and air pollution conference 2014

Lou Y, Zhang C, Zheng Y, Xie X, Wang W, Huang Y (2009) Map-matching for low-sampling-rate GPS trajectories. In: Conference proceedings: ACM SIGSPATIAL GIS 2009, Seattle, WA

Marquette A, Rosenstiel W (2012) Learning infotainment component for an improved eco-routing and eco-driving performance. In: 19th ITS world congress, Vienna, Austria

Messelodi S, Modena C, Zanin M, De Natale F, Granelli F, Betterle E, Guarise A (2009) Intelligent extended floating car data collection. Expert Syst Appl 36:4213–4227

Müller M (2007) Information retrieval for music and motion. Springer Verlag, Berlin/Heidelberg

Petitjean F, Gancarski P (2012) Summarizing a set of time series by averaging: from steiner sequence to compact multiple alignment. Theoret Comput Sci 414(1):76–91

Petitjean F, Ketterlin A, Gancarski P (2011) A global averaging method for dynamic time warping, with applications to clustering. Pattern Recognit 44(3):678–693

Pham ND, Ganti RK, Nangia S, Pongthawornkamol T, Ahmed S, Abdelzaher TF, Heo J, Khan M, Ahmadi H (2009) Fueoogle: a participatory sensing fuel-efficient maps application. Technical Report, Department of Computer Science, University of Illinois, Urbana-Champaign

Smit R, Poelman M, Schrijver J (2008) Improved road traffic emission inventories by adding mean speed distributions. Atmos Environ 42:916–926

Treiber M, Kesting A, Thiemann C (2007) How much does traffic congestion increase fuel consumption and emissions? Applying a traffic emission model to the NGSIM trajectory data. Technical report. In: Submission for the annual meeting of the transportation research board 2008

Van Wee B, Annema JA, Banister D (2013) The transport system and transport policy. Edward Elgar Publishing, Cheltenham/Northampton

Yu LJ, Peng ZR (2012) A better understanding of taxi emissions in Shenzhen, China, based on floating car data. In: Conference proceedings: transportation research board 2013 annual meeting

Zogg JM (2011) GPS und GNSS: Grundlagen der Ortung und Navigation mit Satelliten. ubox AG, Thalwil

Combining Different Data Types for Evaluation of the Soils Passability

Martin Hubáček, Lucie Almášiová, Karel Dejmal and Eva Mertová

Abstract The soils, as one of the basic landscape components, influence directly or indirectly almost all human activities. They have radical influence on a food production, environment formation, water regime and many other aspects. Besides that, they are one of the dominant landscape components, which influences significantly passability of the off-road movement. The knowledge of this influence is necessary especially for farmers, foresters, soldiers and rescuers. Farmers and foresters have an opportunity to postpone their work when the conditions are inconvenient. But rescuers and soldiers must accomplish their tasks immediately. So it is very important for them to know how they can move and which vehicle they can use in the required area. The big amount of data is necessary to evaluate the soil. Except for own soil databases there are mainly meteorological information, because weather has significant influence on the behaviour of the particular soil. Precipitations are the most fundamental of all the meteorological components or phenomena. Except for weather, there are also other components of landscape influencing the soil characteristics, such as vegetation, drainage and especially relief, which affect significantly drain and accumulation of water in soils. The main goal of the undergoing investigation is to predict the influence of soils on vehicle movement in dependence on soil characteristics, weather information and modelled drain conditions. The actual modelling employs digital soil database, detailed elevation models, radar images of precipitations, meteorological information from stationary stations and penetrometric measurement realized in different soil conditions. This modelling will be implemented to the cross-country movement model,

M. Hubáček (✉) · L. Almášiová · K. Dejmal · E. Mertová
Department of Military Geography and Meteorology, Faculty of Military Technologies, University of Defence, Kounicova 65, 662 10 Brno, Czech Republic
e-mail: martin.hubacek@unob.cz

L. Almášiová
e-mail: lucie.almasiova@unob.cz

K. Dejmal
e-mail: karel.dejmal@unob.cz

E. Mertová
e-mail: eva.mertova@unob.cz

I. Ivan et al. (eds.), *The Rise of Big Spatial Data*, Lecture Notes in Geoinformation and Cartography, DOI 10.1007/978-3-319-45123-7_6

which caused improvement to the older version of the model of cross-country movement and makes the model more accurate.

Keywords Geographic data · Meteorological data · Soils · Precipitation · DEM · Vehicle movement

1 Introduction

Soil is the outermost layer of the Earth crust. It represents a part of landscape, where its other components are meeting (Monkhouse 1970; Collins 1998). Soils have a very important role in the civilization development and influence both directly and indirectly most of human activities. In ancient times, the soil fertility influenced an existence of the human civilization in a particular area (Hillel 1991). Today, in time of the globalization and technological progress, it is possible to colonize in fact any part of the Earth.

Nevertheless, the soil will affect human activities at each location. One of these activities is movement. The man transforms landscape and creates various traffic infrastructures to make movement easier. But there are cases, when choosing an off-road movement for various activities is necessary. That kind of activities are for example agricultural and forest works. Most of them take place in an open countryside where agriculture and forest machinery moves not only on unpaved and forest roads, but also on fields, meadows and in forests. In such cases, the machinery movement is significantly influenced by soil conditions, which depends considerably on meteorological conditions. It is possible to postpone this type of works until suitable conditions occur, therefore to avoid situations, when vehicles sink or move with large difficulty throughout the area. But there are situations, when vehicles must go to the field regardless of any conditions, such as during military or non-military crisis situations. The aim of the research is to refine and modify existing simple and not very accurate algorithm for calculating soils passability. The new algorithm will be subsequently implemented into a comprehensive model for calculating of cross-country movement.

2 Initial Assumptions

Every vehicle movement is tied to the Earth's surface. The movement possibilities are affected by the whole range of parameters. The technical specifications of a vehicle, terrain, where vehicle moves, weather at the location and also driver′s skills of the vehicle belong among the most basic parameters (Rybanský and Vala 2009; Rybansky et al. 2015; Cibulova and Sobotkova 2006; Cibulova et al. 2015). The most important terrain components are relief, soil, drainage and vegetation cover, which influence movement during the off-road driving. The influence of terrain relief, drainage and vegetation can be relatively well parametrized and evaluated (Manual 1998; Rybansky 2009). For assessing the effect of the terrain there are known technical parameters of vehicles, such as maximal climbing capability, maximal transversal inclination, maximum width of trench, maximum climbing capability up to rigid step and others. Then it is possible to evaluate area of interest for a particular type or category of vehicles and to determine impassable terrain or terrain passable with restrictions using precise digital model of the terrain.

The similar attitude can be used for vegetation and drainage. The following parameters can be assessed: type of trees, spacing between stems, stem diameter, tree height, depth and width of rivers, flow speed, characteristics of bottom and others. Based on this knowledge, it is possible to make decision when planning vehicle movement through a particular area. However, such parametrization is more difficult in case of soils. It is especially caused the following:

- less quality data displaying soils,
- considerable dependence of soil behaviour on meteorological conditions,
- usually unclear boundaries between different soil types,
- affecting soils by water infiltration from rivers and lakes and by accumulation of precipitations in low lying locations.

The problem of the influence of soils on vehicle movement is relatively complex and cannot be solved using a simple analysis like in case of other landscape components. In addition to soil data and vehicle parameters, it is necessary to consider also other phenomena in calculation. Principally, it is the influence of meteorological conditions, especially precipitations, as well as terrain relief, which affects drain conditions in a given area. The research team at the Department of Military Geography and Meteorology has been dealing with problems of cross-country movement for a relatively long time.

In addressing the issue of cross-country movement, the requirement for solving the influence of the dominant soils on movement of military and rescue vehicles in the territory of the Czech Republic was considered. Total of 12 locations representing main soil types and soil textures for the entire territory of the Czech Republic was selected in collaboration with soil scientists from the Mendel University in Brno. The soil sampling was carried out within these locations for the comparison with data sources of soils (Specialized military database of soils—SMDoS) being used in the Army. Also the penetrometric measurements were performed for dry, humid and wet seasons according to the NATO classification (Manual 1994).

These penetrometric measurements showed that these soils are passable or passable with restrictions in most of the year regardless of the weather conditions (Hubacek et al. 2014). Any of the selected locations were impassable for monitored categories and vehicle types. The problem emerging from the comparison of soil analysis and values shown in the SMDoS was that this database is not convenient for evaluating soils from the point of view of its cross-country movement with

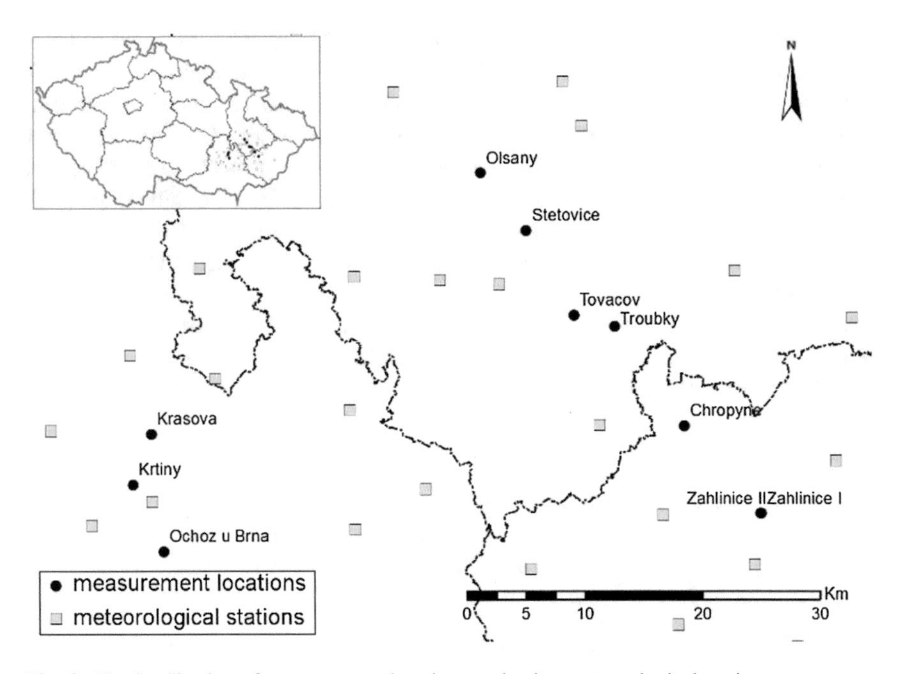

Fig. 1 The localization of measurement locations and using meteorological stations

respect to its quality, which comes out of the way of the classification and of the scale original maps used for creation of SMDoS (Technical Report 2000).

Considering these results, the second phase has been started in 2014. This was focused on soils, which have prerequisite for significantly greater impact on vehicle movement and which change their own characteristics significantly according to the amount of water. During this phase it was chosen nine appropriate locations (Fig. 1) situated in the area of south and middle Moravia. The locations were selected with regard to the time accessibility from Brno. It is suitable for opportunity to organize measurements flexibly with respect to meteorological situation. The next task of this phase was searching for convenient digital data describing soils.

3 Tested Area and Used Data Sources

The area of penetrometric measurements is relatively extensive. It occupies an area of approximately 3000 km^2 and it is relatively varied in terms of orography. There it is possible to find lowlands, such as the Hornomoravský Valley, highlands, such as the Drahanská Upland, or karst area, such as Moravian Karst. There is a variety of soils, such as gley, pseudogley, gleyic, organogenic, but also chernozem or phaeozem occuring on loessic sward. They are mostly sandy-loam and loam soils

Fig. 2 Measurement locations **a** agriculuture land, **b** wasteland (reed-mace)

from the perspective of soil texture. The test locations are situated out of forest areas usually at the agriculture land (meadows, fields) (Fig. 2a) or at wasteland with low vegetation (grass, reed-mace, bush) (Fig. 2b).

In addition to the penetrometric measurements the appropriate data are needed for the analysis. These are the soil data, meteorological data (precipitations) and elevation data. The main information about soils, which are necessary for calculating, are the information about the soil texture and the soil type. Both these attributes are contained in the SMDoS. Comparison of information from the database and information obtained from the soil samples revealed considerable unreliability of data, especially the soil type classification. Therefore, the digital soil map at scale 1:50,000 (DSM 50) was chosen as the main information source and was compared with the SMDoS and the terrain samples. Despite the fact that even this source is not 100 % reliable in terms of drawing accuracy, it is still noticeably more accurate in comparison with the SMDoS (Hubacek et al. 2015a). One disadvantage of DSM 50 is missing information about soil texture (soil granularity). This information is thus taken from SMDoS, where reliability of soil textures was more than twice higher to taken soil samples in comparison with information about soil type. Currently, there is no other suitable digital data source of required information from the territory of the Czech Republic. The compilation manuscript drafts from the time of the Complex soil mapping (Prax et al. 2008; Research Institute for Soil and Water Conservation 2015) could serve as the alternative source, however these materials exist in an analogue form only and they do not have any digital equivalent, be it the vector form or the raster form. Employing these manuscripts would require extensive amount of the manual work in terms of vectorization. The territory of the Czech Republic is situated on more than 16,000 map sheets and vectorization would thus one thousand hours of work.

Precise mapping of soil boundaries is very difficult. Border areas of soils in principle are not as sharp as in other geographical objects. In most cases, there is a gradual change in soil properties, which reflected the influence of soil on movement of vehicles. For this reason, in the calculation expressed by the uncertainty boundaries soils using fuzzy logic tools.

The main source of meteorological data is represented by data from meteorological stations of the Czech Hydrometeorological Institute. There are total of 39 stations of different categories in the area of interest. Varying amount of meteorological elements is collected at the stations according to station category (i.e. meteorological, climatic, precipitation gauged). The only precipitations, which are available from all stations of the area, are included to the processing in a case of the first investigation of the research currently. It is possible to use compound radar images from the radars Skalky u Protivanova and Brdy-Praha in addition to data from individual meteorological station. The compound radar images are being acquired every 5 min. They are then post-processed to generate the compound information of the total estimates of the precipitations of the duration of 1 h. The pixel's information of the compound radar image gives an intensity of precipitation in millimetres per hour on the base of the total estimates of the precipitations. The values of radar image pixels represent the intensity of precipitations according to the classes in Table 1.

The last source of data is represented by the digital elevation model (DEM). Several DEM types of various accuracy exist in the Czech Republic. Currently, the most accurate elevation model is the DEM 4, which was created using the Airborne Laser Scanning method. This model contains elevation points in a grid of 5 by 5 m and its accuracy is expressed by 0.3 m mean error in the open terrain and 1.0 m mean error in the areas covered by vegetation (Brazdil et al. 2012). The results of

Table 1 The classes of radar intensity of the precipitations

Class	Intensity of precipitations (mm/h)	Class	Intensity of precipitations (mm/h)
0	0	8	3.65
1	0.6	9	6.48
2	0.12	10	11.53
3	0.21	11	20.50
4	0.36	12	36.46
5	0.65	13	64.84
6	1.15	14	115.31
7	2.05		

independent accuracy testing show that its accuracy is usually twice than is declared, despite systematic errors, which model contains (Silhavy and Cada 2015; Hubacek et al. 2015b). The most accurate available model is the most appropriate data source, because during own analysis cross-country movement the drainage ratio and the capabilities of water accumulation in landscape are monitored.

4 Methodology

A big variety of geographic data is being employed during solving of this problem. They have different structure and they are primarily in different coordinate systems. Therefore, it is necessary to transform all data to the same coordinate system before the processing. The SMDoS and DEM 4 data produced by the Military Geographic and Hydrometeorological Office (MGHMO) use the WGS84 coordinate system and the UTM projection. The DSM uses the S-JTSK national coordinate system of the Czech Republic. The data transformation between these coordinate systems is described very well and the most of data analyst tools can execute transformation easily.

Meteorological data from meteorological stations represent the discrete point set. Their coordinate system can be transformed into the selected system. The situation with radar image is a little bit complicated, even though it is possible to execute data transformation only with knowledge of the projection parameters of the compound radar image. In this case, it relates to a set of data (report), which are transformed to raster image of the size 256 × 356 pixels, where one pixel has size 2 × 2 km. The images are in gnomonic projection in universal position, where the middle of projection is situated at the Prague Libuš station, which has geographic coordinates 14.45° east longitude and 50.01° north latitude.

After transforming data into the same coordinate system, it is necessary to execute the following steps:

- to analyse precipitations in previous season;
- to calculate drainage ratio and accumulation polygons at the area using DEM data;
- to express uncertainty in location of soil polygons using fuzzy logic tools;
- to estimate the water amount in soil on the basis of overlap function with usage precipitation and elevation data;
- to calculate bearing capacity of soils.

The first three tasks are currently tested and verified. They proceed from own experiments as well as from generally known principles of modelling of geographic reality. The estimate of the water amount in soil is problem, to which the main attention is paid at the moment and its solution will noticeably make calculation of cross-country movement more accurate. Current calculation of cross-country movement is based on the application of previously implemented procedures using the SMDoS database with using newly collected precipitation data, relief characteristics with regard to its ability to drain water, and the DSM soil types.

4.1 Precipitation Modelling

Although there are 39 meteorological stations in the monitoring area and its vicinity, where data are collected, their distribution is not optimal with respect to locations of penetrometric measurements (Table 2). It is evident that minimum distance between the measurement location and the nearest meteorological station is 2.2 km. Maximum distance is 9.3 km and the average distance is 5.7 km. Therefore it is not possible to consider the precipitation value at the nearest station as coincident with the precipitations amount falling down at location of measurement for calibration of calculation (Salek 2012). It is necessary to compute the precipitation amount for the locations of the measurement. Except for values at the locations of the measurement, it is necessary to have the precipitation values from the entire area. To extrapolate data to entire area, it is possible to use some of the interpolation algorithms, such as Spline, IDW, Kriging and so on (Childs 2004; Myers 1994). However, the precipitation course in the real world has usually the arrangement not corresponding to any commonly used interpolation (Fig. 3).

Therefore different method was chosen for gaining the overview of precipitation amount over the selected area and that is usage radar data and their rectification due to data from meteorological stations. The rectification of radar data is executed because radar data give only estimates of precipitations. The attenuation in the precipitations is very important phenomenon whose value depends mostly on precipitation intensity. In case of heavy precipitations it could become considerably underestimated due to noticeable shading of distant precipitations (Salek 2012; Andersson and Ivarsson 1991; Almasiova 2015).

Table 2 Distance between the measurement location and the nearest meteorological station

Measurement locations	Meteorological stations	Distance (km)
Zahlinice I	Kvasice	4.3
Zahlinice II	Kvasice	4.3
Troubky	Kojetin	8.3
Tovacov	Kralice na Hane	6.8
Stetovice	Kralice na Hane	5.0
Krasova	Bukovina	5.7
Olsany	Kralice na Hane	9.3
Krtiny	Bukovina	2.2
Ochoz u Brna	Bukovina	4.2
Chropyne	Kojetin	7.2
Mean		5.7

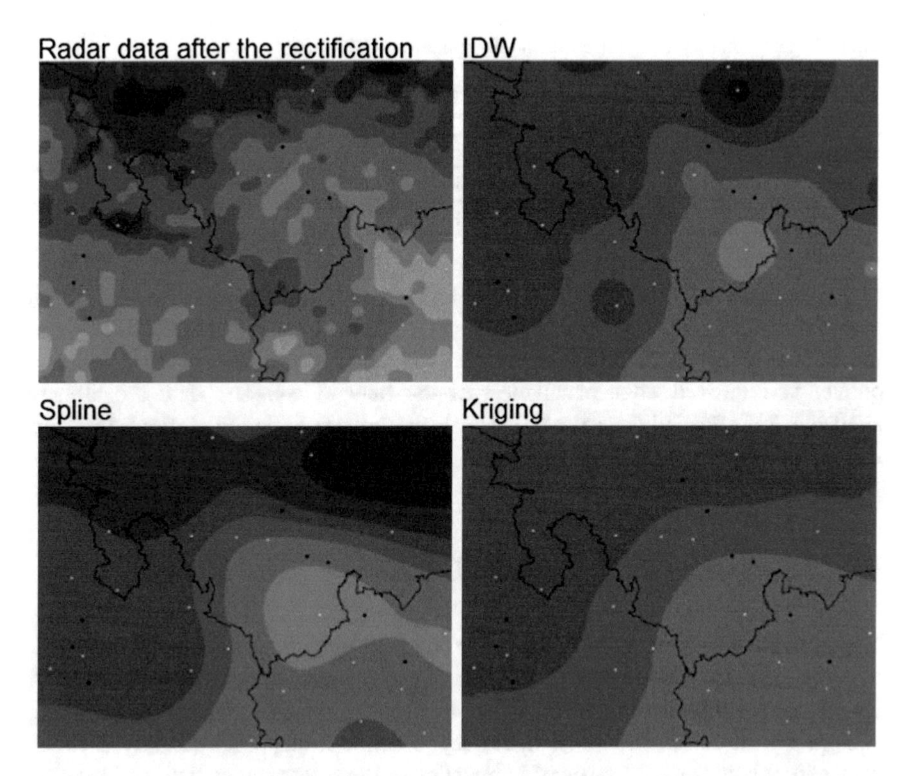

Fig. 3 The comparison of the precipitation distribution to different interpolation algorithms

There were determined errors on the basis of the comparison between the values of 24 h precipitation totals at meteorological stations to the 24 h precipitation totals gaining from radar data during the second half of March 2015. These were used afterwards for executing the radar image rectification and its accuracy improvement in the way that values express real precipitation total. The mentioned season was chosen on the basis of data availability, precipitation intensity and the term of the executed penetrometric measurement in the terrain. This method is not optimal, but it seems to be more accurate. To verify the method, the four precipitation different days we chosen, when precipitation totals were within the interval of 5–20 mm at least at half of the monitoring stations. In this case only half of stations regularly located at the whole area were used for rectification of the radar data. The second half was used as control points. The precipitation estimates were created from the same data by interpolation methods. The results from these control stations are better in comparison to the interpolation methods such as IDW, Spline and Kriging, which were used for generating continuous distribution of the precipitations. The mean error and RMSE for each method are showed in Table 3.

The noises represent a problem of radar data and it has different causes—both meteorological and non-meteorological. The most common non-meteorological error is the interference with other source, especially with Wi-Fi networks.

Table 3 The accuracy comparison of the interpolation algorithms

Interpolation algorithm	Mean error (mm)	RMSE (mm)
Radar data after the rectification	1.7	0.9
Spline	2.2	1.7
IDW	2.0	1.8
Kriging	2.0	1.9

The right band effect, influence of droplet size, rise and evaporation of droplets below cloudiness belong among often meteorological errors. Reduction of these noises is problem, which has not been solved yet. It could be expected, that data quality is improved after installation of the new radars. The data are already available, but currently we do not use data from precipitation stations for their rectification.

4.2 Expressing Uncertainty in Location of Soil Polygon

The expression of a border between the two soil polygons is generally quite difficult in comparison to other geographical phenomena. It is caused by two factors. First of them is the creation method of the most of soil maps, which is built of the sampling soils probe. In this method, it is necessary to define a shape of boundary between each soil polygons on the basis of knowledge of the soils characteristics in location of the soil probe, of the landscape configuration and of the knowledge of the soil forming processes. This method, however, shows a considerable uncertainty in definition of soils polygons. The second factor, which affects the soil characteristics, is the change soil types from one to the other, which can be more or less continuous. It is possible to say, that boundaries of the two adjacent soil polygons are significantly uncertain and that soil characteristics will be overlapping. It is necessary to express this uncertainty in location of soil polygons for modelling purposes. The fuzzy logic tools are thus employed (Ross 2009). Fuzzy sets are characterized by the membership function or the characteristic function. This is a functional dependency, where each element of the universal set has one exactly specific value of a degree of belonging to the fuzzy set. However, it is necessary that the functional values are just within the interval <0; 1>. Similarly to mathematical functions, fuzzy function can be linear, sinusoidal, discrete and continuous or in the shape of the Gaussian curve (Kainz 2008; Hofmann et al. 2013). Linear function of distance was selected for fuzzification of transition of individual soil types.

It is very important to choose the correct overlapping distance especially for the calculation of the uncertainty in location of soil polygons, which have different shape and size. It was taken 13 soil probes in Zastávka-Rosice terrain profile for boundary determination. There was found out the soil parameters by analysis in the laboratory. Afterward, they were compared in location in the soil maps. The three distances were defined on the basis of this analysis and of the particular soil polygons size in database. They are applied for the creation of the uncertainty layer of soil boundaries. The distances are the following: 250 m for large polygons, 125 m for medium-size polygons and 10 m for very narrow polygons adjoining to the rivers. The influence of uncertainty can be expressed for each soil area as follows (formula 1 sample for 250 m the distance):

$$\mu_S(x) = \begin{cases} 1, & x \leq -250 \\ \frac{x-250}{-500}, \ldots & -250 < x < 250 \\ 0, & x \geq 250 \end{cases}$$

where:

μs the degree of conformity of the type soils with the reality;
x distance from the interface of soils in database.

These modified soil layers serve as the additional data source for final modelling of soil impact to vehicle movement.

4.3 Calculating Drainage Ratio

Hydrological modelling tools, which enable the calculation of drainage ratio at a given area, are currently intergraded in the most of GIS tools. Their workings principle is very well described in a whole range of publications (Beven 1996; Vieux 2001). Their usage ranges from modelling of drainage ratio in drainage basin, flash floods simulation, searching slopes in danger of landslide, to calculation of erosion or pollution spreading by surface or underground waters (Kubíček et al. 2011; Gritzner et al. 2001; Vranka and Svatonova 2006). The final goal is usage of these tools to making precipitation-drainage ratio more accurate in the monitoring area.

The layers of slope, flow direction and flow accumulation are generated from DEM data. The last two generated layers are not used in the simplified form of the calculation. They are expected to enter the phase of the calculation of the water amount in soil, of the determination of the ratio between precipitation amount, which soaks into the ground and which flows away. Currently only the layer of slope is used in calculation. It is classified for two basic values, which come from the definition of the plain terrain of the NATO standards. When slope exceeds 5° value, it is expected, that drainage speed on the surface in noticeably higher than in case of plain surface. From this reason there is lower probability of absorption at these locations than on plain surface. The slope layer is classified in the way, that plain surface has value 1 and terrain with slopes exceeding limit has value 0.8. This value determination was counted on the basis penetrometric measurements of the soil resistance in season of recent rainy days. This value is determined by only two different soil types measurement currently. In future, it is expected to improve the accuracy of that value or to define more values in dependence on different soil characteristics.

4.4 Soil Passability Modelling

The bearing capacity of soils in most cases depends on meteorological conditions, especially on amount of precipitations, which influences soil saturation by water. This soil characteristic is defined particularly by vertical soil permeability, which is basically a function of soil texture (granularity). Light soils are good permeable, heavy soils are permeable with restrictions or impermeable. Further important characteristic, that affects the bearing capacity, is drainage speed depending on slope. The lower bearing capacity occurs at soils with permanent high water table or at soils, where water accumulates or where drainage is difficult. These locations are determined by the terrain relief which is tied to the presence of specific soil types (Novak et al. 1991). These attributes and dependencies emphasize the proposed approach to the passability modelling.

The problem of its adequate implementation is a correct expression of soil saturation and amount of precipitation, which flows away from the area without any distinct impact on soils. Until satisfying solution of the problem is reached, modified process determined during the SMDoS formation for soil impact modelling on vehicle passability will be applied. This procedure was designed by experts from the Research Institute of Amelioration and Soil Protection depending on the knowledge of the behaviour of soils. Influence of soil was generalized regardless of the specific type of vehicles that should be in the area (User Manual 2000). It is possible to create a map of soil influence on the movement of vehicles (Fig. 4). Individual soils polygons depending on the parameter soil type and soil texture represent the following parameters:

- GO in all weather conditions (light grey);
- SLOW GO in wet season (grey);
- NO GO in wet season (dark grey);
- NO GO throughout the year (black).

The wet season is defined as follows: rainfall in a liquid state greater than 40 mm during 3 days in the period from October to April and rainfall greater than 70 mm during 3 days in the period from May to September.

This relatively simple process is modified and it is applied in the following way with combining created data layers:

1. Creation of layer from the combination of data from precipitation-measured stations and from meteorological radars;
2. Calculation of the area slope and reclassification into defined categories;
3. Creation of modified precipitation layer by multiplying precipitation layer with reclassified slope;

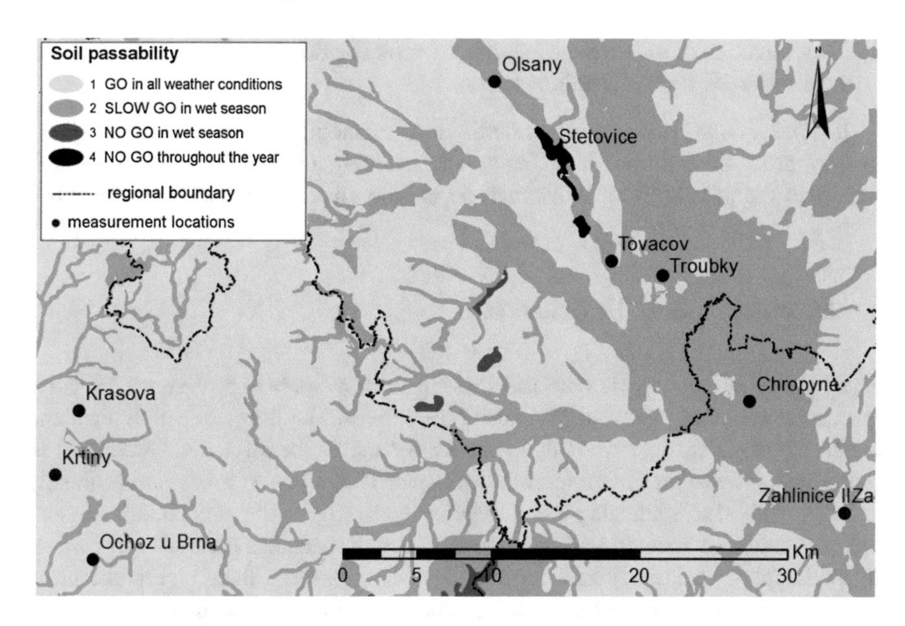

Fig. 4 The map section of soil passability created by the original algorithm for all vehicle types

4. Adaptation of soil texture and soil type layers using fuzzy logic;
5. Calculation of passability using the algorithm defined for SMDoS;

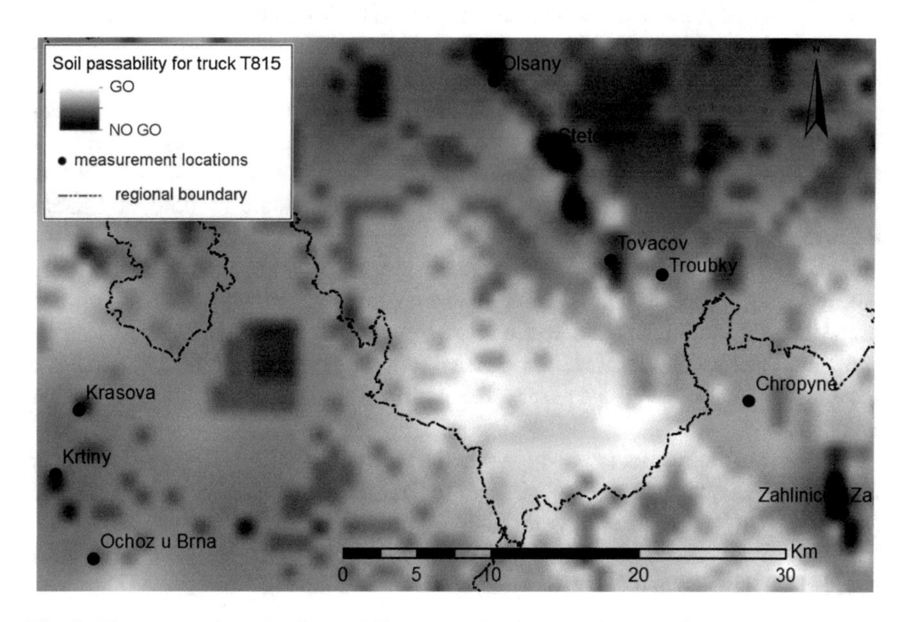

Fig. 5 The map section of soil passability created by the new algorithm for truck T815

6. Multiplication of final soil passability layer by coefficients of a particular vehicle type collected during terrain research.

It is possible to create a soil passability map by modifying the original algorithm and by processing further input data. The resulting map could express the soil impact to the possibility of movement depending on used vehicle type (Fig. 5).

5 Discussion and Conclusion

The creation of geographic data and making them available are very useful activities. It enables their wide usage in many disciplines. The usage of pedological and meteorological data together with accurate relief model for soil impact modelling to the vehicle movement can serve as the example. The proposed process uses already existing method of evaluating soil polygons depending on the precipitation amount and it extends the method with new possibilities. It evaluates precipitation amount in landscape by relatively reliable method and uses more than 1 year executing penetrometric measurement of bearing capacity at localities with soils which are passable with restrictions. Although the process is not capable of considering ratio between absorption and rain precipitations drainage and other meteorological influences as a temperature, time of sunshine, wind speed and others, the proposed solution is more accurate than the present method. Its improvement depends on a way of determination of precipitation distribution in landscape and combination various pedologic data including expression uncertainty in location of soil polygons. Modelled values of passability are not 100 % reliable, but their accuracy reaches approx. 65–87 % of values measured in terrain depending on soil characteristics in the area of measuring and on vehicle type.

Results that were achieved, are influenced to a considerable extent by climatically non-typical year 2015, when the precipitations in the Czech Republic did not reach even one half of annual mean at some places. Almost half of locations, where measurements were performed, are still insufficiently saturated (Global Change Research Institute of the Czech Academy of Sciences 2015). It is expected on the basis of recent autumn and winter months course, that situation will be improved during spring season. It will enable further measurement and checking of proposed procedures. It is planned to select and verify other locations with regard to soils showing lower reliability of results, besides continuation in measurement at present localities. The next step is to solve the problem of absorption and precipitation drainage.

At present, the results are being applied to the complex model of cross-country movement, which is being developed by the Department of Military Geography and Meteorology (Rybansky and Vala 2009; Rybansky et al. 2015; Rybansky 2009; Hubacek et al. 2015a, b; Hofmann et al. 2014). After its completion, this model could be implemented to the systems of provision of the geographic and hydrometeorological support for the benefit of Czech Republic Army's units and the agencies of the Integrated Rescue System.

References

Almasiova L (2015) A relationship between radar reflectivity and rainfalls recorded by the ground gauges. In: International conference on military technologies (ICMT), IEEE, Brno, May 2015, pp 285–288

Andersson T, Ivarsson K (1991) A model for probability nowcasts of accumulated precipitation using radar. J Appl Meteorol 30:135–141

Beven KJ (1996) A discussion of distributed hydrological modelling. Springer, Netherlands, pp 255–278

Brazdil K et al (2012) Technical report to the digital terrain model DTM 4th generation. Pardubice, Dobruska, Czech Republic (in Czech)

Childs C (2004) Interpolating surfaces in ArcGIS spatial analyst. ArcUser, July–September, pp 32–35

Cibulova K, Sobotkova S (2006) Different ways of judging trafficability. Adv Military Technol 1 (2):77–88

Cibulova K, Hejmal Z, Vala M (2015) The influence of the tires on the trafficability. In: International conference on military technologies (ICMT), IEEE, Brno, May 2015, pp 1–4

Collins JM (1998) Military geography for professionals and the public. Potomac Books Inc, Sterling

Global Change Research Institute of the Czech Academy of Sciences (2015) http://www.intersucho.cz. Accessed 12 Dec 2015 (in Czech)

Gritzner ML et al (2001) Assessing landslide potential using GIS, soil wetness modeling and topographic attributes, Payette River, Idaho. Geomorphology 37(1):149–165

Hillel D (1991) Out of the earth: civilization and the life of the soil. University of California Press, Berkeley

Hofmann A et al (2014) Creation of models for calculation of coefficients of terrain passability. Qual Quant 49(4):1679–1691.

Hofmann A, Mayerova S, Talhofer V (2013) Usage of fuzzy spatial theory for modelling of terrain passability. Adv Fuzzy Syst 2013:1–7

Hubacek M et al (2014) The soil trafficability measurement in the Czech republic for military and civil use. In: 18th international conference of the ISTVS, September 2014, Seoul

Hubacek M et al (2015a) Assessing quality of soil maps and possibilities of their use for computing vehicle mobility. In: 23rd central European geographic conference, Masaryk University, Brno, October 2015 (in print)

Hubacek M et al (2015b) Accuracy of the new generation elevation models. In: International conference on military technologies (ICMT), IEEE, Brno, May 2015, pp 1–6

Kainz W (2008) Fuzzy logic and GIS. University of Vienna, Vienna

Kubíček P et al (2011) Flood management and geoinformation support within the emergency cycle (EU example). In: Environmental software systems. Frameworks of environment, Berlin, pp 77–86

Manual (1994) Field. Manual 5-430 planning and design of roads, airfields and heliports in the theatre operations. Washington, DC, USA

Manual (1998) Field. Manual 5-33, Terrain analysis. Washington, DC, USA

Monkhouse FJ (1970) Principles of physical geography. American Elsevier, New York

Myers DE (1994) Spatial interpolation: an overview. Geoderma 62(1):17–28

Novak P et al (1991) Synteticka pudni mapa Ceske republiky 1:200000. VÚMOP Praha (in Czech)

Prax A, Hrasko J, Nemecek J (2008) The importance of comprehensive agricultural soil survey in the former Czechoslovakia and processing the result thereof. In: 2008 soil in modern information society: 1st conference of the Czech Society of Soil Science and Societas Pedologica Slovaca. Bratislava, pp 22–28

Research Institute for Soil and Water Conservation (2015) http://wakpp.vumop.cz/. Accessed 12 Dec 2015 (in Czech)

Ross TJ (2009) Fuzzy logic with engineering applications. John Wiley & Sons, London

Rybansky M (2009) The cross-country movement—the impact and evaluation of geographic factors. CERM, Brno

Rybansky M, Vala M (2009) Analysis of relief impact on transport during crisis situations. Morav Geogr Rep 17(3):19–26

Rybansky M et al (2015) Modelling of cross-country transport in raster format. Environ Earth Sci 74(10):7049–7058

Salek M (2012) Analysis of rainfall intensities using very dense network measurements and radar information for the Brno area during the period 2003–2009. Meteorol Z 21:29–35

Silhavy J, Cada V (2015) New automatic accuracy evaluation of altimetry data: DTM 5G compared with ZABAGED® altimetry. Lecture notes in geoinformation and cartography. Springer, Berlin, pp 225–236

Technical Report (2000) UDB Pudy. Vojenský zeměpisný ústav Praha. Czech Republic (in Czech)

User Manual (2000) UDB Pudy. Vojenský zeměpisný ústav, Praha, Czech Republic (in Czech)

Vieux BE (2001) Distributed hydrologic modeling using GIS. Springer, Netherlands, pp 1–17

Vranka P, Svatonova H (2006) Continuous soil lost modeling in the Haraska watershed—an application of 4D digital landscape model. Morav Geogr Rep 14(1):38–45

Sparse Big Data Problem. A Case Study of Czech Graffiti Crimes

Jiří Horák, Igor Ivan, Tomáš Inspektor and Jan Tesla

Abstract Sparse data sets may be considered as a one of the issues of big data generating extremely uneven frequency distribution. To deal with this issue, special methods must be applied. The study is focused on the Czech graffiti crimes and selected factors (property offences, buildings, flats, garages, educational facilities, and gambling clubs) which may influence the graffiti crimes occurrence. For regression analysis decision trees with the exhaustive CHAID growing method were applied. Grid models with 100, 500 and 1000 m cells were tested. The model of 1 km grid was evaluated as the best. The most influencing factors are the occurrence of secondary schools and gambling devices enhanced for several territorial units. The results of the decision tree for 1 km grid are validated using alternative models of data aggregation —aggregation around the randomly selected building and randomly distributed points.

Keywords Big data · Graffiti · Sparse data · Data mining · Crime · Multidimensional modelling · Decision tree

1 Introduction

The big data faces various problems that are usually not as evident as the high volume of available data. One of the specific issues which causes severe problems in data processing and analysis occurs in a situation of uneven data distribution,

J. Horák (✉) · I. Ivan · T. Inspektor · J. Tesla
Faculty of Mining and Geology, Institute of Geoinformatics,
VŠB-Technical University of Ostrava, 17. Listopadu 15/2172,
708 33 Ostrava-Poruba, Czech Republic
e-mail: jiri.horak@vsb.cz

I. Ivan
e-mail: igor.ivan@vsb.cz

T. Inspektor
e-mail: tomas.inspektor@vsb.cz

J. Tesla
e-mail: jan.tesla@vsb.cz

I. Ivan et al. (eds.), *The Rise of Big Spatial Data*, Lecture Notes
in Geoinformation and Cartography, DOI 10.1007/978-3-319-45123-7_7

especially in cases of extremely left skewed distribution (it means a large portion of very small or even zero data).

Sparse data matrices can be effectively stored and managed in multidimensional databases. The multidimensional database is usually understood as a part of Business Intelligence (Horák et al. 2015). Business intelligence (BI) mainly refers to computer-based techniques for identifying, extracting and analysing business data. They are abstracted using Extract, Transform, Load (ETL) tools into a data warehouse (Badard et al. 2012). The ETL processes include selection, harmonisation and aggregation processes. In case of crime events several important harmonisation procedures have to be provided—smart procedures to calculate referential time of event from time interval based on the type of event, length, beginning and ending of the interval; classification of facilities, buildings, age classification etc. (Horák et al. 2016; Loshin 2012).

Aggregated data is stored in fact tables. Data is organised and aggregated according to dimensions. Dimensions usually contain hierarchical structure which can be stored as an explicit (set of normalised tables) or implicit (one table with de-normalised solution) hierarchy (Horák and Horáková 2007). Each element of the hierarchy can be used for grouping facts.

A data warehouse integrates data from multiple data sources. Data warehouse and multidimensional modelling may be applied for different purposes not only for business strategies. The public sector usually repeatedly integrates data from regularly provided data sources and needs effective tools for repeated spatiotemporal analysis of large volume of incremental data. One of such application can be the database for crime prevention, developed for the Ministry of Interior of the Czech Republic (Horák et al. 2016).

In our database, two dimensions are shared among all fact tables—geographical and temporal dimensions. They enable to link fact tables together. The location is considered as a cornerstone of the database and is expressed in each fact table by two geographical dimensions—the administrative dimension and geometrical dimension (square grid). The administrative dimension identifies the administrative units. The basic unit is the part of a municipality which seems to be an appropriate compromise between data availability, privacy protection and the required highest detail. The elementary unit for the geometrical dimension is 100×100 m cell, arbitrarily aggregated. It is consistent with the 4th level of the scale system for public authority where 10 km window and 100 m grids for communes and urban districts are recommended (Bacler 2014). The basic structure of the multidimensional database and structure of the fact table about crimes is described in detail in Horák et al. (2015).

To explore issues and possible solutions of analysing high volume sparse dataset we selected evidence of graffiti crimes in the Czech Republic and explored influences of several factors to evaluate the possibility of improved predictions. Graffiti is a common phenomenon. Often it is represented by simple patterns, signatures (tags) from young people. These people call themselves graffiti artists and consider it as a lifestyle. The graffiti can be seen from several perspectives—as an art (Lachmann 1988), as a reflection of societal customs and attitudes (Stocker et al.

1972), a method to attain notice or fame (Lachmann 1988), as a form of political statement (Ferrell 1995), or as territorial markers (Ley and Cybriwsky 1974).

Unlike sprayers themselves, most of the society considers graffiti as an expression of vandalism as well as the Czech law. The act of vandalism expressed as graffiti is included in §228 par.2 Act No. 40/2009 as an independent fact of a criminal damage offense. In literature and (negative) consciousness of people, the concepts of vandalism and graffiti are often confused. Nevertheless, there is no direct evidence that those who create graffiti, carried out vandalism simultaneously (Thompson et al. 2012). Also Wilson and Healy (1987) have not found statistically significant relationship between graffiti and vandalism in trains. However, what and under which conditions the difference between graffiti and vandalism forms, remains unclear (Halsey and Young 2002). Cohen (1969) distinguished five main motivations for graffiti and vandalism—gaining or sustaining membership within a 'deviant' group through anti-social acts that reinforce group membership, self-esteem, self-expression, to disrupt the order of authority, enjoyment and the rush associated with the illegality of the behaviour. Similarly, Bandaranaike (2001) identified following main reasons for graffitist's behaviour from questionnaires in Australia—they are asserting identity, in defiance of societal norms, reacting to a heartfelt feeling (need for love or company; anguish of abuse, discrimination), larrikinism and seeking adventure, the adrenaline rush and pleasures in risk taking and association with fad and fashion (imitate peers in other urban environments).

Graffiti is associated with a young age and often linked to school also. Most often mentioned reason is "boring" at school and lack of opportunities for self-expression. This leads to the breaking the rules and disrupting the system by youth (Thompson et al. 2012; Iveson 2007; McCormick 2003). "The Broken Windows" theory (Wilson and Keeling 1982) says that the presence of vandalism, no matter how small, means that unorganised social environment normalises and even supports other disorders. Similarly "contagion theory" principle (Armitage 2002) can be applied. It means the distribution of graffiti should be clustered.

According to Tygart (1988), vandalism peaks in seventh grade of primary school and progressively decreases with further successfully completed year. From a socio-economic perspective, the typical school vandal comes from middle economy class with low income (Howard 1978) or was fired from the school (Yankelovich 1975 in Goldstein 1997). Schools are not only a place of concentration of young people, but the campus is also a suitable environment for graffiti. Offenders are attracted to such public places where their exhibition may have a wide audience, i.e. a form of exhibitionism (Buck et al. 2003). Also other factors should be taken into account. According to Gibbons (2004), graffiti comes out as one actor which is consistently correlated with measures of fear of crime, neighbourhood decline, property crimes and escalating crime rates.

Spatial analyses are usually made at the level of individual cities. Popular techniques include creating a square grid and comparison of multiple factors using regression models. E.g. Megler et al. (2014) reported 59,000 records of graffiti within 2 years from San Francisco. No temporal pattern in the data set was recognised. Thus, all data was aggregated to square grid 381 m. Finally, they applied a GWR model which explained more than 2/3 of the variation of graffiti records.

Grid models generated from multidimensional databases may provide an effective solution for sparse big data. The aim of the study was to verify results of grid models in regression analysis, compare them with alternative models and evaluate the stability of significant outputs.

2 Data Sources

The evidence of graffiti comes from the crime register of the Police of the Czech Republic (PCR). Data was exported from centralised information system for event recording (ETR, Evidence of Criminal Proceedings) which contains detail information about each crime commission. All events in the CR with classification as a criminal act "sprejerství" (§228 par.2 Act No. 40/2009) in the period 1.1.2014–10.12.2014 were selected. Thus, there are only events registered by PCR as serious events (suspicion of crime act) and we have no evidence of other graffiti manifestation. It is obvious that the number of all graffiti occurrences is much higher but it is almost impossible to evaluate the missing part. Marešová (2011) estimates the number of unregistered (latent) crimes of vandalism in the range of hundreds to thousands per year. Less serious damages are more frequent. Nevertheless, according to German outputs (Marešová 2011), it is valid for the urban environment that main typical features of latent crime are well overlapped with features of registered crime. Thus, the relationships between registered graffiti crime occurrence and influencing factors should be similar as for the latent or total crime. The relationships are also influenced by other factors like different level of tolerance to graffiti (issued on a different level of reporting), differences in mapping (different frequency of visiting places due to the different number of people or officers), the occurrence of police stations (i.e. Megler et al. 2014), etc.

Three types of locations are distinguished in ETR—a place of commitment, place of effect and place of reporting. Only first two types were selected (and the second represents only 4 %). The same time interval and type of locations were applied for the selection of records about property offences.

Other data sources contain referential and potentially explanatory information:

- Register of schools (Min. of Education, Youth and Sports) offers contact information and classification of education level and type (http://stistko.uiv.cz/registr/vybskolrn.asp). Data was exported on 6/2015 and represents valid records about registered educational facilities in the school year 2014/15.
- Register of gambling devices contains registration ID, addresses and description of placement for all technical lottery devices approved by the Min. of Finance to 10/2014.
- Register of Statistical Units and Buildings (RSO, Czech Statistical Office) contains information about address points and buildings with an indication of

the type of usage, construction classification and number of flats. Used data comes from the version valid to 1.7.2014. Finally number of buildings, flats and garages were processed.

Data from all registers except RSO was harmonised. In the case of the crime register several types of control procedures were applied (integrity constraints, validity check of the time range, geographical range, the validity of codes in the database) (Horák et al. 2015). Address descriptions in all registers were harmonised (e.g. name of the municipality, municipal part, street, numbers).

Further, geocoding of registers of schools and gambling devices were performed using a specially developed SW Geocoder (Fojtík et al. 2016). The quality of geocoding has to be assured on the satisfactory level. The minimal limit of the success rate is 85 % which should not cause a systematic spatial error (Ratcliffe 2004 in Andresen and Malleson 2013). The success rates for our data sets reached almost 100 % but the quality (positional errors) is variable. The quality of geocoding of Geocoder SW was compared with results of geocoding API of web search engines (Fojtík et al. 2016) with a positive evaluation. In total, 1511 graffiti crimes, 26,046 schools and 6342 gambling clubs were selected in the Czech Republic.

3 Grid Evaluation

To explore problems of high volume sparse data, we analyse relationships among the graffiti crime acts and the occurrence of property offences or locations of selected objects like buildings, flats, garages, schools, gambling clubs.

To verify factors with higher influence (they are more concentrated around graffiti crimes), the distribution of explored type of objects was analysed using an aggregation of selected objects around graffiti crimes within increasing distance (Fig. 1). It is clear the secondary schools and gambling clubs are much more concentrated around graffiti crimes than lower levels of education facilities.

The problem is such calculation is highly time-consuming because all distances between crime placement and the objects have to be calculated, select the closest one and a share from the total enumerated. The main issues are that such way we can model and explore the influence of only one factor at once (no combination of factors), no non-numerical factors (predictors) (including regional differences) can be included, and no statistical significance of the results is provided.

The main idea was to utilise square grids for analysis of relationships (dependencies) and compare results with alternative methods of data aggregation. Usage of grid analysis (incl. pre-processing) is fast and flexible. Point data is aggregated directly to cells according to its coordinates. All events and objects are aggregated independently and no enumeration of spatial relationships between them is required (the advantage of the implicit topology of the raster model). These features facilitate

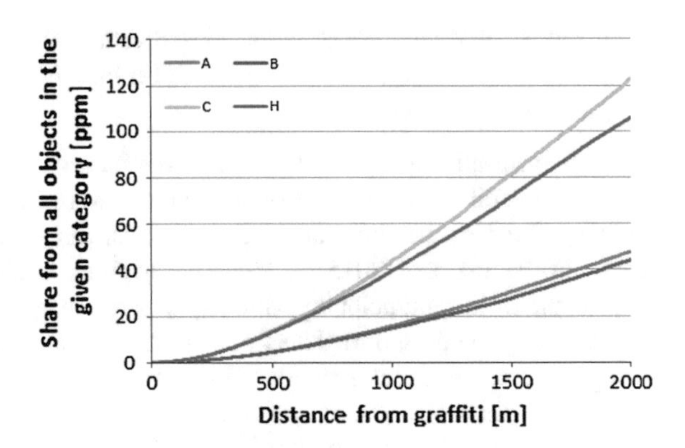

Fig. 1 The relative share of kindergartens (*A*), primary schools (*B*), secondary schools (*C*), gambling clubs (*H*) within an increasing distance from graffiti crimes

spatial analysis and provide great benefits especially for processing of high data volumes.

The harmonised data was aggregated to square grids and stored in a multidimensional database. The multidimensional database effectively stores such data due to the specific structure containing only non-zero counts. However, when we expand data set to cover the whole country we obtain massive data sets with rare occurrences of investigated events. Such data can be classified as a sparse big data. **The main question is how to deal with such data**.

Due to the small number of events we did not explore the temporal pattern. Finally, three spatial aggregation levels were applied—grid 100 m (basic units), 500 m and 1000 m. The grid 100 m in the Czech Republic contains 7,423,937 cells (the graffiti crimes occur only in 1189 cells, total sum is 1511), the grid 500 m includes 297,134 cells (the graffiti crimes occur only in 803 cells), and finally the number of cells in 1 km grid is 80,131 (only 578 "positive" cells). The data volume (above 7 mil. records) is too big for processing various statistics methods in software like SPSS (e.g. insufficient memory, too big data volume); in these cases, the random selection of 10 % of data was applied.

The frequency distribution of gridded graffiti crimes is extremely curved and skewed (with a very high portion of NULL cells), obviously more with higher spatial resolution (Fig. 2). The power distribution is highly skewed with a long tail (Brown and Liebovitch 2010) and follows a straight line in log-log graphs. Shapes of histograms indicate that their middle part can follow a power law distribution, more with more coarse spatial resolution, but both tails of the distribution are more extreme than the power law distribution. Such extreme type of distribution makes impossible to use many of typically used statistical and spatial methods (e.g. geographically weighted regression) requiring the normal distribution (Table 1).

The other variables aggregated to cells (selection from the multidimensional database) are number of property offences, buildings (all), flats, garages, kindergartens, primary schools, secondary schools, gambling clubs and gambling devices.

Due to such extreme imbalance between cells with and without graffiti crimes, the proximity measures between variables are explored only for 1 km grid.

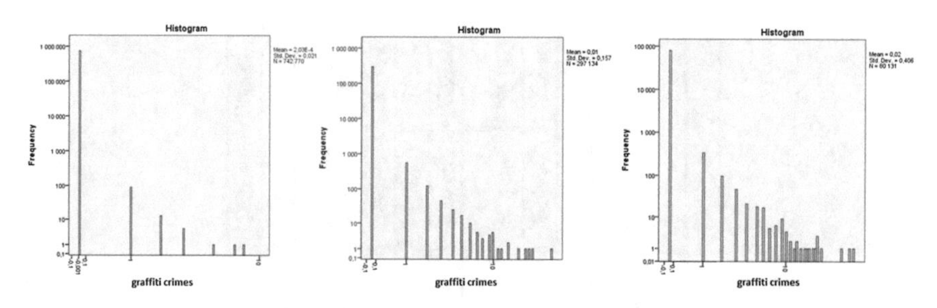

Fig. 2 Histograms of gridded graffiti data sets (100, 500, 1000 m; log-log scales)

Table 1 Statistical characteristics of gridded graffiti data sets

Grid (m)	N	Share from all data (%)	Maximum	Mean	Variance	Skewness	Kurtosis
100	742,770	10	8	0.0002	0.000	193	55,027
500	297,134	100	34	0.01	0.025	88	12,590
1000	80,131	100	45	0.02	0.165	55	4470

The partial correlation cannot be evaluated because it requires a normal distribution. The calculation of Spearman's coefficient of correlation was impossible for all data because the volume of the data matrix is out of the SPSS capabilities and thus, 10 % random sample was used. The results (Table 2) show the most important dependencies between graffiti and secondary schools (0.268), gambling clubs or gambling devices (0.251 and 0.250 resp.), followed by primary schools (0.207), garages (0.192) and kindergartens (0.181). The correlation with property crimes is surprisingly low.

Due to the fact of small numbers of graffiti and several other factors, also a proximity matrix for categorised (binary) variables was calculated using Phi 4-point correlation for the full data set (Table 3). This index is a binary analogue of the Pearson correlation coefficient and it has a range of −1 to 1 (SPSS 2007). The differences are better expressed than Spearman's rho in the previous table. The occurrence of graffiti crimes shows the highest proximity to secondary schools (0.307) and gambling clubs (0.306), followed by primary schools (0.206), garages (0.195) and kindergartens (0.189). It well corresponds to the distance analysis (Fig. 1).

For evaluation of factors influencing graffiti crimes, decision trees were selected. The decision tree procedure creates a tree-based classification model. It classifies cases into groups or predicts values of a dependent (target) variable based on values of independent (predictor) variables. The decision tree in SPSS uses various growing methods (SPSS 2007):

- CHAID (Chi-squared Automatic Interaction Detection) chooses the independent variable that has the strongest interaction with the dependent variable. Categories of each predictor are merged if they are not significantly different on the dependent variable.

Table 2 Spearman's coefficients of correlation between explored variables

	Correlations									
Type = spearman's rho, Statistics = correlation coefficient										
	Graffiti oc.	Property of.	Gambling	Gamblmach	Kindergarten	Primary	Secondary	Building occ.	Flat occur.	Garage
Graffiti oc.	1.000	.170**	.251**	.250**	.181**	.207**	.268**	.126**	.130**	.192**
Property of.	.170**	1.000	.315**	.315**	.352**	.339**	.225**	.563**	.554**	.338**
Gambling oc.	.251**	.315**	1.000	1.000**	.340**	.352**	.375**	.240**	.239**	.274**
Gamblmach	.250**	.315**	1.000**	1.000	.339**	.352**	.375**	.240**	.239**	.273**
Kindergarten	.181**	.352**	.340**	.339**	1.000	.713**	.269**	.354**	.360**	.255**
Primary sch.	.207**	.339**	.352**	.352**	.713**	1.000	.325**	.318**	.325**	.284**
Secondary s.	.268**	.225**	.375**	.375**	.269**	.325**	1.000	.167**	.171**	.264**
Building occ	.126**	.563**	.240**	.240**	.354**	.318**	.167**	1.000	.946**	.301**
Flat occur.	.130**	.554**	.239**	.239**	.360**	.325**	.171**	.946**	1.000	.295**
Garage occ.	.192**	.338**	.274**	.273**	.255**	.284**	.264**	.301**	.295**	1.000

**Correlation is significant at the 0.01 level (2-tailed)

Table 3 Phi 4-point correlation between explored variables

	Proximity matrix								
	Fourfold point correlation								
	Graffiti oc.	Property off	Gambling oc.	Kindergar oc.	Primary sch.	Second sch.	Building oc.	Flat occur.	Garage oc.
Graffiti oc.	1.000	.145	.306	.189	.206	.307	.065	.072	.195
Property off	.145	1.000	.274	.330	.317	.192	.354	.366	.269
Gambling oc.	.306	.274	1.000	.343	.377	.396	.121	.133	.262
Kindergar. oc.	.189	.330	.343	1.000	.725	.269	.187	.211	.252
Primary sch.	.206	.317	.377	.725	1.000	.325	.167	.188	.254
Second sch.	.307	.192	.396	.269	.325	1.000	.082	.092	.221
Building oc.	.065	.354	.121	.187	.167	.082	1.000	.882	.155
Flat occur.	.072	.366	.133	.211	.188	.092	.882	1.000	.171
Garage oc.	.195	.269	.262	.252	.254	.221	.155	.171	1.000

This is a similarity matrix

- Exhaustive CHAID is a modification of CHAID that examines all possible splits for each predictor.
- Classification and Regression Trees (CRT) splits the data into segments that are as homogeneous as possible on the dependent variable. A terminal node in which all cases have the same value for the dependent variable is homogeneous, "pure" node.
- Quick, Unbiased, Efficient Statistical Tree (QUEST) is a method that is fast and avoids other methods' bias in favour of predictors with many categories. QUEST can be specified only if the dependent variable is nominal.

CHAID and exhaustive CHAID enables multiway node splitting of trees while CRT and QUEST only binary. The main advantages of decision trees are—no assumptions about the type of distribution, the data population can be heterogeneous, the effect of predictors can be nonlinear (Hendl 2006), data can be measured on different scales (Alkhasawneh et al. 2014) and it is also possible to predict values of the dependent variable.

It is recommended to designating one or more categories as target categories. The selected categories are treated as the categories of primary interest in the analysis and enable to use some classification rule options and gains-related output. It has no effect on the tree model, risk estimate, or misclassification results (SPSS 2007). In our case, the category "yes" for graffiti crimes was selected of primary interest.

In all models, a category (occurrence) of graffiti crimes was the dependent variable and independent variables were property offences, gambling clubs, kindergarten, primary schools, secondary schools, all buildings, all flats, garages and name of district.

The testing showed the usage of scale variables as useless, due to prevalent dividing trees in nodes according to the values 0 and 1. Therefore, we reclassified all numerical scale variables into binary values (0, 1) for the occurrence of given phenomena in a cell. The best results (approved by validation) are obtained using exhaustive CHAID method of tree growing. Significance levels for splitting nodes and merging categories are both 0.05. The maximal depth of trees for CHAID and exhaustive CHAID is three. The results for three different spatial resolutions (100, 500 and 1000 m) indicated best characteristics for a model with 1000 m grid (Tables 4 and 5). Larger grid cells were not explored due to the limited capacity of full comparison with alternative aggregation models and decreasing of model performance found for 2000 m grid (88 %).

It seems that model of 100 m is less probable according to 0 correct predictions, low maximal response but also the provided selection of predictors where secondary schools are absent and the maximal influence is given to property offences.

The best results are reached by 1000 m grid model. The most influencing variable was the occurrence of secondary schools, followed by the districts and the occurrence of gambling. Other variables may have a local effect (garages, property offences and primary schools).

Table 4 Parameters of classification trees for prediction of graffiti crime occurrence in various grids

Grid (m)	List of significant independent variables	Nodes	Percent correct prediction "yes" (%)	Max. response (prediction % for yes)	1st variable	2nd variable	3rd variable
100	Property offences, district, all flats, garages, gambling clubs, kindergartens, all buildings, primary schools	158	0	18	Property offences = 1	District = Brno-město	Gambling clubs = 1
500	Gambling clubs, district, property offences, garages, all flats, primary schools, kindergartens, secondary schools	160	6.6	58.2	Gambling clubs = 1	District = Brno-město	
1000	Secondary schools, district, property offences, primary schools, garages, kindergartens, gambling clubs, all flats	171	26.6	71.1	Secondary schools = 1	District = Písek, Brno-město, Jeseník, Prostějov, Ostrava-město, Beroun	

Table 5 Classification results for three grid models

Model (m)	Observed no – predicted no	Observed no – predicted yes	Observed yes – predicted no	Observed yes – predicted yes	% Correct for yes	Overall % correct
100	7,422,764	0	1,173	0	0	100
500	296,294	38	749	53	6.6	99.7
1000	79,454	99	424	154	26.6	99.3

Table 6 Classification results for three models

Model	Observed no – predicted no	Observed no – predicted yes	Observed yes – predicted no	Observed yes – predicted yes	% Correct for yes	Overall % correct
Grid1km2	79,454	99	424	154	26.6	99.3
RB1km2	78,992	1642	3876	3310	46.1	93.7
RP1km2	79,508	41	528	54	9.3	99.3

The whole tree cannot be visualised due to a large number of nodes (171). The detail is in the Fig. 3. Each statistically significant splitting of the tree is documented by the name of variable (predictor) with adjusted *p*-value, the values of Chi-square test and degrees of freedom. Each tree branch is headed by the value of this variable valid for the whole branch (i.e. district is one of the following names: Zlín, Přerov, Hodonín, Česká Lípa, České Budějovice, Praha-západ; the occurrence of graffiti club is 1). The box describes the node as a data group defined by this tree branch. It contains the identifier of the node, frequency of No and Yes answers in absolute and relative values and the total share of this group to the whole dataset again in absolute and relative values.

The highest positive response in graffiti crimes prediction was reached for the following combination:

- 71 %: (secondary schools occur.) AND (districts: Brno-město BM, Ostrava-město OT, Prostějov PV, Písek PI, Jeseník JE, Beroun BE)
- 56 %: (secondary schools occur.) AND (gambling clubs occur.) AND (districts: Praha AB, Olomouc OL, Šumperk SU, Karviná KA, Kroměříž KM, Znojmo ZN, Blansko BK, Třebíč TR, Jihlava JI, Havlíčkův Brod HB, Litoměřice LT, Uherské Hradiště UH, Žďár nad Sázavou ZR).
- 47 %: (secondary schools missing) AND (district = Brno-město BM) AND (garages occur.)
- 32 %: (secondary schools missing) AND (district = Praha AB) AND (gambling devices occur.)

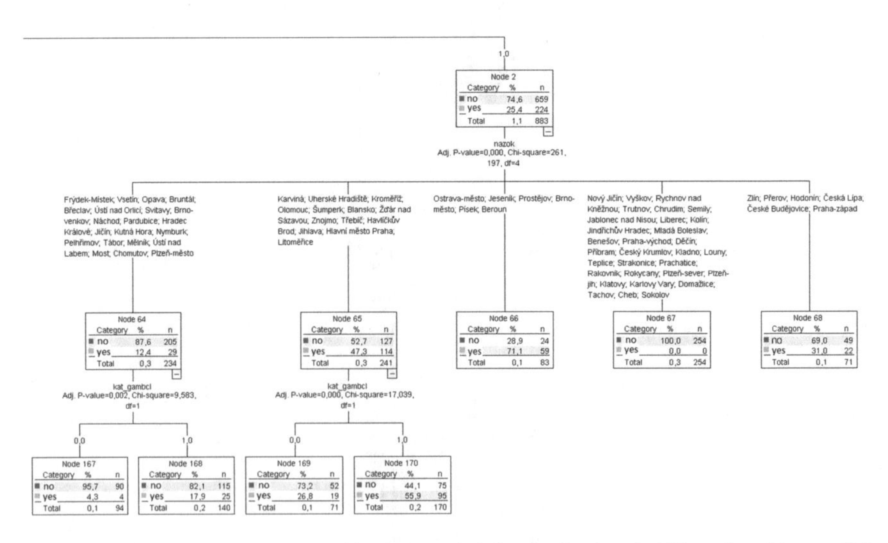

Fig. 3 Detail of decision tree for grid 1 km around the maximal probability of positive graffiti crime results (node 66)

4 Comparison of Three Different Models of Aggregation

To independently verify if the results of the selected grid model are relevant, we performed alternative ways of aggregations and build other data sets for decision trees.

4.1 Aggregation Around Randomly Selected Buildings

The random sample of 85,589 buildings from 2,765,726 was created using mod operation from the meaningless artificial identifier of buildings in RSO. The sample size is the approximation of the number of cells for 1 km grid. The uniformity of distribution in the CR was checked and evaluated as satisfactory.

The aggregation was performed for several distances; finally for this study we use aggregation within the distance 564.19 m which provides a circle with the same area as the square 1 $\times$ 1 km. Furthermore, the model is labelled as RB1km2. The aggregation of buildings around each selected building was quite slow and without optimisation, the calculation takes 2 weeks (PC AMD 8320 8 cores, 8 GB RAM).

4.2 Aggregation Around Randomly Generated Points

The set of randomly distributed points was generated in ArcGIS using a function "create random points". The number of points (80,131) is the same as cells in 1 km grid, the distance for aggregation was 564.19 m to assure the same aggregation area as other two methods. Furthermore, the model is labelled as RP1km2.

4.3 Decision Trees

The variables and the settings for the decision tree method are the same as in the above analysis. The example displaying a part of the decision tree for the RB1km2 model is given in Fig. 4. The maximal probability of graffiti crime incidence is 94.8 % (91 occurrences) for data groups with a positive occurrence of secondary schools in the Prostějov (PV) district.

The resulted trees look differently showing different lists of independent variables, in a different order, different selected districts (Table 7). The RP model seems to be weaker according to the low probability for yes target category (Table 6). The random distribution of points in the country leads to slightly higher distances to the location of crimes or descriptors, but it does seem to fully explain the worse behaviour of this model.

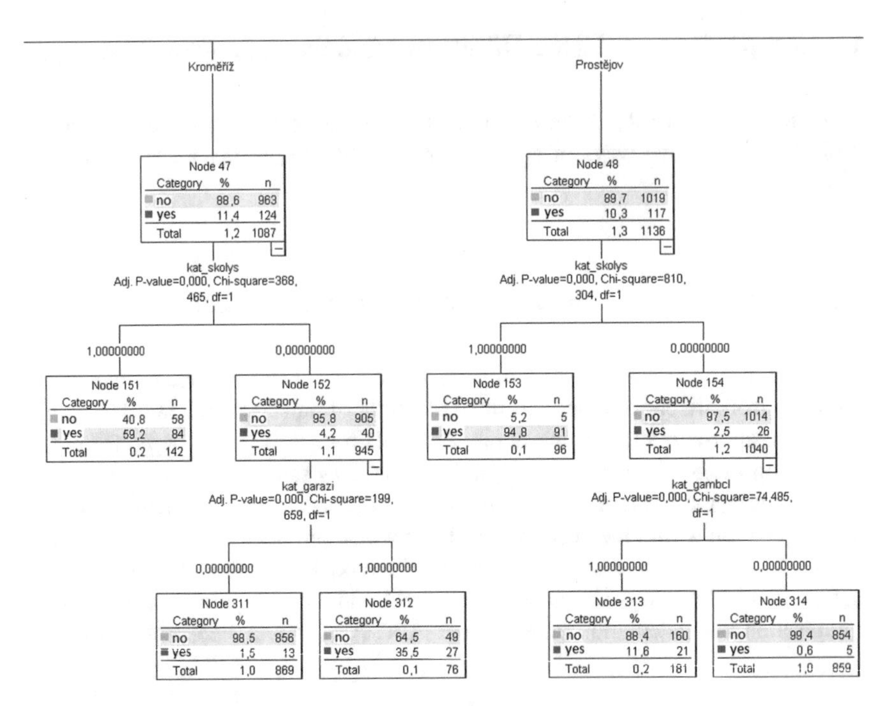

Fig. 4 Detail of decision tree for RB1km2 around the maximal probability of positive graffiti crime results (node 153)

Further, the differences between models were evaluated in more details. SQL WHERE conditions were written for all nodes in the classification trees with probability p for yes higher than 1 %. In the next step, districts were selected for those models where the positive occurrence of secondary schools (secondary = 1) is the part of SQL WHERE condition. The similar selection was made for positive occurrences of gambling clubs. The outputs follow.

4.4 *Influence of Secondary Schools Occurrence*

The overlay of results for grid 1km and RB1km2 for $p > 1$ % reaches 47.8 % (22 from 46 districts). All non-overlayed districts (occurrence only in one model) are provided by the grid model. Thus, the grid model provides more indications. The overlay of results for grid1 km and RP1km2 for $p > 1$ % is 28.6 % (22 from 77 districts). All three models coincide in 22 following districts: Praha (AB), Beroun (BE), České Budějovice (CB), Písek (PI), Litoměřice (LT), Havlíčkův Brod (HB), Jihlava (JI), Třebíč (TR), Žďár nad Sázavou (ZR), Blansko (BK), Brno-venkov

Table 7 Parameters of decision trees for prediction of graffiti crime occurrence in three models

Aggreg	List of independent variables	Percent correct prediction "yes" (%)	Max. response (prediction % for yes)	1st variable	2nd variable	3rd variable
Grid1km2	Secondary schools, district, property offences, primary schools, garages, kindergartens, gambling clubs, all flats	26.6	71.1	Secondary schools = 1	District = Písek, Brno-město, Jeseník, Prostějov, Ostrava-město, Beroun	
RB1km2	District, gambling clubs, secondary schools, garages, primary schools, kindergartens, property offences	46.1	94.8	District = Prostějov	Secondary schools = 1	
RP1km2	Gambling clubs, secondary schools, property offences, district	9.3	56.8	Gambling clubs = 1	Secondary schools = 1	District = Přerov, Tachov, Děčín, Ústí nad Labem, Jablonec nad Nisou, Liberec, Jičín, Plzeň-jih, Brno-město, Plzeň-město, Opava, Benešov, Kladno, Mělník, Nymburk, Praha-východ, Praha-západ, Pardubice

(BO), Břeclav (BV), Hodonín (HO), Znojmo (ZN), Olomouc (OL), Prostějov (PV), Přerov (PR), Šumperk (SU), Kroměříž (KM), Uherské Hradiště (UH), Zlín (ZL) and Karviná (KA).

The overlay of results for grid 1 km and RB1km2 for $p \geq 20$ % reaches 74 % (20 from 27 districts). The majority of partly indicated districts (occurrence only in one model) are provided by the grid model except Břeclav (BV). All three models approve 11 following districts: Beroun (BE), Písek (PI), Litoměřice (LT), Jihlava (JI), Třebíč (TR), Znojmo (ZN), Olomouc (OL), Přerov (PR), Šumperk (SU), Zlín (ZL) and Karviná (KA). Increasing p limit to 30 % does not show any change in district assignments.

The overlay of results for grid1 km and RB1km2 for $p \geq 40$ % is slightly decreased to 70 % (16 from 23 districts). Districts České Budějovice (CB), Hodonín (HO), Přerov (PR) and Zlín (ZL) are declared only for RB model, while Brno-město (BM), Jeseník (JE) and Ostrava-město (OT) are only in the grid model. The p limit of 40 % seems to be too high for grid models—four missing districts are indicated in the grid model but with lower p. The model RP1km2 provides totally different results, the coincidence with grid models is in one district (Brno-město BM) and with the RB1km2 model also in one district (Přerov PR).

4.5 Influence of the Gambling Devices Occurrence

The overlay of results for grid1 km and RB1km2 for $p > 1$ % is 41.9 % (18 from 43 districts). Majority of partly indicated districts (occurrence only in one model) is provided by the grid model, and only following districts are indicated only by RB1km2 model: České Budějovice (CB), Strakonice (ST), Plzeň-jih (PJ), Brno-město (BO), Hodonín (HO), Jeseník (JE), Přerov (PR), Zlín (ZL), Ostrava-město (OT). All three models coincide in 18 districts (23.4 % from all indicated districts)—Hlavní město Praha (AB), Tábor (TA), Litoměřice (LT), Most (MO), Havlíčkův Brod (HB), Jihlava (JI), Pelhřimov (PE), Třebíč (TR), Žďár nad Sázavou (ZR), Blansko (BK), Brno-venkov (BO), Břeclav (BV), Znojmo (ZN), Olomouc (OL), Šumperk (SU), Kroměříž (KM), Uherské Hradiště (UH) and Karviná (KA).

Increasing p limit to 20 % provides better coincidence with the RB model (48.1 %, 13 from 27 districts). All partly indicated districts [České Budějovice (CB), Brno-město (BM), Hodonín (HO), Přerov (PR), Zlín (ZL)] are generated by the RB model; no surplus districts are indicated by the grid model. Coincidence of three models is found only in 9 districts [14.5 %; Litoměřice (LT), Jihlava (JI), Třebíč (TR), Žďár nad Sázavou (ZR), Blansko (BK), Znojmo (ZN), Olomouc (OL), Šumperk (SU) and Karviná (KA)].

4.6 *Simultaneous Combination of Both Secondary Schools and Gambling Devices Occurrences*

The overlay of results for grid1 km and RB1km2 for $p > 1$ % is 39.4 % (15 from 38 districts). The majority of partly indicated districts (occurrence only in one model) is provided by the grid model, and only following districts are indicated only by RB1km2 model: České Budějovice (CB), Hodonín (HO), Přerov (PR), Zlín (ZL). All three models coincide in 15 districts (19.7 % from all indicated districts)—Hlavní město Praha (AB), Litoměřice (LT), Havlíčkův Brod (HB), Jihlava (JI), Třebíč (TR), Žďár nad Sázavou (ZR), Blansko (BK), Brno-venkov (BO), Břeclav (BV), Znojmo (ZN), Olomouc (OL), Šumperk (SU), Kroměříž (KM), Uherské Hradiště (UH) and Karviná (KA).

Increasing p limit to 20 % improves the overlay. The coincidence reaches 68.4 % (13 from 19 districts). All partly indicated districts (occurrence only in one model) is provided by the RB model and the grid model was unable to discover them. All three models agree only seven districts (only 12.1 %)—Litoměřice (LT), Jihlava (JI), Třebíč (TR), Znojmo (ZN), Olomouc (OL), Šumperk (SU) and Karviná (KA). The level of coincidence of two main factors for p above 30 % is portrayed in following figures (Figs. 5, 6).

The majority of selected districts are the same in both figures. Overall, the model of aggregation around randomly selected buildings (RB1km2) provides stronger

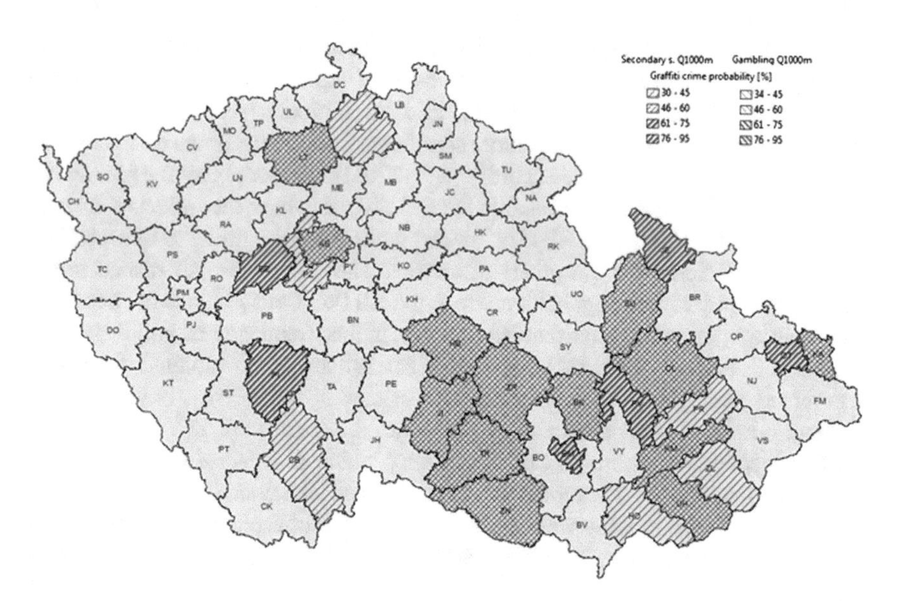

Fig. 5 Probability of the influence of secondary school occurrence and gambling device occurrence on graffiti crime occurrence by grid1km2 model

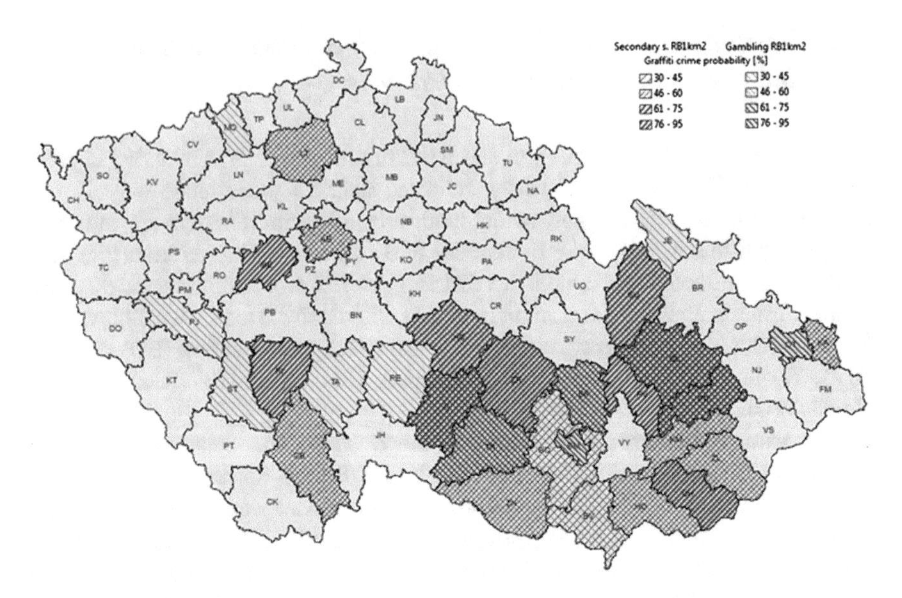

Fig. 6 Probability of the influence of secondary school occurrence and gambling device occurrence on graffiti crime occurrence by random building1km2 model

relationships than grid1km2. It is the reason why several districts (MO, PJ, ST, TA, BO, BV) are added to the selection in this model (while only PZ and CL are missing). More interesting is the fact that three districts (Jeseník JE, Ostrava-město OT, Brno-město BM) show opposite dominance of explored factors in both models.

In these districts, according to the grid1km2 model, the influence of the occurrence of secondary school is more significant than the occurrence of gambling clubs while the opposite is true according to the RB1km2 model (but the probability is higher for grid1km2). Additionally, several districts show the influence of both factors with $p > 30\ \%$ in the RB1km2 model while the only influence of secondary schools is significant for the grid1km2 model. The differences are related to the different method of data aggregation where the RB1km2 model tends to strengthen occurrences in an urban environment due to a higher frequency of buildings.

The selection of districts with significant influence of explored factors is determined mainly by the count of graffiti crimes in the district; obviously less frequent crimes cannot be evident as a significant relationship. The higher occurrence is reported in eastern part of our country which is reflected in the figures.

Despite the evident differences between obtained classification trees, the coincidence in indicating places with significant influence of secondary schools and/or gambling clubs occurrence is on the satisfactory level and it strengthens our trust to these models.

5 Conclusion

The grid models are easy to create using the multidimensional modelling and mainly they facilitate to combine various factors together. Other models grouping objects around point or building are extremely time-consuming and this fact is disqualifying them for operational analytical usage. The study highlighted the differences between these models and indicated the suitability of grid based regression models.

The study focused on the graffiti crimes and selected factors (property offences, buildings, flats, garages, educational facilities, and gambling clubs) which may influence the graffiti crimes occurrence. For regression analysis decision trees with the exhaustive CHAID growing method were selected.

Three different spatial resolutions of the grid were tested (100 m, 500 m, 1 km). The model of 1 km grid was evaluated as the best. The most influencing factors are occurrence of secondary schools and gambling devices, especially in following districts: Brno-město (BM), Ostrava-město (OT), Prostějov (PV), Písek (PI), Jeseník (JE), Beroun (BE), Praha (AB), Olomouc (OL), Šumperk (SU), Karviná (KA), Kroměříž (KM), Znojmo (ZN), Blansko (BK), Třebíč (TR), Jihlava (JI), Havlíčkův Brod (HB), Litoměřice (LT), Uherské Hradiště (UH), and Žďár nad Sázavou (ZR).

To verify the results of the decision tree for grid1km2, alternative aggregation models were built and analysed in a similar way. The comparison approves the satisfactory level of a coincidence especially for probability above the limit of 20 %.

The utilisation of decision trees for analysis of sparse big data should be further studied and verified using other data sets. The sensitive issues are mainly the ability to detect statistically significant relationships of investigated phenomena reported in sparse data and the stability of results using different aggregation approaches.

Acknowledgments Data is provided by the courtesy of the Czech Statistical Office, Police of the Czech Republic, Czech Ministry of Finance. The research is supported by the research of the Czech Ministry of Interior, project VF20142015034 "Geoinformatics as a tool to support integrated activities of safety and emergency units".

References

Act No. 40/2009, Zákon trestní zákoník (Penal Code). In: Collection of acts 9. 2. 2009. ISSN 1211-1244

Alkhasawneh MS, Ngah UK, Tay LT et al (2014) Modeling and testing landslide hazard using decision tree. J Appl Math. doi:10.1155/2014/929768 Article ID 929768

Andresen MA, Malleson N (2013) Spatial heterogeneity in crime analysis. In: Leitner M (ed) Crime modeling and mapping using geospatial technologies. Geotechnologie and Environmnet 8, Springer, Dordrecht. doi: 10.1007/978-94-007-4997-9_1

Armitage R (2002) Tackling anti-social behaviour: what really works. NACRO, London

Bandaranaike S (2001) Graffiti: a culture of aggression or assertion? The character, impact and prevention of crime in regional Australia. Australian Institute of Criminology, Townsville

Bacler LH (2014) EFGS and integration of geography and statistics. In: Proceedings of EFGS Krakow. 22–24 October 2014
Badard T, Kadillak M, Percivall G, Ramage S, Reed C, Sanderson M, Singh R, Sharma J, Vaillancourt L (eds) (2012) Geospatial business intelligence (GeoBI). OGC White paper. Ref. Number OGC 09-044r3
Brown C, Liebovitch L (2010) Fractal analysis. SAGE. 2010, 165 s. ISBN 978-4129-7165-2
Buck AJ, Hakim S, Swanson Ch, Rattner A (2003) Vandalism of vending machines: factors that attract professionals and amateurs. J Crim Justice 31(1):85–95. doi:10.1016/S0047-2352(02)00201-5
Cohen S (1969) Hooligans, vandals and the community: a study of social reaction to juvenile delinquency. PhD thesis, The London School of Economics and Political Science (LSE)
Ferrell J (1995) Urban graffiti: crime, control, and resistance. Youth Soc 27(1):73–92
Fojtík D, Horák J, Orlíková L, Kocich D, Inspektor T (2016) Smart geocoding of objects. In: Proceedings of ICCC 2016, Tatranská Lomnica
Gibbons S (2004) The costs of urban property crime. Econ J 114(499):F441–F463. doi:10.1111/j.1468-0297.2004.00254.x
Goldstein AP (1997) Controlling vandalism: The person-environment duet. In: Conoley J, Goldstein A (eds) School violence intervention: a practical handbook. Guilford Press, New York, pp 290–321
Halsey M, Young A (2002) The meanings of graffiti and municipal administration. Australian. 35 (2):165–186. doi:10.1375/acri.35.2.165 ISSN 0004-8658
Hendl J (2006) Přehled statistických metod zpracování dat. Portál, Praha. ISBN 80-7367-123-9
Horák J, Horáková B (2007) Datové sklady a využití datové struktury typu hvězda pro prostorová data. In: Proceedings of GIS Ostrava 2007. Ostrava, 28–31 January 2007. ISSN 1213-2454
Horák J, Ivan I, Drozdová M, Horáková B, Bala P (2016) Multidimensional database for crime prevention. In: Proceedings of ICCC 2016, Tatranská Lomnica
Horák J, Ivan I, Horáková B (2015) OLAP for heterogeneous socio-economic data—the challenge of integration, analysis and crime prevention: a Czech case study. In: Proceedings of European forum for geography and statistics, Vienna, 10–12 November 2015
Howard ER (1978) School discipline desk book. Parker, West Nyack, NY
Iveson K (2007) Publics and the city, vol xii. Blackwell, Oxford. ISBN 978-140-5127-301
Lachmann R (1988) Graffiti as career and ideology. Am J Sociol 94(2):229–250
Ley D, Cybriwsky R (1974) Urban graffiti as territorial markers. Ann Assoc Am Geogr 64(4):491–505
Loshin D (2012) Business intelligence: the savvy managers guide. MORGAN Kaufmann, Newnes, USA, p 370
Marešová A (2011) Resortní statistiky – základní zdroj informací o kriminalitě v České republice. Vybrané metody kriminologického výzkumu. Institut pro kriminologii a sociální prevenci, Praha. ISBN 978-80-7338-110-3. Available at http://www.ok.cz/iksp/docs/385.pdf
McCormick J (2003) "Drag me to the Asylum": disguising and asserting identities in an urban school. Urban Rev 35(2):111–128
Megler V, Banis D, Chang H (2014) Spatial analysis of graffiti in San Francisco. Appl Geogr 54:63–73. doi:10.1016/j.apgeog.2014.06.031
SPSS (2007) PASW statistics 18 command syntax reference. Documentation
Stocker TL, Dutcher LW, Hargrove SM, Cook EA (1972) Social analysis of graffiti. J Am Folklore 85(338):356–366. doi:10.2307/539324
Tygart C (1988) Public school vandalism: toward a synthesis of theories and transition to paradigm analysis. Adolescence 23:187–199
Thompson K, Offler N, Hirsch L, Every D, Thomas MJ, Dawson D (2012) From broken windows to a renovated research agenda: a review of the literature on vandalism and graffiti in the rail industry. Transp Res Part A Policy Pract 46(8):1280–1290. doi:10.1016/j.tra.2012.04.002. ISSN09658564
Wilson JQ, Keeling GL (1982) Broken windows. Atl Mon 249(3):29–38
Wilson P, Healy P (1987) Research brief: graffiti and vandalism on public transport. Australian Institute of Criminology, Woden, ACT. ISBN 06-421-1868-X

Towards Better 3D Model Accuracy with Spherical Photogrammetry

Handoko Pramulyo, Agung Budi Harto, Saptomo Handoro Mertotaroeno and Arnadi Murtiyoso

Abstract Spherical photogrammetry (SP) is an efficient method in 3D model acquisition. The SP method is fast, low cost, and the resulting of 3D model is reliable. SP has many advantages, but it also has many disadvantages. Since it uses panoramic images as the basic information, significant distortion within the panoramic image causes the accuracy of the 3D model to be less accurate. These distortions need to be corrected and one way to correct and reduce the distortion effect is by performing camera calibration through laboratory calibration and self-calibration. In previous research, self-calibration actually had been performed without laboratory calibration, but there is still miss stitching and alignment in the high resolution panoramic image (Brown and Lowe in Int J Comput Vision 74 (1):59–73, 2007). Laboratory calibration and self-calibration are procedures which aims to determine the lens distortion parameters, such as radial and tangential distortion, as well as the interior camera parameters, such as focal length and the principle point of photos. All parameters are used to correct the distorted images. Thus, this research tries to perform laboratory calibration together with self-calibration. Different scenarios are made, and the result shows that the laboratory calibration and self-calibration corrects panoramic image distortion and lowers the standard deviation of the panoramic image stitching process. Although the laboratory calibration and self-calibration shows better results in the panoramic

H. Pramulyo (✉) · A.B. Harto · S.H. Mertotaroeno
Geodesy and Geomatics, Faculty of Earth Science and Technology, Bandung Institute of Technology, Ganesha No. 10, 40132 Bandung, West Java, Indonesia
e-mail: handoko.pramulyo@yahoo.co.id

A.B. Harto
e-mail: agung@gd.itb.ac.id

S.H. Mertotaroeno
e-mail: saptomohm@gmail.com

A. Murtiyoso
Photogrammetry and Geomatics Group, ICube Laboratory UMR 7357, INSA Strasbourg, Strasbourg, France
e-mail: arnadi.murtiyoso@insa-strasbourg.fr

I. Ivan et al. (eds.), *The Rise of Big Spatial Data*, Lecture Notes in Geoinformation and Cartography, DOI 10.1007/978-3-319-45123-7_8

image stitching process, the corrected panoramic image does not affect the 3D model reconstruction accuracy significantly. The accuracy is rather affected by the photogrammetric network design, which significantly influences the panoramic image projection distortion.

Keywords Distortion · Feature point · Panoramic image · Structure from motion

1 Introduction

Spherical photogrammetry (SP) is a photogrammetry method that uses panoramic images as the basic information (Fangi 2015). A panoramic image presents a wide field-of-view (FOV) of a scene that can reach 360o × 180o with high resolution (Peleg et al. 2001; d'Annibale et al. 2013; Fangi 2015). Creating a panoramic image could be done by acquiring individual photos sequences with any kind of camera and overlap all the photos sequences by a certain percentage (depends on the FOV of the camera) while rotating the camera around the "no-parallax" point of the lens by using a specific tripod and panoramic head (Kwiatek and Tokarczyk 2015).

Spherical photogrammetry is also used for accurate 3D model reconstruction (d'Annibale et al. 2013). The surveying time is fast and effective, low cost, and the result's accuracy is satisfactory (Fangi 2015), yet at the same time, this method still has many problems. The problem such as camera motion, scene motion (Brown and Lowe 2003; Sangle et al. 2011), illumination changes, and especially camera lens distortion (Shum and Szeliski 2000; Brown and Lowe 2007) affects the accuracy of the 3D model.

In earlier researches, camera lens distortion shows no significant error in the panoramic images (Brown and Lowe 2007). A good panoramic image is subjectively defined as an image with a natural mosaic result (Chung-Ching et al. 2015). At least two steps are required to create panoramic image, image stitching and image alignment (Feng et al. 2008). Stitching is a feature-matching method using feature point extraction. Scale Invariant Feature Transformation (SIFT) is among the most widely used in feature matching (Feng et al. 2008; Sangle et al. 2011; Kapur and Baregar 2013; Xiaohui et al. 2013). After that, search of the correspondence between each image pair by estimating a homography using RANSAC (random sample consensus) (Brown and Lowe 2007; Xiaohui et al. 2013) is performed. Finally, the alignment process begins, that is the process which blend the images in a seamless manner, taking care to deal with potential problems such as blurring (Sieberth et al. 2014), ghosting, etc. (Feng et al. 2008).

The purpose of those two steps is to remove miss stitching and aligning, causing artefacts in panoramic image (Feng et al. 2008), but it does not guarantee that the camera distortion will be removed (geometry specification quality of the panoramic image is not achieved). Panoramic image is good in many ways, but it has lots of distortion (Shum and Szeliski 2000). Brown and Lowe (2007) has tested the stitching process under moderate amounts of radial distortion. Actually, panoramic image recognition and approximate alignment is robust to distortion (Cannelle et al. 2010) but there still can be found noticeable artefacts in the panoramic image. These phenomena prove that the distortion correction during the stitching process is not good enough to create a good panoramic image.

Thus, this research tries to correct the distortion error by doing a camera calibration through laboratory calibration. The laboratory calibration aims to correct the error in panoramic images. The corrected panoramic image will show better data quality and better 3D model data accuracy.

2 Camera Calibration

Camera calibration is a must procedure in order to satisfy the metric requirements of photogrammetric measurement (Remondino and Fraser 2006) and to correct geometric error in an image (Brown and Lowe 2007). Camera calibration is a procedure to find the lens distortion parameter and the interior orientation parameters (Tommaselli et al. 2012).

2.1 Camera Calibration Parameter

The value of camera calibration parameter, such as radial distortion, tangential distortion, and the interior orientation parameter, such as principle distance and principle point of auto-collimation (Tommaselli et al. 2012) could be determined. Here are the explanations of the parameter:

- Radial distortion—The radial distortions cause the image to be distorted in such a way that the target seems closer or farther from the principal point. Gaussian radial distortion describes the magnitude of the radial distortion when the nominal principal distance as a basis for calculations is used, so the magnitude of these distortions varies with radial distance and may change with focusing (Remondino and Fraser 2006). It is usually expressed as a polynomial function of the radial distance from the point of symmetry (Wolf 1983).
- Tangential distortion—Tangential distortion is caused by imperfection of the camera lens manufacturing. The camera lens is ideally aligned in a collinear manner with the optical axis in the lens system. The displacement of the lens causes the geometric displacement of images known as tangential distortion (Remondino and Fraser 2006).
- Principal distance (focal length)—The principal distance is the perpendicular distance between the image plane and the perspective centre of the lens system. There is a difference between aerial and close range photogrammetry. The aerial photogrammetry uses infinite camera lens focus, however in close range

photogrammetry, a fixed camera lens focus (focusable) is used, and it is the principal distance which must be determined.

- Principal point—The principal point is the location of an image on the image plane formed by the direct axial ray passing through the centre of the lens system.

2.2 Calibration Method

The lens and camera system (Canon Powershot SX40HS is used) with laboratory camera calibration method is calibrated (with chessboard or calibration grid pattern) to determine the lens distortion and camera interior orientation. At least 15 images of the calibration grid pattern (see Fig. 1) from each of the pattern's four sides are taken before and after the surveying process.

In this research, MATLAB® (The MathWorks, Inc.) is used to compute and process the laboratory calibration. The program is used to determine camera calibration parameters and to correct the distortion in an image (see Fig. 2). The camera

Fig. 1 Photographs used for the of camera calibration of a chessboard pattern

Fig. 2 Superimposition of the distorted image and the corrected image

calibration parameters itself are f, s, cx, cy as interior orientation, k_1, k_2 as radial distortion, and p_1, p_2 as tangential distortion.

3 Spherical Photogrammetry

Spherical photogrammetry (SP) has been tested as a method to acquire 3D model reconstruction, and the result over a few years is satisfactory (Fangi 2015). SP only uses a digital camera to perform the survey, thus it is very simple and affordable. SP could be an alternative or solution when an expensive instrumentation like a terrestrial laser scanner is not practical to use (d'Annibale et al. 2013), but it has to be noted that terrestrial laser scanner produces point clouds with higher accuracy than SP. Depending on the application, SP may not be sufficient.

SP uses panoramic images as the basic information to reconstruct a 3D model (see Fig. 3). In this study, Autopano (Kolor) is used only to create the panoramic images, as it cannot reconstruct the 3D model. Thus, Agisoft Photoscan (Agisoft LLC.) was used to reconstruct the 3D model, which will be described in the 3D model reconstruction section. Autopano (Kolor) generally creates panoramic

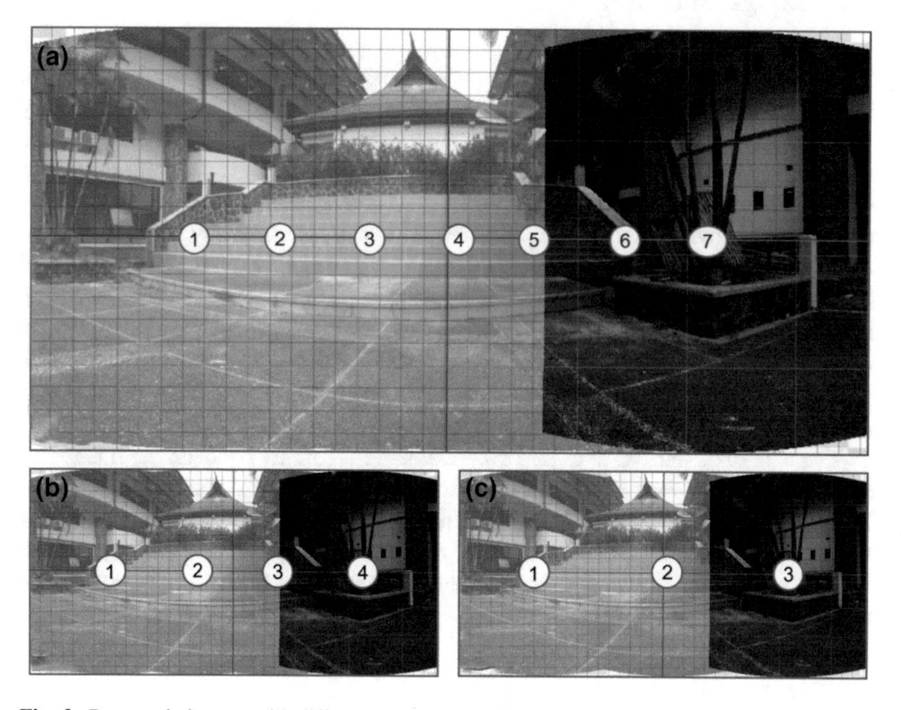

Fig. 3 Panoramic images with different percentages of image overlap. **a** Seven images registered with 73 % overlap. **b** Four images registered with 46 % overlap. **c** Three images registered with 19 % overlap

images in a few simple steps (yet the algorithm is complex); these steps are feature matching, image matching, bundle adjustment, and image blending together with a rendering step.

3.1 Feature Matching

The first step to stitch an image is to extract feature points in each images using Scale Invariant Feature Transformation (SIFT). SIFT is a feature matching algorithm which is invariant to rotation and scale change (Brown and Lowe 2003, 2007). SIFT algorithm not only establishes correspondences between points, but it also uses invariant local feature matching, which allows features to be matched under arbitrary orientation change between the two images. David Lowe proposed SIFT in 2007 with a more robust algorithm, which allows it to be invariant to rotation, scale, and change in 3D viewpoint and illumination. This method has been successfully used for panoramic image stitching (Sangle et al. 2011).

3.2 Image Matching

In photogrammetry, point extraction in an image has the same principle as the triangulation method. The aim of it is to find or extract points automatically. They are two extraction point techniques in photogrammetry, direct and feature based matching (Atkinson 1998; Mikhail et al. 2001). Direct based matching compares each pixel window between images, it is a time-consuming process, which is complex and invariant to rotation and scale, but it automatically finds the correspondence between the extracted points in images. In contrast, feature based matching cannot do the same. Feature based (for example SIFT) requires only one step to extract the feature points, but no correspondence between the images is established just yet. To finish the image matching process, matching the feature points by using Random Sample Consensus (RANSAC) is needed (Brown and Lowe 2007).

The idea of RANSAC is to find compatible homography between the images by selecting certain feature points (Brown and Lowe 2003, 2007). The basic concept of RANSAC is statistical sampling approach, which detects the inliers and outliers; if the compared feature point between the images matching represent the homograph and the ratio is within the inliers, the match is selected, otherwise the outliers are removed (Sangle et al. 2011).

3.3 *Bundle Adjustment*

Bundle adjustment is the critical step for optimizing the quality of stitched panorama. Since image matching generates a lot of accumulated error, the error is minimized by a certain amount using bundle adjustment. Bundle adjustment is also used to solve all the camera's interior parameter and the lens distortion (Pérez et al. 2011; Tommaselli et al. 2012). Knowing the interior parameter and the lens distortion allows us to perform robust panorama recognition and to better estimate the alignment for high-resolution panorama without noticeable artefact and excellent geometry.

3.4 *Image Blending and Rendering*

In commercial and free software nowadays, bundle adjustment is not the end of the step to create spherical panorama. The next step is to blend the image, so there are no colour intensity differences between images. There is always the possibility to capture images in different lighting condition, with a colour intensity differences error all over the image. Normalizing it by a function could be done, which is the sum of gain normalized intensity errors for all overlapping pixels (Brown and Lowe 2007). After blending the image, the last step is to render the image to create the spherical panorama.

4 3D Model Reconstruction

This section describes the 3D model reconstruction data processing. One way of 3D model reconstruction is to retrieve the 3D information from 2D images. Here we used Agisoft Photoscan (Agisoft LLC.) to obtain a 3D model. In photogrammetry, the techniques to acquire 3D models are divided into two categories. They are techniques to retrieve the depth of a surface in 3D space, as shape form shading and shape from focus, and they are techniques to retrieve the full reconstruction of volumetric object, such as shape from stereo, shape from motion, shape from silhouettes, and spherical photogrammetry. All of these techniques use the triangulation concept as the basic principle of the image-based modeling, except for silhouettes and shading techniques. In this research, SP is used to reconstruct 3D model, which is actually similar to shape/structure from motion (SfM) techniques (d'Annibale et al. 2013). Basic SfM uses a sequence of photographs or images as the basic information to reconstruct 3D model, but spherical photogrammetry which is also called panoramic SfM uses panoramic images as the basic information to reconstruct 3D model instead (Cohen et al. 2012).

Fig. 4 Dense point cloud of 3D model generated by using Agisoft Photoscan (Agisoft LLC.)

4.1 Measurements and Reconstruction

Image based modeling (IBM) also called photometric modeling is the method of reconstructing 3D models from images and then mapping them onto the surface of the model. The modeling software reconstructs the geometry by identifying sets of common points from two or more source images (Cohen et al. 2012). The correlation between those points is taken into consideration to form the edges and surfaces of the 3D object. To reconstruct a complete 3D model, the following problems need to be solved: projective reconstruction, sparse depth estimation, and dense depth estimation (see Fig. 4).

4.2 Structure from Motion (SfM)

Structure from motion (SfM) is an automated process of calculating camera orientations and the 3D positions of the tie points by analysing an image sequence (Kwiatek and Tokarczyk 2015). SfM includes a process where point correspondences along the stereo images are identified. This research uses panoramic images as the basic information. Thus, the image orientation should be provided by using the SIFT method (Brown and Lowe 2007). Afterwards, the algorithm will estimate the fundamental matrix of each stereo pair with the use of outliers and inliers detected by RANSAC. The software computes the orientation (projection) matrix for every image and optimizes it by doing bundle adjustment for the whole block.

4.3 Meshing and Texturing

The last steps in the reconstruction of a 3D model are meshing and texturing (Kwiatek and Tokarczyk 2015). There are different techniques of creating a reality based 3D model. The classifications of the three main techniques are volumetric representation (voxels), surface representation (polygonal mesh), and point patching techniques. Volumetric representation is commonly used in scientific visualization, computer graphics and computer vision. The disadvantage of this technique is the need for a huge amount of memory and processing, but it gives a better representation for the complex objects.

The surface representation or the polygonal meshing is a flexible way to represent 3D measurements accurately, providing an optimal surface description (Remondino and El-Hakim 2006). Furthermore, texturing is the process where the software maps the image colour onto a 3D surface. The image consists of red, green, blue (RGB) values per pixel, and it has corresponding image coordinates for each vertex of a triangle on the 3D surface by knowing the parameters of the interior and exterior orientation of the images.

5 Network Geometry Design

Understanding 3D model reconstruction is important, but there is a more important aspect that influences the outcome of 3D model reconstruction of the object, which is network geometry design (see Fig. 5) (Wolf 1983; Atkinson 1998; Saadat-Seresht et al. 2004). The aim of network geometry design is to satisfy the network quality requirement at minimum cost (Olague 2002).

Fangi (2015) proposed an orientation procedure, which is also a way to optimize the network geometry design called trinocular model. There is also an orientation procedure called binocular model. It is a model where two panoramas as one model are having common three or more tie points. The model is formed by coplanarity and linked together in a block adjustment. However, the trinocular model is a combination of three different panoramas to form a unique model. It has the advantage of being much more robust in comparison to binocular vision in the sense that the trinocular model is likely to have a smaller error than any of the three composing binocular models. It contains less model deformation. The mutual comparison of the three intersecting binocular models validates the model coordinates.

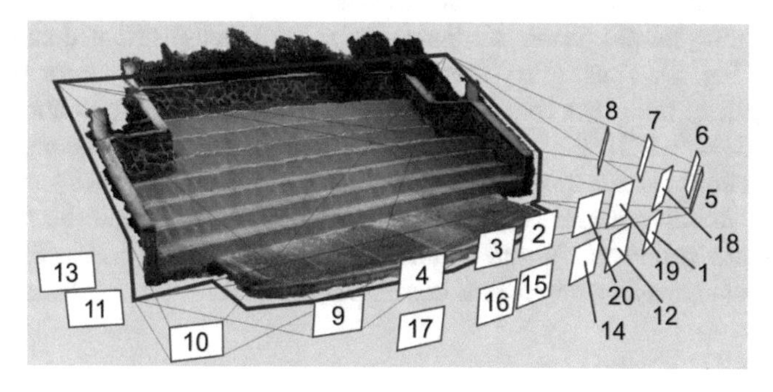

Fig. 5 Illustration of the photogrammetric network design

6 Results and Discussion

In this research, three different scenarios of panoramic image stitching were done. The first scenario stitches panoramic images with 73 % image overlap, the second scenario stitches panoramic images with 46 % image overlap, and the third scenario stitches panoramic images with 19 % image overlap. Each scenario has four different projects, the first project stitches the panoramic image without adding laboratory calibration nor self-calibration, the second project stitches panoramic image only by including laboratory calibration, the third project stitches panoramic image only by including self-calibration, and the fourth project stitches panoramic image by including both laboratory calibration and self-calibration. The entire scenario shows different panoramic image stitching standard deviation (SD) and the fourth project in every scenario show better SD results than the others (see Fig. 6). Although the software has a robust ability to recognize and align the distorted panoramic images, laboratory calibration could still be useful to determine better camera lens distortion parameters by the self-calibration process and lower the SD. However, it has to be noted that the improvement is insignificant (see Fig. 7).

From Fig. 6, we conclude that the bigger the image overlap (such as image overlap in the first scenario with 73 % image overlap), the lower the SD, but only insignificant. In addition, it is obvious that the calibration did not give satisfying results if the image set has a small overlap.

Furthermore, four different scenarios of 3D model reconstruction were done. In these scenarios, control points were measured on clear points of interest and distributed evenly in the 3D model to support the geometry for bundle adjustment (see Fig. 8).

The aim of the 3D model reconstruction scenario and the field surveying procedure is to help to understand the difference between the proposed scenario accuracy. We noticed that the 3D model reconstruction in the scenario with laboratory calibration and self-calibration (the calibration process within the panoramic image stitching) shows the smallest total error (error 4, Table 1). However, the total error of this scenario is not significantly different from the other three scenarios. In our study, both laboratory calibration and self-calibration do not significantly affect the accuracy of the 3D model reconstruction.

In this research, 20 panoramic images were with well-distributed camera positions (see Fig. 5). Thus, every scenario produces a satisfactory error of ± 3 cm. However, there are some control points with greater errors, for example, control points 1, 5, 6, 10, 21, 22, and 23 (see Table 1). These control points lie in the panoramic image area with greater projection distortion effect, which is causing the 3D model to suffer a great deal of deformation. The error could be reduced by adding more panoramic images with different orientations and adding farther camera stations while using longer camera focal length.

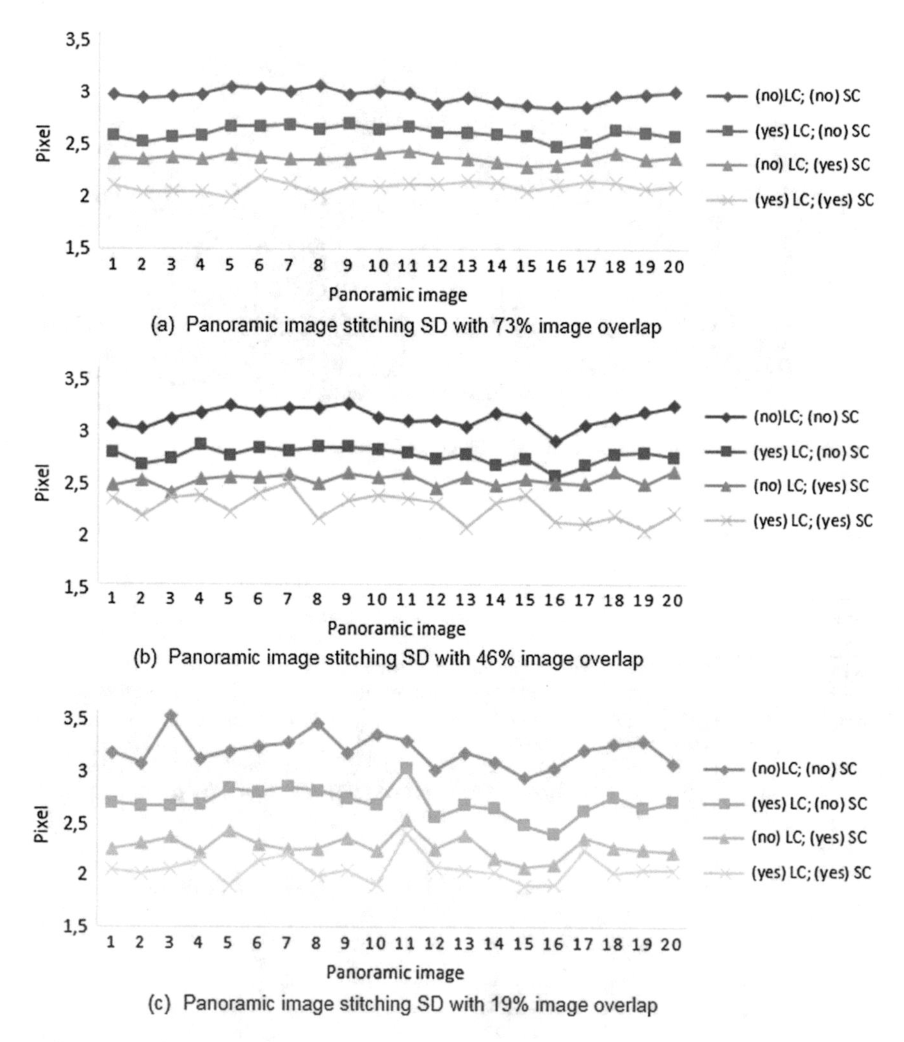

Fig. 6 Standard deviation (SD) results of panoramic image stitching. In the graphic's legend, LC denotes lab calibration and SC denotes self-calibration

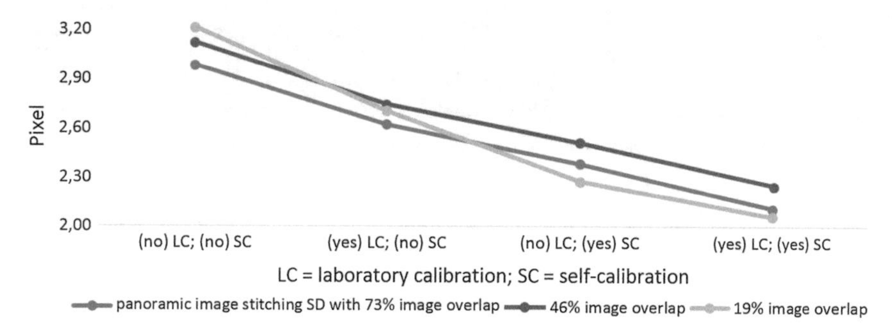

Fig. 7 The average of standard deviation (SD) results of panoramic image stitching

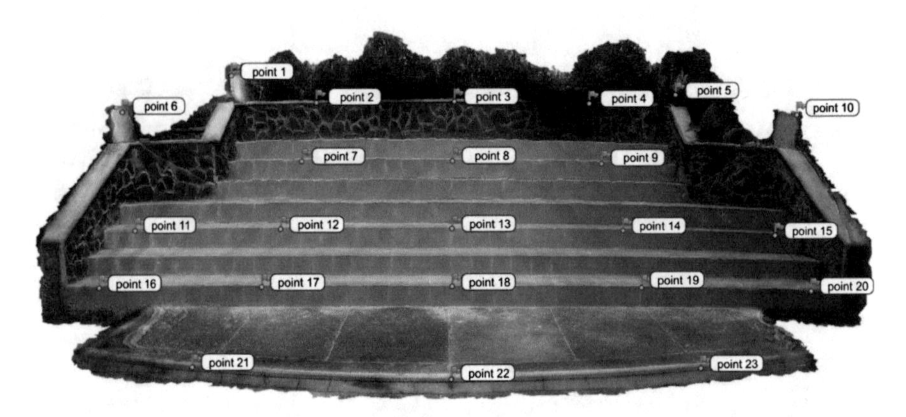

Fig. 8 Complete 3D model (textured) with 23 control points

Table 1 Errors of the control points produced by SP (panoramic images with 73 % were used) in Agisoft Photoscan (Agisoft LLC.)

Control point	X (m)	Y (m)	Z (m)	First project error (m)	Second project error (m)	Third project error (m)	Fourth project error (m)	Sparklines
point 1*	-4.38	11.96	1.98	0.086112	0.089216	0.082551	0.082127	
point 2	-2.71	11.90	1.42	0.041679	0.040805	0.03838	0.03587	
point 3	0.02	11.92	1.42	0.027144	0.026737	0.025431	0.031587	
point 4	2.67	11.93	1.43	0.024902	0.021447	0.019861	0.017641	
point 5*	4.15	11.34	1.65	0.090184	0.085083	0.080888	0.076264	
point 6*	-5.67	9.95	1.36	0.124098	0.132413	0.127184	0.132624	
point 7	-2.75	10.52	0.40	0.009906	0.007918	0.007591	0.016355	
point 8	-0.01	10.54	0.39	0.028652	0.026482	0.026152	0.024391	
point 9	2.69	10.55	0.39	0.017333	0.02384	0.025219	0.023772	
point 10*	5.79	10.00	1.42	0.186258	0.182203	0.17728	0.170228	
point 11	-5.10	8.73	-0.50	0.03789	0.037883	0.039439	0.036146	
point 12	-2.74	8.74	-0.50	0.016391	0.021018	0.018972	0.022017	
point 13	0.00	8.75	-0.50	0.040856	0.041075	0.036686	0.035309	
point 14	2.70	8.77	-0.51	0.027601	0.02857	0.028191	0.030479	
point 15	5.10	8.78	-0.51	0.041759	0.037294	0.037344	0.033726	
point 16	-5.09	7.53	-1.10	0.026412	0.029208	0.028595	0.035182	
point 17	-2.73	7.54	-1.11	0.019819	0.019606	0.022944	0.018136	
point 18	0.01	7.56	-1.11	0.031676	0.028128	0.025826	0.025966	
point 19	2.71	7.57	-1.10	0.022048	0.018784	0.017038	0.016243	
point 20	5.11	7.58	-1.11	0.048049	0.050058	0.050487	0.045977	
point 21*	-2.92	5.35	-1.44	0.069606	0.082089	0.085816	0.093201	
point 22*	-0.01	5.03	-1.45	0.049648	0.060702	0.066196	0.069158	
point 23*	2.76	5.36	-1.43	0.07877	0.088922	0.089847	0.088656	
Total error				0.030701	0.030445	0.029783	0.029311	

Points with (*) were not used in total error calculation process

7 Conclusion

We conclude that no matter which laboratory calibration and self-calibration scenario, there is only insignificant difference. Image overlap while stitching the panoramic image makes the difference. To achieve better panoramic images, large percentage of image overlap is recommended.

References

Atkinson K (1998) Close range photogrammetry and machine vision. Whittles Publishing, Scotland

Brown M, Lowe DG (2003) Recognising panoramas. ICCV, Nice, pp 1218–1225

Brown M, Lowe DG (2007) Automatic panoramic image stitching using invariant features. Int J Comput Vision 74(1):59–73

Cannelle B, Paparoditis N, Tournaire O (2010) Panorama-based camera calibration. International archives of photogrammetry, remote sensing and the spatial sciences. vol XXXVIII, Part 3A, pp 73–78

Chung-Ching L, Pankanti SU, Ramamurthy KN, Aravkin AY (2015) Adaptive as-natural-as-possible image stitching. In: Computer vision and pattern recognition (CVPR), IEEE, pp 1155—1163

Cohen A, Zach C, Sinha SN, Pollefeys M (2012) Discovering and exploiting 3D symmetries in structure from motion. In: IEEE computer society, pp 1514–1521

d'Annibale E, Tassetti A, Malinverni E (2013) From panoramic photos to a low-cost photogrammetric workflow for cultural heritage 3D documentation. International archives of the photogrammetry, remote sensing and spatial information sciences, vol XL-5/W2, pp 213–218

Fangi G (2015) Towards an easier orientation for spherical photogrammetry. International archives of the photogrammetry, remote sensing and spatial information sciences, vol XL-5/W4, pp 279–283

Feng L, Yu-hen H, Gleicher ML (2008) Discovering panoramas in web videos. In: 16th ACM international conference on multimedia, pp 329–338

Kapur J, Baregar AJ (2013) Security using image processing. Int J Manag Inf Technol 5(2):13–21

Kwiatek K, Tokarczyk R (2015) Immersive photogrammetry in 3D modelling. Geomat Environ Eng 9(2):51–62

Mikhail EM, Bethel JS, McGlone JC (2001) Introduction of modern photogrammetry. Wiley, New York

Olague G (2002) Automated photogrammetric network design. Photogramm Eng Remote Sens 68 (5):423–431

Peleg S, Ben-Ezra M, Pritch Y (2001) Omnistereo: panoramic stereo imaging. IEEE Trans Pattern Anal Mach Intell 23(3):279–290

Pérez M, Agüera F, Carvajal F (2011) Digital camera calibration using images taken from an unmanned aerial vehicle. International archives of the photogrammetry, remote sensing and spatial information sciences, vol XXXVIII-1/C22, pp 1–5

Remondino F, El-Hakim S (2006) Image-based 3D modelling: a review. Photogram Rec 21 (115):269–291

Remondino F, Fraser C (2006) Digital camera calibration methods: considerations and comparisons. International Archives of Photogrammetry, Remote Sensing and the Spatial Sciences, vol XXXVI, Part 5, pp 266–272

Saadat-Seresht M, Samdzadegana F, Azizi A, Hahn M (2004) Camera placement for network design in vision metrology. International archives of the photogrammetry, remote sensing and spatial information sciences, vol XXXV-B5, pp 105–109

Sangle P, Kutty K, Patil A (2011) A novel approach for generation of panoramic view. Int J Comput Sci Inf Technol 2(2):804–807

Shum HY, Szeliski R (2000) Construction of panoramic. Int J Comput Vision 36(2):101–130

Sieberth T, Wackrow R, Chandler JH (2014) Influence of blur on feature matching and a geometric approach for photogrammetric deblurring. International archives of the photogrammetry, remote sensing and spatial information sciences, vol XL-3, pp 321–326

Tommaselli AMG, Junior JM, Telles SSS (2012) Camera calibration using straight lines: assessment of a model based on plane equivalence. Photogramm J Finl 23(1):1–11

Wolf PR (1983) Elements of photogrammetry. McGraw-Hill, Madison

Xiaohui W, Kehe W, Shengzhuang W (2013) Research on panoramic image registration approach based on spherical model. Int J Signal Process Image Process Pattern Recognit 6(6):297–308

Surveying of Open Pit Mine Using Low-Cost Aerial Photogrammetry

Ľudovít Kovanič, Peter Blišťan, Vladislava Zelizňaková and Jana Palková

Abstract For the geodetic survey of the terrain surface, the use of conventional surveying methods and instruments is common. New technologies, such as e.g. UAVs and their combination with a digital camera, bring new opportunities also in the documentation of Earth's surface. This combination of technologies allows the low-cost digital photogrammetry use to document the Earth's surface in relation to mining activities. The aim of the research presented in this paper is to analyse the accuracy of the digital elevation model obtained using low-cost UAV photogrammetry. The surface mine Jastrabá (Slovakia) was chosen as the test area. The mine has morphologically dissected surface and is thus suitable for verifying the use of UAV photogrammetry to capture fairly intricate details on the surface. The use of UAV in a photogrammetric survey of the mine brings new opportunities for the creation of documentation because with this technology we can measure the entire surface in detail, create orthophoto maps of the entire area and document inaccessible parts of the area such as sludge dumps, steep slopes, etc.

Keywords Low-cost UAV · Photogrammetry · Point cloud

Ľ. Kovanič (✉) · P. Blišťan · V. Zelizňaková · J. Palková
Institute of Geodesy, Cartography and Geographical Information Systems,
Faculty of Mining, Ecology, Process Control and Geotechnology,
Technical University of Košice, Park Komenského 19, 04200 Košice,
Slovak Republic
e-mail: ludovit.kovanic.2@tuke.sk

P. Blišťan
e-mail: peter.blistan@tuke.sk

V. Zelizňaková
e-mail: vladislava.zeliznakova@tuke.sk

J. Palková
e-mail: jana.palkova@tuke.sk

I. Ivan et al. (eds.), *The Rise of Big Spatial Data*, Lecture Notes
in Geoinformation and Cartography, DOI 10.1007/978-3-319-45123-7_9

1 Introduction

During the process of minerals extraction by surface mining, we often see requirements to document the surface in order to track the mining process. The surveying of the quarry is usually carried out regularly depending on the method of extraction. The terrain and objects in the mine are surveyed, subsequently, a map of the surface is prepared and usually also a 3D model of the mine is constructed, which is used to calculate the volume of mined material. Surveying of the quarry is usually performed using conventional surveying methods and instruments. This process is in the case of large cast mines laborious and tedious. The result of geodetic measurements, carried out with the goal to document the mined surface, is therefore a set of points with coordinate X, Y, Z. The quality of realization of the surface in the DEM directly depends on the accuracy of the devices and the number of measurement points. To achieve an accurate and detailed surface, it is necessary to survey a greater number of points on the surface, what is time-consuming. The first option for documentation of the surface is to use photogrammetry, with special software which enables creating a point cloud from acquired photos. The modern trend is the use of unmanned aerial vehicles—UAV (Zhang and Elaksher 2011). Photogrammetry from a UAV is an inexpensive and relatively accurate method to capture the extraction and document the process of mining.

The purpose of this article is to present the usability of UAVs for the documentation of surface mines. Currently, commercially available and inexpensive UAV carriers exist on the market, and they are an interesting solution to reduce the cost of the documentation process. Their main drawback is often the lower quality of compact cameras. They are used thanks to their low weight in cheap UAVs. In our research, we aimed to create the digital surface model of the quarry, which was obtained photogrammetrically using low-cost UAV and by traditional geodetic methods represented by the total station (TS). This issue is addressed in several papers (Fritz et al. 2013; Niranjan et al. 2007) and brings interesting results.

2 Study Area

Area of interest selected for testing the quality of digital surface models of mine obtained by photogrammetry from UAV and by traditional surveying techniques using TS was surface mine in the area Jastrabá—near Žiar nad Hronom (Slovakia) (Fig. 1). The surface in this mine is unpaved and morphologically rugged, which complicates the surveying. In the direction of the terrain slope, there are numerous scratches created by running water in heavy rainfall.

Quarry Jastrabá was particularly chosen with the goal to test the precision of aerial photogrammetry utilizing the low-cost UAV for documenting mines to create a detailed and precise enough surface model.

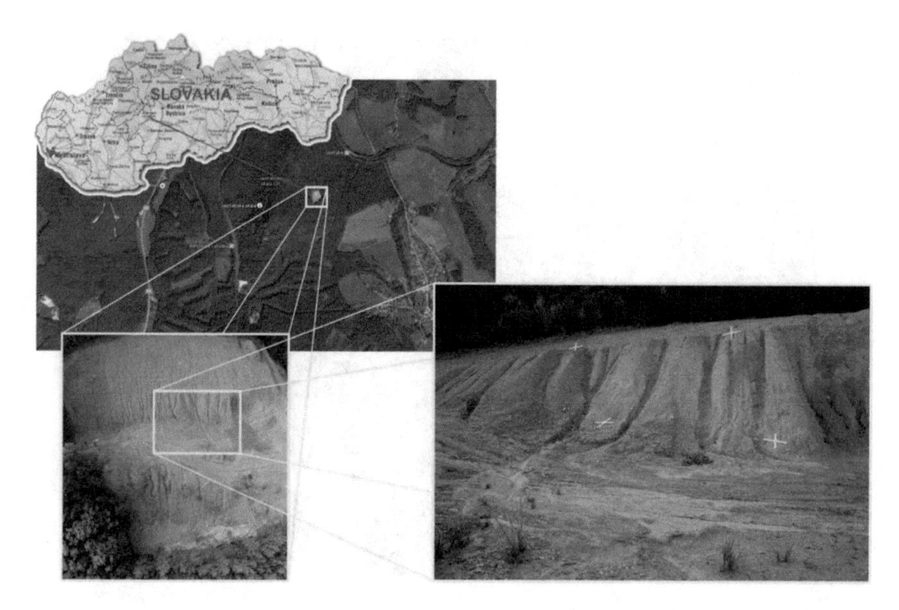

Fig. 1 Area of interest

3 Methods, Equipments and Application

Testing area—surface mine Jastrabá was surveyed terrestrially using Leica TS 02 total station (Leica FlexLine TS 02 2015) and photogrammetrically using UAV DJI Phantom 2 Vision+ with a 14 megapixels FC200 camera. Points obtained by terrestrial surveying and photogrammetric acquisition were subsequently processed into a DEM.

3.1 Total Station

Altogether 439 points composing the basic skeleton of the terrain and 10 ground control points (GCP)—photogrammetric targets were surveyed by TS. The connection of the measurement was performed through the mining field of points previously surveyed by the GNSS—RTK method in the coordinate system S-JTSK (the unified trigonometric cadastral network) and height system Bpv (Balt after settlement). These coordinate systems are obligatory in the Slovak Republic. On the Fig. 2 the measured points are marked by a flag and GCP points by a circle.

Fig. 2 Model of the area, detailed points and control points

3.2 UAV Photogrammetry

The flight was carried out in six strips in the height of 35 m above the average height of the terrain (Fig. 3). Manual handling of the UAV was applied. Altogether 135 images were captured. Ten GCPs with coordinates acquired terrestrially using the total station for location and registration of the model were used. Results of the

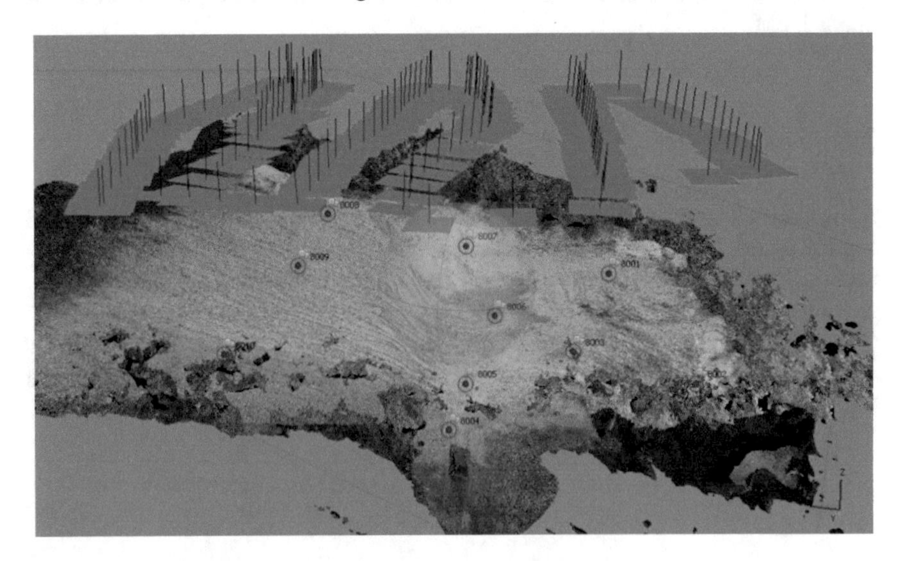

Fig. 3 Model of the area with positions of images and control points

photogrammetric data processing show the following characteristics of accuracy. Ground resolution is 0.01 m/pixel. Mean error in the position of the measurement point is 1.96 pixels. The mean average error on ground control points is 0.043 m, which corresponds to 0.86 pixels. Resolution of the point cloud is 0.0403 m/pixel and the density of points is 613 points/m^2. Longitudinal and lateral overlay of the images 80 and 60 % was achieved.

4 Results and Discussion

The final result of the aerial imagery processing in Agisoft PhotoScan® software was a cloud of points that creates the 3D model of the surface. Subsequently, this point cloud was processed in Trimble RealWorks® software. The same software for processing of data obtained by direct geodetic measurements with TS was chosen. The reduced point cloud (2,324,715 points) was used to continuous surface model creation—mesh. Equally, the mesh surface using the points obtained by terrestrial methods of measurement using the TS was created.

To assess the accuracy of the determination on control points was calculated the root mean squared error—RMSE (Wu et al. 2012). RMSE value for the control points is 4.79 cm.

4.1 Analysis of the Accuracy of the DEM Derived from the Photogrammetric Processing

Digital elevation models derived from terrestrial measurements and photogrammetric data were merged, compared and analyzed in Trimble RealWorks® software. As a reference the model created from the points obtained by terrestrial measurements using the TS was selected. Heights of the points as the distances between single surfaces were compared. The result of comparison of the two surface models is a hypsometric expression of the difference between the two surfaces. We divided the points based on their deviation from the reference surface in the specified intervals Fig. 4. Table 1 shows the absolute and percentage proportion of points based on their deviation from the reference surface in the specified intervals. The maximum value of the difference between the two compared surfaces is 834 mm. This extreme value can be easily identified and assessed.

Figure 4 shows that the largest deviations of the generated surfaces were detected in locations of terrain furrows, in the potholes and on the vertical walls of the quarry. As a result, we can only confirm that usability of the total station to document such complex morphological shapes and detailed documentation of steep and uneven walls of mines is highly limited (Uysal et al. 2015; Stöcker et al. 2015). However specific morphological shapes on the surface model e.g. potholes caused

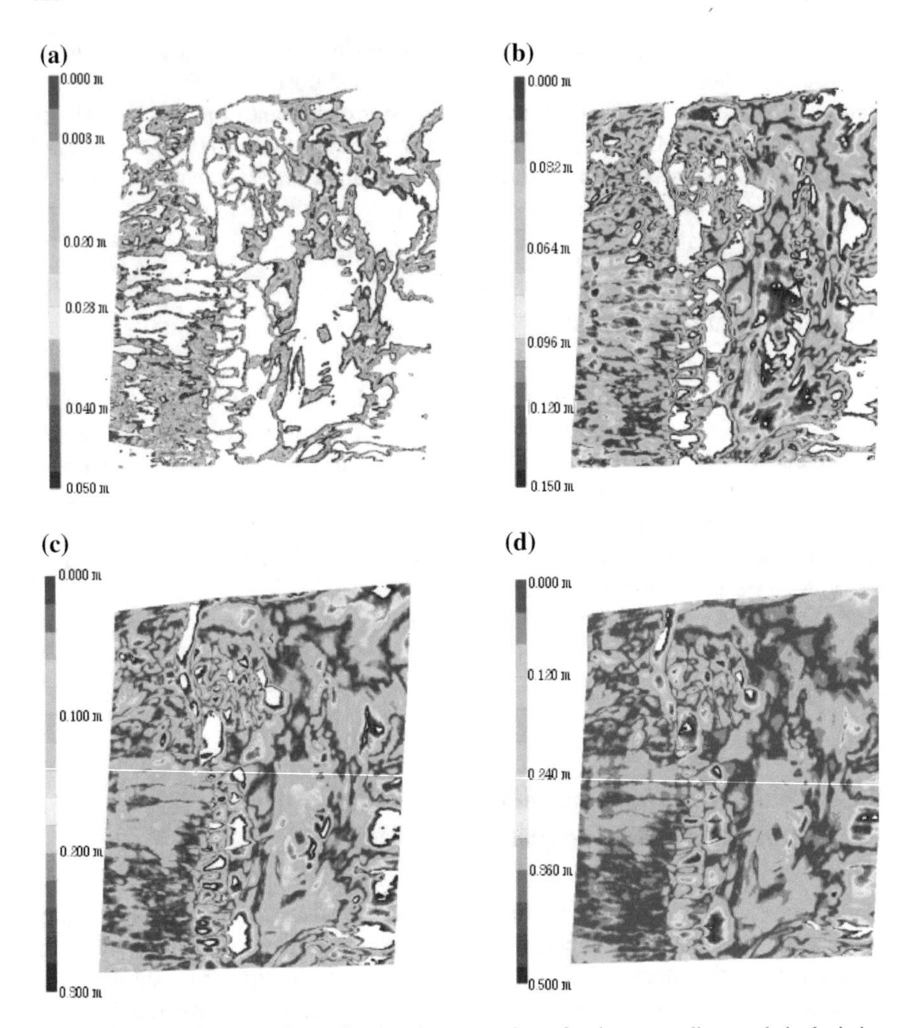

Fig. 4 Hypsometric expression reflecting the proportion of points according to their deviation from the reference surface in the defined intervals: **a** deviation up to 0.05 m, **b** the deviation up to 0.15 m, **c** deviation up to 0.30 m and **d** deviation up to 0.50 m

Table 1 Calculation of the proportion of points according to their deviation from the reference surface and the division into intervals

Deviation from the reference surface (m)	Percentage ratio of points (%)	Absolute number of points
Up to 0.050	44.48	783,267
Up to 0.150	82.56	1,761,053
Up to 0.300	93.09	2,133,158
Up to 0.500	98.57	2,291,569
Up to 0.834	100.00	2,324,715

by water erosion on the top of the vertical wall, are well captured photogrammetrically. These potholes appearing as a deviation of the two surfaces are well identifiable also in Fig. 4c. Points at those parts of the quarry then appear as the areas with the greatest mutual deviations of the compared models.

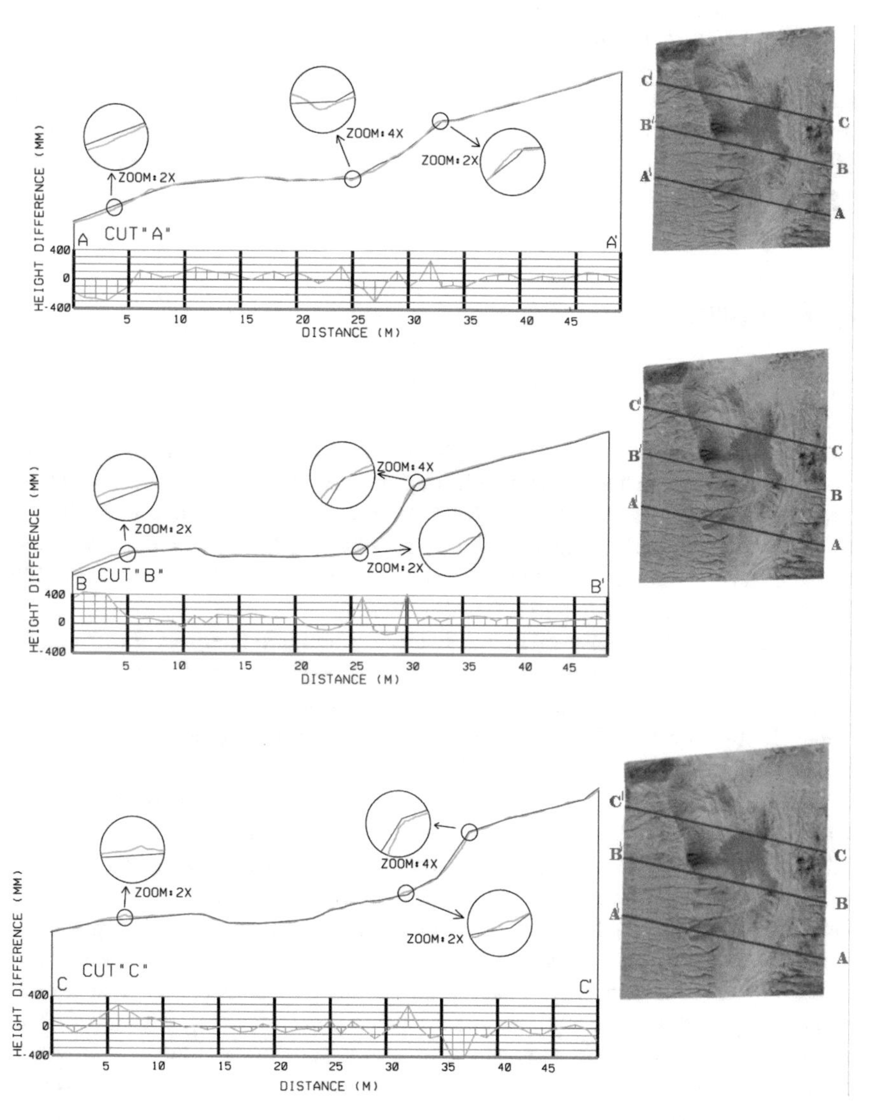

Fig. 5 Cross-sections through modelled surfaces with vertical distances between the surfaces obtained from photogrammetric data and the reference surface

4.2 *DEM Accuracy Evaluation Based on Cross-Sections*

To assess the accuracy of the used geodetic methods cross-sections of modelled surfaces have been created. In these sections, along section lines, the vertical distances between the surface created from photogrammetric data and the reference surface were determined (Kovanič and Németh 2013). These distances of modelled surfaces are presented in the charts under the section line. The red colour represents the reference surface and green colour the evaluated surface (Fig. 5).

At Fig. 5 the biggest differences between these two surfaces are shown in details. They are visible mainly in the areas of the upper and lower edges of the mine walls. The maximum value of deviation to the reference surface in the selected cross-sections achieved ±400 mm. These differences are caused by the low density of points measured by TS compared to the dense point cloud obtained from photogrammetric data processing.

5 Conclusion

The impact of each human activity is changing the Earth's surface, and it is important to identify these changes and then document their size and scope. Such changes also occur in mining for example in surface mines during mining activities. Classical terrestrial methods of documenting objects using total stations are still the most commonly used geodetic methods in surface mining. However, new data collection methods are being introduced, for example LIDAR or the UAV photogrammetry. Their big advantage is especially the speed of data collection, the density of measured points and their affordability.

To assess the suitability of low-cost UAV photogrammetry as an alternative method to conventional terrestrial surveying methods, we conducted research in surface mine Jastrabá (Central Slovakia). The surface of the quarry was documented by two methods listed above to compare the created surface models in terms of accuracy and detail.

It was confirmed that the surface model created by a photogrammetric method using low-cost UAV and the low-cost camera can be considered as a convenient data collection tool for surveying in surface mines. Approximately 45 % of the points on the surface were identified with the value of deviation to the reference surface up to 5 cm. Regarding on detail of the created model we can even say that such models are, thanks to the principle of digital photogrammetry and software processing of aerial imagery, much more detailed in comparison with models created only from terrestrially measured detailed points. The big advantage of the method used is simple data acquisition, time undemanding and fairly comfortable processing of measured data. A better camera would be appropriate to use for increase the accuracy. The method would be more efficient, and the results would improve more faithful surfaces and more accurate results.

Acknowledgments The authors are grateful to the Slovak Research and Development Agency for the support of the Project APVV-0339-12.

References

Fritz A, Kattenborn T, Koch B (2013) UAV-based photogrammetric point clouds—tree stem map-ping in open stands in comparison to terrestrial laser scanner point clouds. In: International archives of the photogrammetry, remote sensing and spatial information sciences, vol XL-1/W2, UAV-g2013, 4–6 September, Rostock, Germany

Kovanič Ľ, Németh Z (2013) Využitie geodetických metód na modelovanie stien povrchových lomov a skalných útvarov a pri ich štruktúrnej analýze: príklad z kameňolomu Sedlice, Mineralia Slovaca 45, pp 69–84, Web ISSN 1338-3523, ISSN 0369-2086

Leica FlexLine TS 02. http://www.geotech.sk/downloads/Totalne-stanice/FlexLine_TS02_Datasheet_en.pdf. Accessed 8 Nov 2015

Niranjan S, Gupta G, Sharma N, Mangal M, Singh V (2007) Initial efforts toward mission-specific imaging surveys from aerial exploring platforms: UAV, Map World Forum, Hyderabad, India

Stöcker C, Eltner A, Karrasch P (2015) Measuring gullies by synergetic application of UAV and close range photogrammetry—a case study from Andalusia, Spain; Catena, vol 132, September 01, pp 1–11

Uysal M, Toprak AS, Polak N (2015) DEM generation with UAV photogrammetry and accuracy analysis in Sahitler hill. Measurement 73:539–543

Wu Y, Zhu J, Zuo T (2012) Investigation on the square root of the linear combination of the square of RMSE and smoothness for the Vondrak filter's evaluation. Geomatics Inf Sci Wuhan Univ 37(10):1212–1214

Zhang C, Elaksher A (2011) An unmanned aerial vehicle based imaging system for 3D measurement of unpaved road surface distresses. J Comput Aided Civ Infrastruct Eng 27(2):118–129

Sentinel-1 Interferometry System in the High-Performance Computing Environment

Milan Lazecky, Fatma Canaslan Comut, Yuxiao Qin and Daniele Perissin

Abstract This paper describes the architecture of the system for monitoring deformations based on interferometrical processing of Sentinel-1 satellite data using specific software installed at the facility of the Czech national supercomputing center (IT4Innovations). The system is ready for Big Data storage and processing and leads to partially autonomous competitive environment for following deformation monitoring projects. The processing chain is prepared to be performed using Sarproz software where the reader of specific Sentinel-1 Terrain Observation with Progressive Scans in azimuth data was newly introduced—the reader implementation is described in the paper. First application of the system has been applied to subsiding area of Konya city in Turkey. The processing was performed using a small fraction of the supercomputer power, i.e. using 23 processors running at 2.5 GHz speed, and led to results estimating subsidence in Konya city during 2014–2015—a subsidence of Konya buildings in the rate of more than 6 cm/year has been evaluated, that is slightly more than the expected values from previous analyzes. The processing strategy of the described system is demonstrated in this case study.

Keywords Satellite radar interferometry · High-performance computing · Deformation monitoring · Sentinel-1 · Konya city subsidence

M. Lazecky (✉)
IT4Innovations, VSB-TUO, 17. listopadu 15, 708 33 Ostrava-Poruba, Czech Republic
e-mail: milan.lazecky@vsb.cz

F. Canaslan Comut
Turkey Disaster and Emergency Directorate (AFAD) of Denizli, Denizli, Turkey
e-mail: fatma.c.comut@afad.gov.tr

Y. Qin · D. Perissin
Lyles School of Civil Engineering, Purdue University, Lafayette, USA
e-mail: qyuxiao@purdue.edu

D. Perissin
e-mail: perissin@purdue.edu

I. Ivan et al. (eds.), *The Rise of Big Spatial Data*, Lecture Notes in Geoinformation and Cartography, DOI 10.1007/978-3-319-45123-7_10

1 Introduction

The IT4Innovations national supercomputing centre is a research institution of the Vysoka Skola Banska-Technical University of Ostrava (VSB-TUO). The first part of the supercomputer, the Anselm supercomputer, was installed into temporary mobile units in May 2013. Its theoretical computing performance was at 94 trillion floating-point operations per second (FLOPS). Supercomputer Salomon was put into operation in July 2015 as the 40th most powerful supercomputer in the world in that time. The Salomon cluster consists of 1008 compute nodes, totaling 24,192 compute cores with 129 TB RAM and giving over 2 peta FLOPS of theoretical peak performance. Each node is a powerful ×86–64 computer, equipped with 24 processing cores and at least 128 GB RAM running CentOS Linux operating system (Palkovič et al. 2015).

Sentinel-1A is a synthetic aperture radar (SAR) satellite launched in April 2014 as part of European Copernicus programme. This satellite generates acquisitions continuously every 12 days for European sites and with a lower revisit frequency for majority of other world areas, with a spatial resolution of around 20 × 4 m and a very large extent of each distributed image (240 × 350 km). This is achieved using a specific image acquisition mode called Terrain Observation by Progressive Scans (TOPS). After the launch of the second SAR satellite Sentinel-1B planned in the late 2016, the acquisition frequency in Europe will be increased into an acquisition per 6 days. Depending on stored polarimetric information, the SAR data are distributed in single look complex (SLC) data in the size of 4 or 8 GB per an acquisition, transforming the current data processing tasks into Big Data processing strategies.

The SAR Interferometry (InSAR) is a modern remote sensing technology allowing both generation of digital elevation models or measuring geophysical parameters, especially deformation in time, within sensitivity of around a millimeter per year for vertical displacements (Hanssen 2001). Various SAR satellite systems tackle with specific problems for a reliable application in local area monitoring praxis—either their spatial and temporal resolution was low (ERS, Envisat) or the data acquisition was performed only by previous often commercial orders (TerraSAR-X, Radarsat-2). Within the open access and the continuous and high spatiotemporal resolution acquisition strategies applied for Sentinel-1 data, scientists talk about a "golden age of InSAR" making a leap from an experimental technique into a practically applicable technology, since there was never before such feasible spatio-temporal coverage of SAR data.

The aim of the paper is to introduce a competitive InSAR environment installed at IT4Innovations, referenced to IT4InSAR, that allows an effective processing of Sentinel-1 data. As the core InSAR processing tool at IT4InSAR, the Sarproz software (Perissin et al. 2011) is used, implementing a flexible environment allowing for a range of various SAR and InSAR processing techniques, including also the most commonly used permanent/persistent scatterers (PS) technique (Ferretti et al. 2000). Lately, the whole processing chain for importing Sentinel-1

TOPS data, their coregistration and InSAR processing has been implemented. This paper describes the implementation of this specific reader and demonstrates the first analysis performed at IT4InSAR in order to detect subsidence of Konya city in Turkey, based on PS processing of Sentinel-1A data. The previous GPS and InSAR studies show a relatively fast subsidence of the city due to groundwater pumping of up to 5 cm/year during 2006–2009 (Üstün et al. 2014).

2 Implementation of Sentinel-1 Reader Into Sarproz Codes

Though Sentinel-1A data are routinely available already since September 2014, there was still only a limited number of software packages in the end of 2015 that were able to perform reading and processing of the standard Sentinel-1 data product —the TOPS. One of the most significant differences that distinguish TOPS mode from typical mode of majority of other SAR satellites (stripmap) is that the antenna will steer from backward to forward in azimuth direction while steering between different subswath in range direction. The steering of antenna will sacrifice the azimuth resolution to a reasonable degree, meanwhile increasing the range illuminating area. Due to this special characteristic of TOPS, the standard processes for generating interferograms between TOPS pairs is different from that of stripmap.

Specifically, there are two important steps that are required by TOPS. Both are originated from the backward-to-forward steering of antenna in azimuth direction. The first step is called deramping. The steering of the antenna introduces an additional quadratic phase term in azimuth direction which doppler frequency exceeds the azimuth pulse repetition frequency (PRF). According to the sampling theorem, in order to resample the slave images during coregistration without aliasing the data, a step named deramping is required to remove this quadratic ramp. After reading the single look complex (SLC) value of TOPS data and before coregistration of interferometric pairs, the quadratic ramp of each image will be calculated and removed. The second step is a sub-pixel coregistration. The appearance of this quadratic term indicates that, in case of even a small misregistration error between master and slave images, there would be a phase ramp in azimuth direction superimposed on the interferogram. It has been commonly acknowledged that an accuracy of 1/1000 pixels of coregistration is needed to ignore the azimuth phase ramp introduced by this quadratic term (Prats et al. 2010). To meet this standard, a general sub-pixel coregistration method is needed. This step is usually done after the initial coarse coregistration that is used for the case of stripmap data. The subpixel coregistration is usually achieved by the spectral diversity method that utilizes the overlapping parts between successive bursts. After the subpixel coregistration of TOPS, the common processing steps for generating interferograms will be identical with the stripmap processing.

3 Implementation of Sarproz InSAR System into IT4InSAR Environment

The Salomon cluster works in the framework provided by CentOS system in connection with a portable batch system (PBS) planner. This makes the environment flexible and very effective especially applying autonomous batch scripts to be performed over a multi-node computing environment. However, license of the commercial Sarproz software doesn't allow multi-node installation and its main functionality is based on an interactive Matlab-based graphical user interface (GUI) that reacts slowly when using direct display forwarding, especially when connected by a slow internet provider. Also for this reason it was decided to prepare an InSAR system into a virtual computer running at only one processing node, offering 24 high speed processors, 128 GB RAM and a shared data storage of 1.3 PB in case of Salomon. This was considered sufficient for the IT4InSAR prototype.

The virtual computer is run using QEMU emulator with a virtual harddrive containing installation of latest Debian testing operating system with necessary applications as Mathworks Matlab, Sarproz, ESA Sentinel Application Platform etc. It allows processing using 23 processors of the node (one processor is left for the flawless work of the virtual computer). The Salomon storage is mapped as a virtual device offering high speed connection with the virtual computer. Internet access is not allowed originally in the Salomon computing nodes for security reasons. A double secure shell (ssh) tunnel was prepared to provide a secure exception and establish an internet connection using the virtual distributed ethernet (VDE) in QEMU via Salomon login server node that offers the Internet connection at ports 22, 80 and 443. The virtual computer is run in a temporary (snapshot) mode, thus it defies potential problems with file structure or unwanted user changes of environment of the system.

Access to the system is provided through open-source X2GO remote desktop framework, based on non-free NoMachine framework. This solution was selected as of the highest performance and flexibility of access from nearly any internet access point. The information exchange is performed by secure SSH connections. Each registered IT4InSAR user will achieve a unique port number for establishing his SSH connection. The IT4Innovations Salomon login server works as a proxy server providing an ssh tunnel to a computing node running QEMU. This node transfers the SSH connection towards the virtual computer. Afterwards, the user is connected to the virtual computer system through X2GO client's SSH threads and works in the environment as it would be installed to his local desktop. The remote desktop offers reentering the system without shutting down the virtual computer itself. The system can additionally be entered also using SSH shell, e.g. for file operations. Figure 1 attempts to basically schematize the configuration of the IT4InSAR system environment at the HPC.

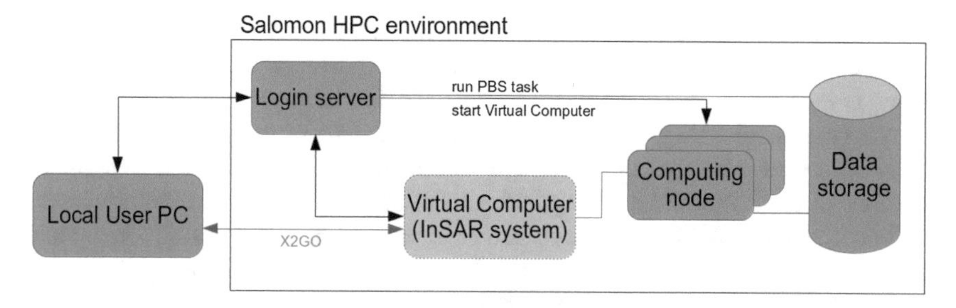

Fig. 1 Scheme of IT4InSAR system configuration as applied at IT4Innovations Salomon HPC environment

4 IT4InSAR System Application for Monitoring Subsidence in Konya City, Turkey

The first area of interest that was processed in the freshly prepared IT4InSAR system was an important subsidence field of Turkish Konya city center and its surroundings located in the biggest closed basin in Turkey and known by significant subsidence. The area has been intensively observed during last years by different engineering techniques such as geological, geodetical and SAR interferometry technologies (Üstün et al. 2014). The city of Konya is known as the hearth of Turkish agriculture and industrial activities. Its population is more than one million (approx. 1,175,000) and the surface area about 39,000 km^2. InSAR has been evaluated as a valuable technique for continuous monitoring of the local subsidence, practically as the only technology that can provide reliable subsidence measurements for a large quantity of points over such large-scale vicinity (Çomut et al. 2015).

The processing chain begins by data download. For purposes of downloading new data as well as updating the existing dataset for new acquisitions, a download script has been prepared based on the batch download script from the official Copernicus Sentinel-1 Data Hub (ESA 2015). With the knowledge of center coordinates and desired relative orbit identifier, the script would download all available SLC data and prepare them for the InSAR processing. In case of Konya closed basin (KCB), 30 Sentinel-1A images have been downloaded ranging 16th October 2014 until 4th November 2015.

From these images, a crop of overlapping area of 80 × 115 km (8150 × 23,500 pixels) from the TOPS Interferometric Wide Swath 3 (IW3) has been prepared, coregistered and stitched. However, two images from the dataset had to be dropped due to coregistration problems. The whole process was finished in several hours;

the performance of the prototype algorithm is planned to be further optimized in order to decrease the processing time of the coregistration and stitching, and to allow coregistration of all the overlapping acquisitions.

Afterwards, the PS processing itself has been performed for a selection of 130,000 points with appropriately high values of the SAR reflection (amplitude) stability through the multiple acquisitions. These points of a high amplitude stability correspond mainly to urban structures, rocks or a bare soil in KCB area. Since the area is urbanized and of a semi-arid character, the number of appropriate points for PS analysis is high. The number of SAR images in the analysis and the density of PS points per a km^2 conforms with the common conditions for a reliable analysis (should be more than 15 SAR images and at least 2–3 points per a km^2) (Hanssen 2001).

A more strict subset of 3000 PS points of a very high amplitude stability forming a more or less regular network throughout the scene was selected in order to perform a spatio-temporal analysis estimating so-called atmospheric phase screen (APS). This represents SAR phase errors due to a variability of atmosphere-caused signal delays, assuming these delays have a spatially correlated character but don't involve any trend in PS point time series (Hanssen 2001). The APS estimated from the network of connections of this subset of points is then interpolated for the whole scene and removed from the whole selection of PS points in the final PS processing. Here, the SAR phase contribution causes within the Sentinel-1 data time series could be distinguished between due to a topography effect (residual heights after the removal of the coarse SRTM digital elevation model) and due to physical changes in PS points (a linear deformation trend in time). The whole process (including estimation of APS) took 2 h that is considered a very short time for such processing.

The resulting estimates of the linear trend of phase changes at the PS points were recomputed into the mean velocity (mm/year) in the satellite line of sight (LOS), that is inclined in 38°–39° from nadir direction in this IW3 area. These results were georeferenced and are presented as colour-coded points in the Google Earth environment as seen at Figs. 2 and 3 (both figures plot only a relevant subset of data over Konya city area affected by a subsidence). The detected subsiding areas visible in Fig. 2 conform with expected locations of subsidence depressions known from previous studies (Çomut et al. 2015).

Figure 3 shows time series of a selected subsiding PS point corresponding to a location of the municipality cadastre building. This building is known to be subsiding and a permanent GPS station is installed on top of the building for a continuous monitoring of subsidence. The Sarproz PS InSAR processing estimates the subsidence trend of −47.5 mm/year in the Sentinel-1 IW3 LOS, with a standard deviation of 2.04 mm/year. The incidence angle at the point corresponds to ~38.3°. Based on (Cumming and Zhang 1999), it is possible to recompute the LOS value into the vertical direction, assuming no horizontal movements of the point.

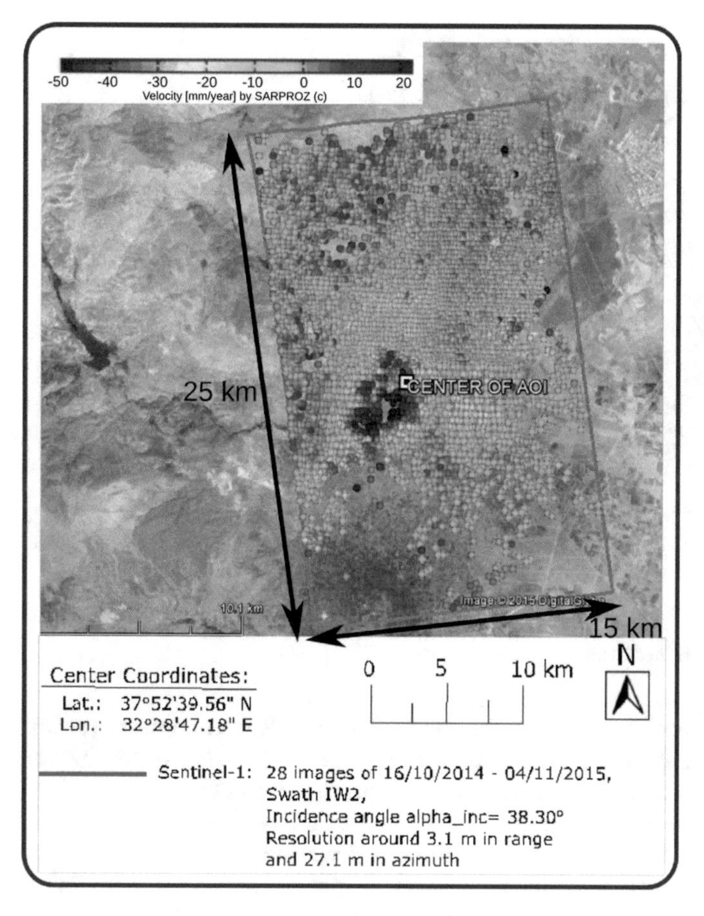

Fig. 2 Mean velocity map of PS points movements in Sentinel-1 IW3 LOS over Konya city area based on semi-automatic PS InSAR processing at IT4InSAR system

This leads to a simplified equation (1) leading to the recomputed value of subsidence of 60.5 ± 2.5 mm/year. This result can be compared with the reference GPS data once made available.

$$d_v = \frac{d_{LOS}}{\cos\theta_{inc}} \tag{1}$$

where d_v is deformation in the vertical direction (neglecting horizontal deformation of the point), d_{LOS} is the original value in LOS direction and θ_{inc} is the incidence angle.

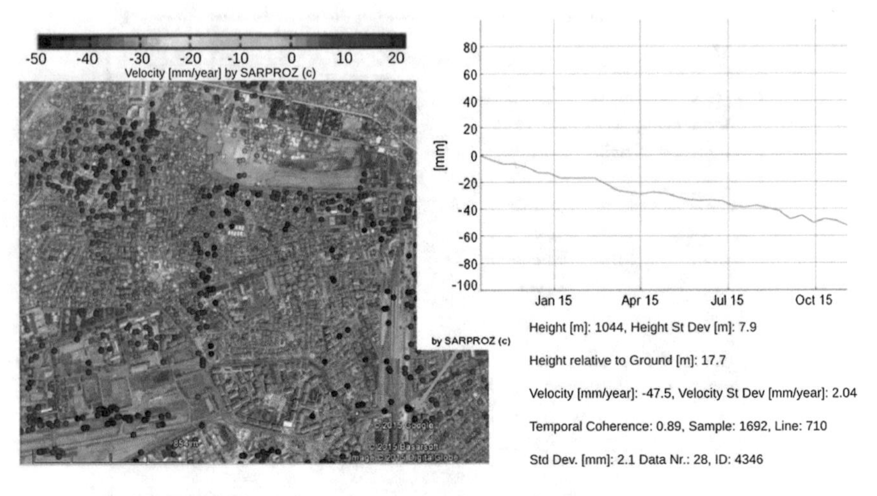

Fig. 3 PS InSAR-based estimation of the mean velocity of movements within 10/2014-11/2015 in the Sentinel-1 IW3 LOS—a zoomed area in the Konya city center with time series plotted for the selected pixel corresponding to a municipal cadastre building

5 Conclusions

A functional environment for satellite-borne SAR interferometry processing in the modern HPC infrastructure has been arranged with a view of routine monitoring of selected sites especially using the modern European Copernicus Sentinel-1 satellite SAR system. This system can be used as a basis for projects with a need of a moderate and supervised performance. Especially the expert supervision over the processing chain is often necessary in the majority of current InSAR-focused projects. Therefore the modular approach where each authorized user may run the IT4InSAR environment at only one processing node per a moment is not a real limitation. However, activities to provide the power of other nodes within the virtual computer can be expected in future in order to extend the performance of the system. A full automation of the system can be planned once it will be possible to batch script the coregistration and stitching operations of the Sentinel-1 TOPS data.

The first processing results over Konya city shows a good performance of Sentinel-1 data for monitoring subsidence. Though at the moment of preparation of this work, there was no data for comparison of results available to the team, the results show very similar outputs compared to the previous studies using older GPS and InSAR data of ERS, Envisat and Alos satellites (Üstün et al. 2014; Çomut et al. 2015). In addition to those, more PS points could be evaluated by Sentinel-1 processing thanks to the high revisit rate and other advanced characteristics of Sentinel-1 indicating also a high reliability in the estimation of velocity rates of the movement of points.

Acknowledgments This work was supported by the European Regional Development Fund in the IT4Innovations Centre of Excellence Project (CZ.1.05/1.1.00/02.0070), The Ministry of Education, Youth and Sports from the National Programme of Sustainability (NPU II) Project "IT4Innovations excellence in science—LQ1602" and the TUBITAK in Turkey (Contract No. 110Y121, Project Name: Monitoring of land subsidence in Konya Closed Basin using geodetic methods and investigate its causes). This work was supported by The Ministry of Education, Youth and Sports from the National Programme of Sustainability (NPU II) project "IT4Innovations excellence in science—LQ1602" and from the Large Infrastructures for Research, Experimental Development and Innovations project "IT4Innovations National Supercomputing Center—LM2015070". Authors are grateful to Dr. Daniele Perissin for offering a free license of Sarproz software for the testing purposes.

References

Çomut CF, Üstün A, Lazecky M, Aref MM (2015) Multi band InSAR analysis of subsidence development based on the long period time series. In: ISPRS/SMPR 2015, Kish Island, Iran; Nov 2015. doi:10.5194/isprsarchives-XL-1-W5-115-2015

Cumming I, Zhang J (1999) Measuring the 3-D flow of the Lowell Glacier with InSAR. In: Proceedings of ESA Fringe'99 SAR workshop. Liége, Belgium, 10–12 Nov 1999, pp 9

ESA (2015) Copernicus SENTINELS SCIENTIFIC DATA Hub: 5 APIs And Batch Scripting. https://scihub.copernicus.eu/userguide/5APIsAndBatchScripting

Ferretti A, Prati C, Rocca F (2000) Nonlinear subsidence rate estimation using permanent scatterers in differential SAR interferometry. IEEE Trans Geosci Remote Sens 38(5):2202–2212

Hanssen R F (2001) Radar interferometry: data interpretation and error analysis. Kluwer Academic, Dordrecht, 328 pp. ISBN 0-7923-6945-9

Palkovič M et al (2015) IT4Innovations Newsletter 01/2015. Ostrava, 11 p. <czech language>. http://www.it4i.cz/wp-content/uploads/2015/10/newsletter-2015-01.pdf

Perissin D, Wang Z, Wang T (2011) The SARPROZ InSAR tool for urban subsidence/manmade structure stability monitoring in China. In: Proceedings of 34th international symposium on remote sensing of environment. Sydney, Australia

Prats P, Marotti L, Wollstadt S, Scheiber R (2010) TOPS interferometry with TerraSAR-X, in synthetic aperture radar (EUSAR). In: 8th European conference on remote sensing, pp 1–4, 7–10 June 2010

Üstün A, Tusat E, Yalvaç S, Özkan Eren Y, Özdemir A, Bildirici Ö, Üstüntas T, Kirtiloglu O, Mesutoglu M, Doganalp S, Canaslan F, Abbak R, Avsar N, Simsek F (2014) Land subsidence in Konya Closed Basin and its spatio-temporal detection by GPS and DInSAR. Environ Earth Sci. doi:10.1007/s12665-014-3890-5

Modelling Karst Landscape with Massive Airborne and Terrestrial Laser Scanning Data

Jaroslav Hofierka, Michal Gallay, Ján Kaňuk and Ján Šašak

Abstract Recent developments in data collection methods such as laser scanning technology contributed to the rise of massive datasets representing the landscapes with an unprecedented accuracy. However, processing of the massive datasets poses new challenges. Current geographic information system softwares have a limited capability to effectively handle such data. Therefore, specialised software must be used in order to further analyse the data in a geographic information system. This study describes the key methodological aspects of processing massive laser scanning datasets with a set of software tools and demonstrates the benefits of integrating the data in a geographic information system. The presented case study represents a complex karst landscape in the south-eastern part of Slovakia.

Keywords Laser scanning · Caves · Digital terrain modelling · LiDAR · Geographic information system

1 Introduction

Digital landscape modelling, as a domain of the geographic information science, is a natural and integral part of its research with numerous applications in geosciences. The main assumption for modelling the landscape is that objects and phenomena existing in the landscape sphere and also their mutual interaction can be quantified and localised. Acquisition and processing of such spatially localised data on landscape became widely accessible in the last decade. This fact relates to the development of modern technologies allowing for a remarkable effective recording of geodata with a high level of detail and accuracy as, for example, laser scanning, hyperspectral scanning, digital photogrammetry or automatized recording of climate or hydrological variables by data loggers. Vast amounts of data originate via

J. Hofierka (✉) · M. Gallay · J. Kaňuk · J. Šašak
Institute of Geography, Faculty of Science, Pavol Jozef Šafárik University in Košice, Jesenná 5, 04001 Košice, Slovak Republic
e-mail: jaroslav.hofierka@upjs.sk

I. Ivan et al. (eds.), *The Rise of Big Spatial Data*, Lecture Notes in Geoinformation and Cartography, DOI 10.1007/978-3-319-45123-7_11

these modern data collection technologies for which, however, the software tools applied in the geosciences research are not sufficiently prepared (Bishop et al. 2012). It especially concerns geographic information systems (GIS) which were developed for data processing and analyses to make the research and landscape modelling more effective. Nevertheless, in terms of processing massive datasets, the use of GIS technology in its current state-of-the-art reaches its limitations. Clearly, GIS softwares were not developed for processing massive data (e.g., billions of points), thus they are very slow and ineffective or even unable to process such data. The range of GIS tools for 3D modelling of landscape is still rather limited to visualisation and basic spatial operations. More sophisticated operations, such as reconstruction of 3D surfaces, representation of dynamics of the 3D surfaces and 3D objects are still missing. Thus processing the massive datasets, such as from laser scanning, requires the use of specialised hardware and software solutions.

The growing amount of data is collected by laser scanning and increasingly used in research of the environment. It is widely recognised as a very progressive method of data collection (Vosselman and Maas 2010). This study is focused on the specific aspects of laser scanning data processing using various software tools in order to create 3D models of specific landscape features readily usable by a GIS. The combination of terrestrial laser scanning (TLS) and airborne laser scanning (ALS) within one region provides a unique chance to show a potential of 3D data in environmental studies. We have selected a small area in the Slovak Karst that contains various surface and subsurface landscape features that can be mapped by TLS and ALS. Karst is a specific type of landscape in which the on-going geoecological processes are determined by the distinctive properties of geological setting which result in unique surface and subsurface landforms, climate-hydrological regime, soil properties and vegetation cover (Jakál 1975; Hochmuth 2004). The karst landscape presents a complex system with a marked horizontal and vertical interaction of its components (Bella 2012). Better understanding of the processes and evolution of the karst requires very detailed mapping methods and 3D modelling. For that reason, karst is an appropriate type of landscape to demonstrate applicability and benefits of mapping the surface and subsurface landforms by TLS and ALS and 3D modelling. However, the presented approach, is applicable also to other types of landscape.

The aim of this paper is to describe the methodological pipelines of processing the massive lidar datasets originating from airborne and terrestrial laser scanning with a set of software tools and to demonstrate the benefits of integrating the data in a geographic information system. The methodology is presented using data from a complex karst landscape in the south-eastern Slovakia that includes surface and subsurface landscape features.

2 Study Area

The study area comprises 68 sq. km of the south-western part of the Slovak Karst geomorphological unit near the state border of Slovakia with Hungary (Fig. 1). The area is a part of the National Park of Slovak Karst and the land cover is a mosaic of forests, shrubs, permanent meadows and some arable land. It is precious for the specifics of a mid-latitude karst landscape for which the several landform features are typical such as dolines, blind valleys, sinkholes, karrens. These features are difficult to map in an unexposed land which is typical in this region. Furthermore, no national coverage lidar database exists in Slovakia so far, therefore, the karst area was flown within our custom laser scanning (LiDAR) mission in August 2014.

The area is abundant with many underground forms such as caves of which the Domica cave (Fig. 1, 48°28′40.4″N, 20°28′12.9″E) is the most well-known and also publicly available. Since its discovery in 1926 by Ján Majko, the cave has been widely studied from various points (Nováková 2009; Kováč and Rusek 2012; Papáč et al. 2014; Hochmuth 2014; Svitavská-Svobodová et al. 2015; Mihailović et al. 2015). For that reason, we conducted a TLS survey to capture the cave morphology in high resolution and with high accuracy (Gallay et al. 2015a). The total length of the Domica cave system is around 5400 m however, it continues into

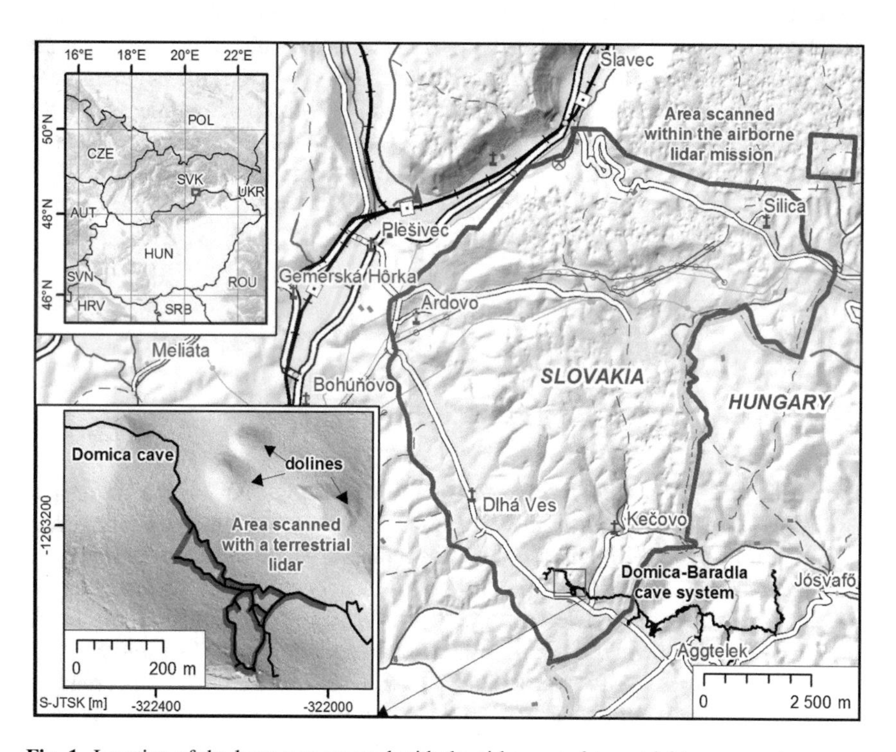

Fig. 1 Location of the karst area mapped with the airborne and terrestrial laser scanning

the Aggtelek Karst region in Hungary as the Baradla cave with a combined length of the whole system of 26,065 m (CAVERBOB 2015). Average annual air temperature in the Domica cave varies between 10 and 11 °C and air humidity is around 95–98 %. The Domica cave was formed by corrosive-erosive processes caused by superficial fluvial water and temporary streams which sank underground at the contact of the Middle Triassic white limestones and the Pontian fluvial-lacustrine gravel-sand-clay sediments. The tectonic framework was generated by south to southeast oriented pressures which caused stress, and the compression was realised in the range of directions of north-south to northwest-southeast (Gaál and Vlček 2011). The oldest (upper) parts of the Domica cave (341 ms a.s.l.) began to form after the uplift of the region above sea level in the Upper Pliocene (Bella et al. 2014) when the current hydrographic network was being established. The cave is a result of the underground flow of the Styx and Domický potok rivers shown by the oval shapes of the corridors and the quantity of allochtone pebbles found in the system. The lower evolution level found by drilling is at 318 m a.s.l. (Droppa 1970).

The landscape surrounding the Domica Cave is one in which superficial and underground processes strongly interact. For example, in the recent history, several rainfall events caused major flooding in the cave which was also induced by inappropriate agricultural practises (Bella 2001; Gaálová et al. 2014). Integration of detailed 3D superficial data and underground 3D data will allow for a detailed morphometric analysis of the karst system and digital modelling of processes forming the landscape to be carried out. Furthermore, this cave is the beginning of a much larger cross-border underground system; therefore, the influence of the Domica Cave area on the formation of the other parts of the system can be assumed. Moreover, the cave is a listed UNESCO Natural World Heritage site and a detailed 3D model derived from laser scanning can help in site management and act as a catalyst for further research and educational initiatives.

3 Laser Scanning Data Acquisition

3.1 Airborne Laser Scanning Data

The ALS dataset originated during a mission flown in August 2014 by Photomap s. r.o., Košice over the area delineated in Fig. 1. The laser measurements were taken by a Leica ALS70-CM scanner capable of multiple target detection and coupled with an onboard GNSS/IMU. There were over 1.99 billion of points recorded across 68 sq. km which yields average point density of 29 points per sq. metre including all returns while it was 21 points per sq. metre for the last returns only.

For the purposes of the research and terrain modelling, it was important to achieve a sufficiently high density of ground returns even under the forest canopy. In the acquired dataset, the minimum ground return spacing in forested areas is

0.5 points per sq. meter. The average density of ground returns across the whole area is 4 points per sq. meter. The accuracy of measurement in open areas is reported at 0.1 m of 1 sigma by the data supplier. The achieved measurement density and accuracy allows for detecting even small building features, power lines, vegetation structure, and terrain forms (Fig. 2). The ALS data are supplemented with RGB and close-range infrared orthoimagery acquired in April 2014. The image data were used to colourize the points.

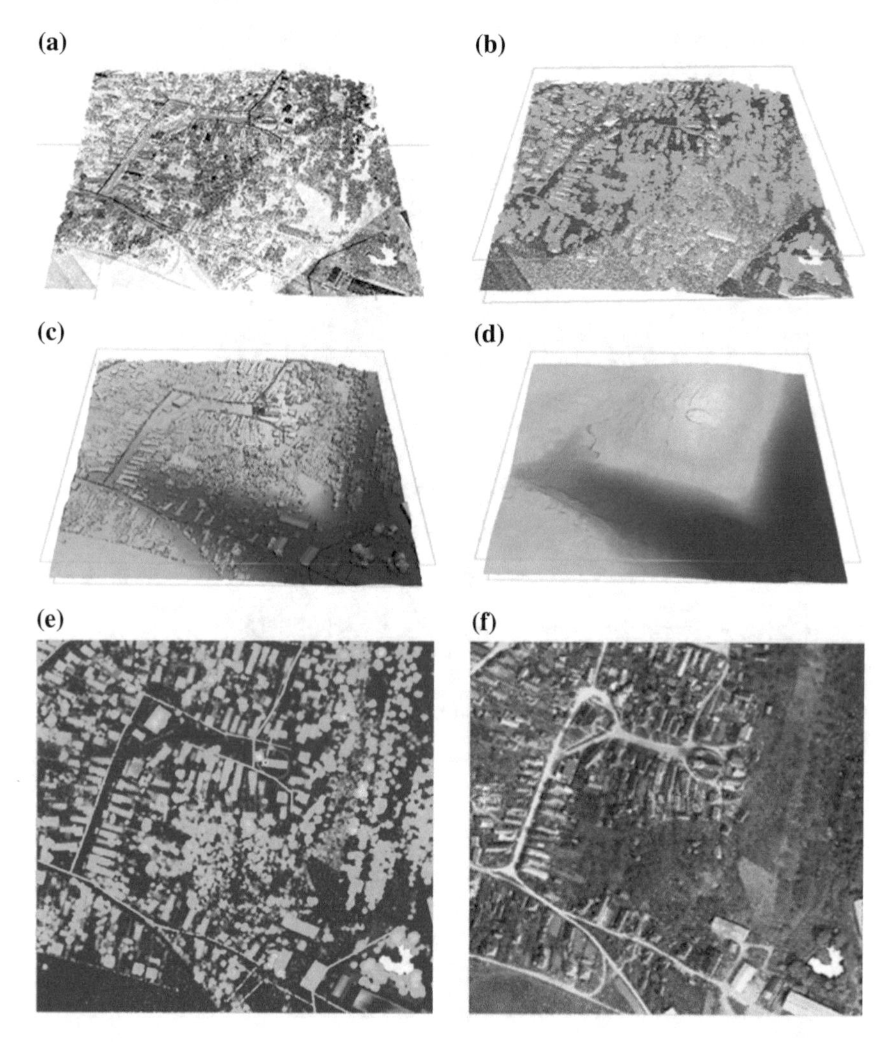

Fig. 2 Sample of the ALS point cloud (500 × 500 m) showing the central part of the Silica village in 3D perspective coloured by **a** intensity, **b** classification (*orange*—roofs, vegetation—*green*, ground—*brown*, unclassified—*grey*), **c** digital land cover surface model, **d** digital terrain model, and in 2D as orthogonal views of **e** digital canopy height model (normalized height with respect to the ground), and **f** colourized point cloud based on natural colour aerial orthoimagery

3.2 Terrestrial Laser Scanning Data

The data used in the presented study were acquired with a terrestrial laser scanner in combination with RTK-GPS surveying within a 5 days mission in March 2014 in the Domica cave, Slovakia. FARO Focus 3D scanner was used to scan around 1500 m of the cave from 328 scanner positions within 40 h in total. The scanning density point spacing was set to 20 mm at 10 m. The scans were oriented relative to each and with respect to the Slovak national coordinate system (S-JTSK) in SCENE8, proprietary software by FARO. The final point cloud contained almost 12 billion of points representing the entire publicly accessible cave and some parts

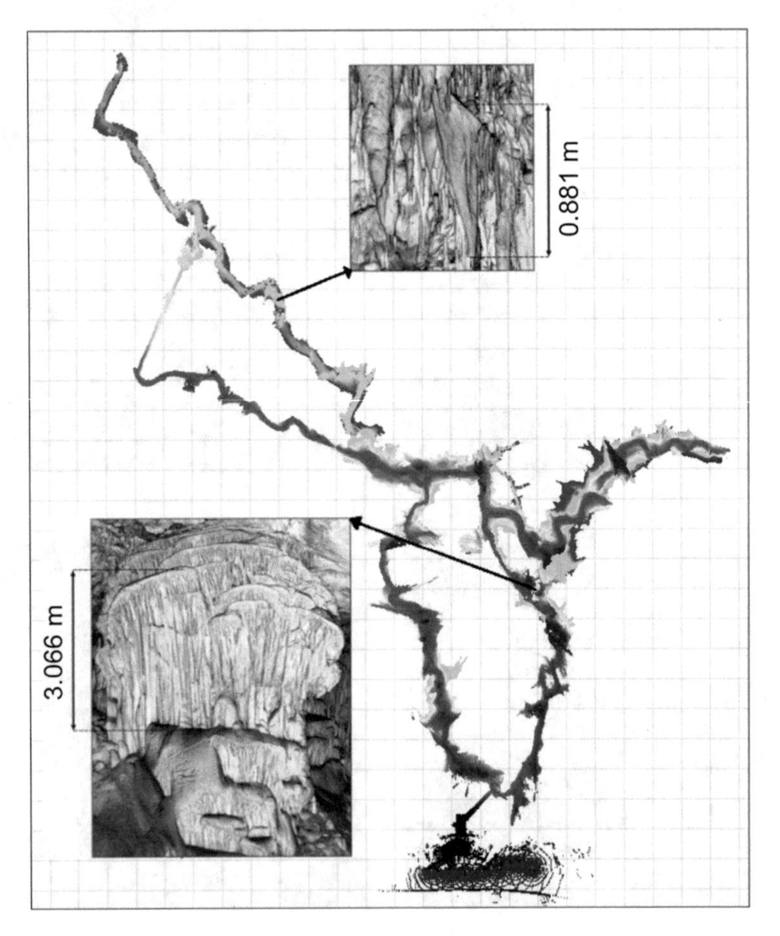

Fig. 3 The footprint of the scanned area of the Domica cave with the outdoor area *coloured* by the altitude above sea level (see the legend for Fig. 5). The grid is aligned with the north direction up and the cells are of 20 × 20 m size. The detailed view in *insets* show point clouds of large and small speleothem forms in full resolution the point cloud *coloured* by the intensity of the reflected laser

inaccessible by public (Fig. 3). The total accuracy of registration of the scans was 4–5 mm. Georeferencing of the registered point cloud in the national grid achieved accuracy of 12 mm measured as the total RTK-GPS positioning error. The technical aspects of this survey is described in (Gallay et al. 2015a). The point cloud provides a very high detail which enables viewing even small geomorphological features such as speleothems (Fig. 3). Visualisation of the point cloud allows for basic measurements of the cave morphometry but it is not applicable for defining morphometric parameters such as curvature or orientation for which a 3D volumetric surface is needed (Gallay et al. 2015b).

4 Processing Methods and Software Tools

The ALS data were supplied as geometrically corrected and partially classified point cloud which was georeferenced in the S-JTSK East North coordinate system and split into data tiles of 500 × 500 m in the LAS 1.2 format for further handling. Processing of the raw ALS data was conducted using the TerraScan proprietary software by TerraSolid on the side of the data supplier, Photomap s.r.o., Košice. The research of the landscape processes requires continuous representation of the terrain and canopy surface therefore further processing was to be done with the ALS data. We opted for the open-source tools which have proven to be very efficient for this kind of massive dataset (Fogl and Moudrý 2015). However, much larger datasets containing hundreds of billions of points seems to be more efficiently managed via database solutions as opposed to file-based solution of the LAStools (van Oosterom et al. 2015). The unlicensed LAStools software package (Isenburg 2014) was used to check and prepare the laser scanning point cloud and to derive the grid-based surface models. LAStools is a suite of executable programmes which can be easily combined into a processing pipelines by batch-scripting. The main advantage of the software is the multicore algorithms providing fast rendering and very efficient processing of millions of points. Part of the tools has an open-source code (LGPL), other part is freely available for non-profit use but the code is closed. All the tools can also be run via a native GUI and are available as LiDAR processing toolboxes for ArcGIS, Quantum GIS, and for ERDAS IMAGINE.

The most important tasks related to the ALS data processing with LAStools involved five steps. Firstly, the original data tiles (500 × 500 m) had to be re-tiled so that additional overlapping buffer zone of 25 m is added to the original tile containing points from the neighbouring tiles (Fig. 4a, b):

lastile -lof file_list1.txt -o "tile.laz" -tile_size 500 -buffer 25 -odir "F:\data\lastile" -olaz.

In some cases, the new tiles had to be shrunk into smaller tiles (100 × 100 m plus the 25 m buffer) to avoid distortion due to the use of unlicensed version of the software.

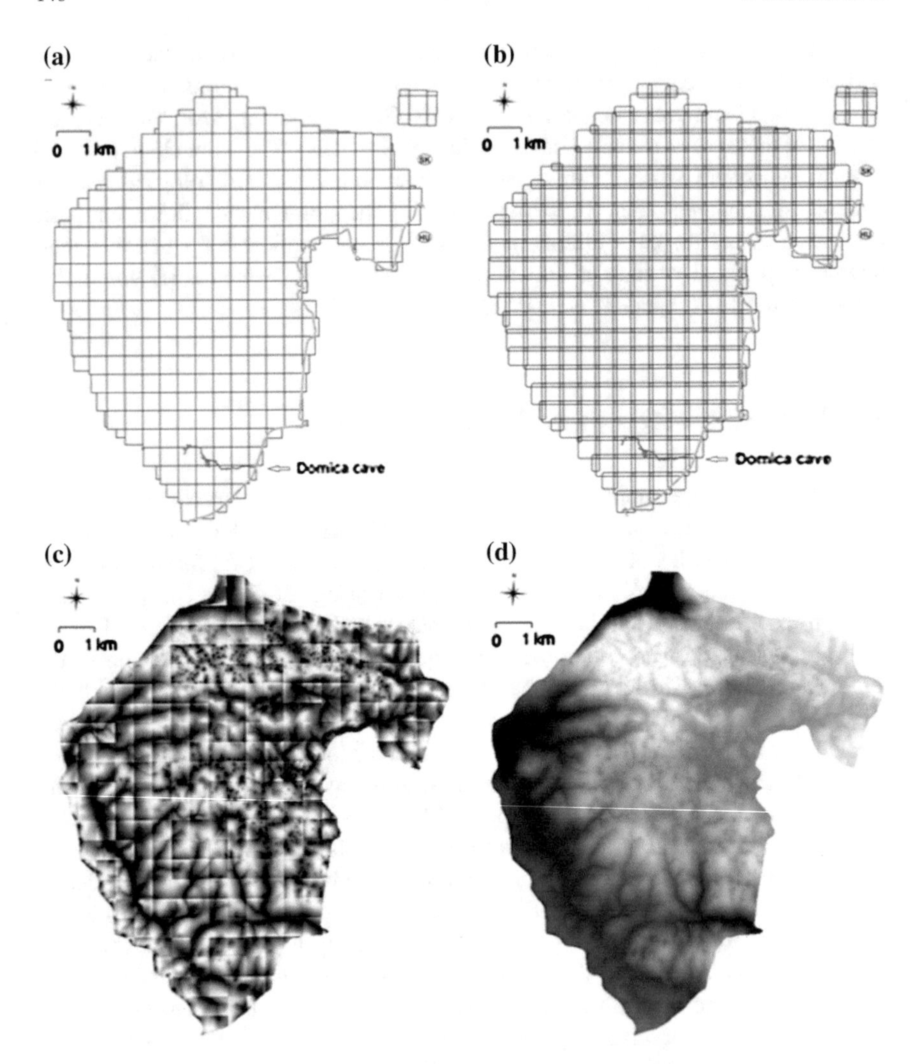

Fig. 4 Segmentation of the originally supplied ALS point dataset into 500 × 500 m tiles (**a**) and after re-tiling (**b**) into tiles having an additional 25-m buffer zone for the sake of creating seamless connection of DTM raster tiles of 500 × 500 m size (**c**). The advantage of a virtual raster dataset is in fast loading and rendering of the DTM raster mosaic as a single file (**d**)

Secondly, data noise was removed. Afterwards the point cloud was classified into ground returns and the remaining points with assigned height above the ground. This dataset was classified into four categories: ground, vegetation, buildings, unclassified:

lasnoise -lof file_list2.txt -step_xy 1 -step_z 0.2 -isolated 3 -remove_noise -odir "F:\data\lastile_denoised" -olaz;
lasground -lof file_list3.txt -town -ultra_fine -odir "F:\data\lasground" -olaz;

lasheight -lof file_list4.txt -drop_below -2 -drop_above 30 -odir "F:\data\lasheight" -olaz;
lasclassify -lof file_list5.txt -odir "F:\data\lasclassify" -olaz;

The third step involved generation of seamless tiles of gridded digital model of terrain height (DTM), digital model of land cover surface height (DSM), digital model of LiDAR intensity (DMI), and digital model of normalized height (DCHM) in the ESRI ASCII grid or TIFF formats. The DSM was generated using the following command as an example:

blast2dem -lof file_list6.txt -elevation -use_tile_bb -odir "F:\data\raster_tiles\dsm" -oasc

The last step concerned re-tiling of the point cloud to remove the data buffers and saving new tiles in the compressed LAZ format as ties of 500 × 500 m:

lastile -lof file_list7.txt -remove_buffer -odir "F:\data\lastile_final" –olaz.

Further processing of the raster tiles involved indexing to create a virtual raster by the gdalbuildvrt utility of the open-source Geospatial Data Abstraction Library (GDAL 2015). We accessed the tool from within the GRASS GIS environment which is also an open-source software providing tools for advanced spatial analysis of geographic data (Neteler and Mitasova 2008). The command:

gdalbuildvrt DTM.vrt F:/data/raster_tiles/dtm/.asc*

builds an index of the raster files contained in the specified folder. The raster tiles can be loaded and viewed directly as the VRT file and then the seamless raster mosaic (Fig. 4c, d) can be saved as a single raster file if desired (e.g., a GeoTIFF). Either way the data can be further used for deriving geomorphometric parameters (Hengl and Reuter 2008). As an example of application, we generated slope angle and slope aspect in GRASS GIS at different levels of scale by the r.param.scale module. The outputs of this procedure were used in hydrological modelling by the r.flow module and r.watershed to analyse concentration of superficial flow and its relation to the underground system of the Domica cave (Fig. 5).

The TLS data collected within the terrestrial laser scanning mission in the Domica cave were registered and georeferenced in the SCENE8 proprietary

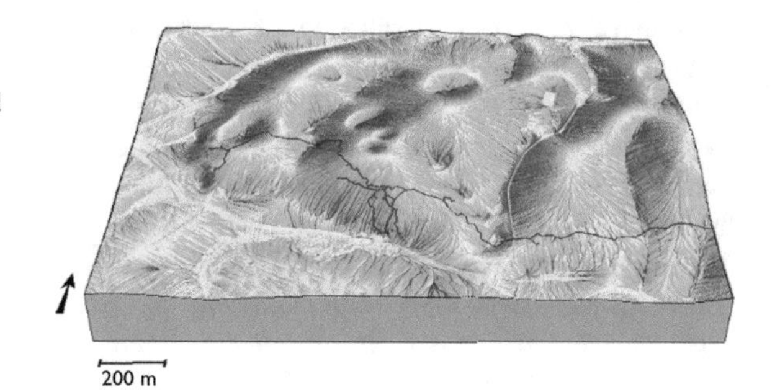

Fig. 5 Overland flow assessment using the r.flow module in GRASS GIS based on digital terrain model derived from airborne laser scanning

software by FARO which is dedicated for processing the data acquired by the FARO scanners. The data are organized within a SCENE project database which is also prepared in the form of a FARO SCENE Webshare Server off-line solution providing product for further use, data visualization and simple measurement by other peer-colleagues. However, more advanced analysis of the cave morphology required production of a continuous representation of the cave surface in the form of a 3D digital model. We opted for the open-source solution provided by Meshlab (Cignoni and Ranzuglia 2014). The first goal in modelling the cave surface was to generate a model of the whole scanned system. For this reason, only a small portion (0.03 %) of the originally acquired laser scanning points were exported from the SCENE software in the PTX format. The format preserves the point normals which are oriented with respect to the scanner position. The normals are needed to reconstruct a model of surface which the points represent. A triangular mesh can be generated using several algorithms. We used the Poisson surface reconstruction approach described in Gallay et al. (2015a, b) in more detail. A similar study was conducted by Silvestre et al. (2013). The surface reconstruction method is based on the work by Kazhdan et al. (2006). With this algorithm, the octree depth is the key input parameter controlling the level of surface detail. The number of faces and vertices comprised in the resulting 3D mesh increases with the higher value of the octree depth. Figure 6 portrays the 3D model derived with the octree depth of 13 from over 3.13 million of input points shown in Fig. 3. The model contains 2.8 million of vertices and 5.6 million of triangular faces. The 3D mesh of the entire cave system is stored in the PLY format and for interoperability with GRASS GIS it was also exported in the DXF format. The mesh in Meshlab allowed for better visualisation and parameterization of the surface by the means of 3D geomorphometry which is demonstrated in Gallay et al. (2015b). Various parameters related to the normals can be computed in Meshlab to study the shape of the cave morphology.

5 Integration in GIS

In the previous sections, we have shown that processing massive datasets from airborne and terrestrial laser scanning requires a complex workflow from basic filtering of raw data to creation of landscape models registered in a national coordinate system. Such models can be fully integrated into a GIS environment with further benefits of available GIS tools for geospatial analysis and modelling. In case of integrating surface and subsurface data, this helps to understand complex landscape processes and evolution. For example, dolines/sinkholes mapped by ALS are clearly visible in Fig. 5. Their shape concentrates rainfall water that infiltrates into the limestone bedrock, later forming specific subsurface geomorphological forms, such as caves mapped by TLS (Fig. 6). To improve spatial context of these features, we integrated their subsequent digital models in an ArcGIS database (Fig. 7). The combination of TLS and ALS data allows for performing other

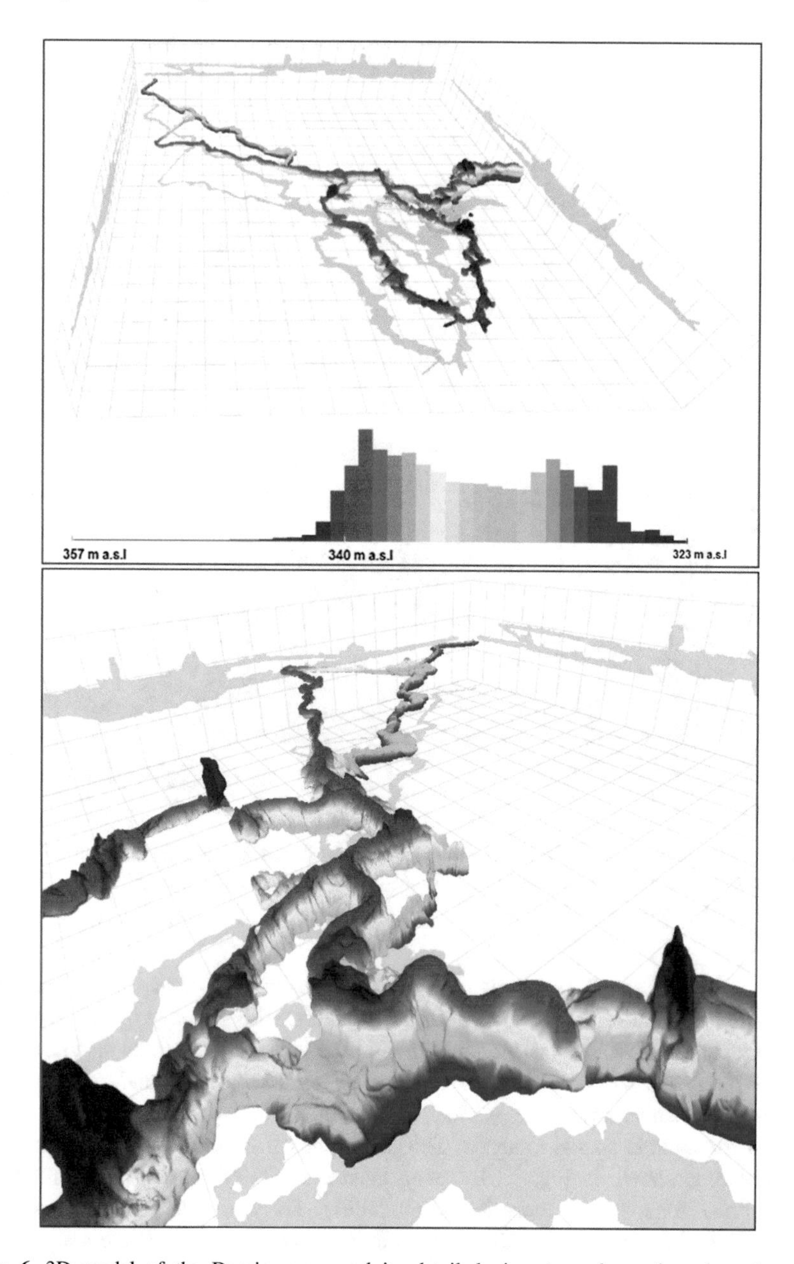

Fig. 6 3D model of the Domica cave and its detailed view towards north and north-west, respectively. The model is orthogonally projected on the XY, YZ, and XZ planes and it is *coloured* by the value of altitude above mean sea level which distribution is shown by the histogram. The size of the background grid cells is 20 m along all three axes

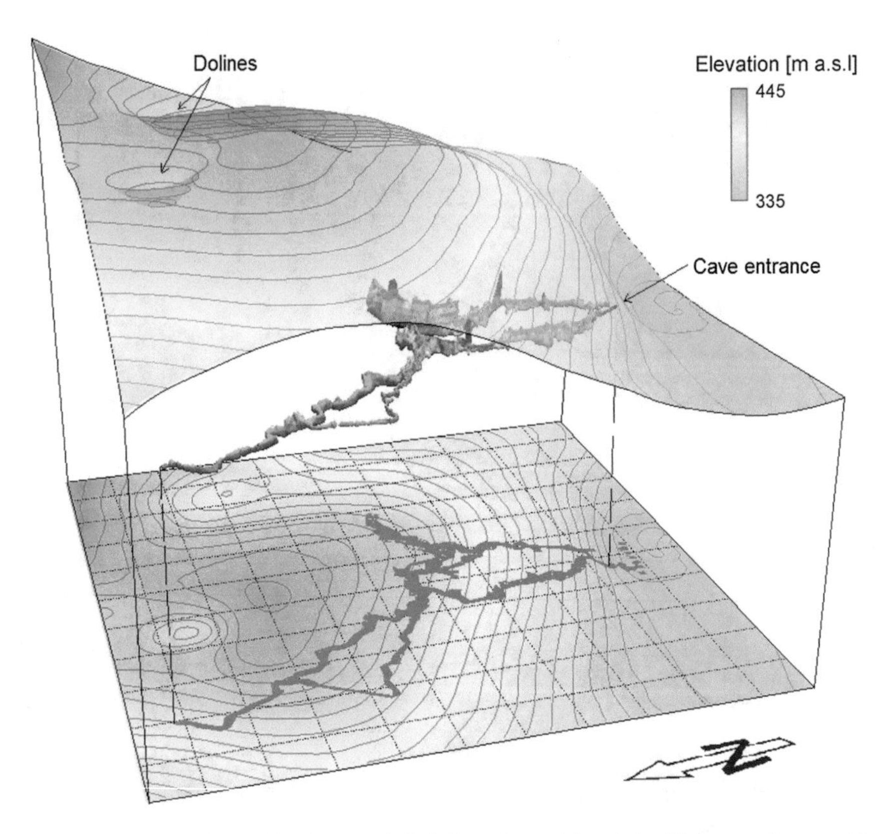

Fig. 7 Perspective view of the integrated digital terrain model and the 3D cave surface model which are also orthogonally projected on the XY plane (adapted from Gallay et al. 2015a). The background grid cell size is 50 × 50 m

analyses in GIS, such as overland flow pattern (Fig. 5) which can be used for analysing the relation of the superficial flow pattern with the cave system location. The interactive 3D visualisation and simple GIS operations such as distance measuring helps to better understand the spatial context of the landscape system. Traditional maps lacking the vertical dimension do not provide sufficient information for reliable spatial analysis. This is obvious from the footprint of the cave and contours shown in Fig. 7. However, current GIS's still have only limited 3D capabilities in geospatial analysis and modelling. Therefore, new 3D tools need to be developed in order to be able to handle 3D surfaces and simulate spatial processes in the 3D domain.

6 Conclusions

Modern data collection methods such as laser scanning produce massive datasets containing billions of points representing 3D landscape features with unprecedented spatial detail and accuracy. Unfortunately, current GIS software are not capable of efficient accessing and processing the raw point cloud data. These datasets need to be processed in a complex workflow starting from basic filtering of the raw point cloud to a full integration of the georeferenced 3D model in a GIS. In this study, we present a methodological pipeline documenting steps in processing airborne and terrestrial laser scanning data using various software tools. The end product of this workflow comprises a 3D landscape model with an acceptable accuracy and level of detail that is integrated in a GIS database. GIS environment provides means to take advantage of further geospatial analysis and modelling to better understand spatial relations and processes forming the real landscape. The applicability of our methodological approach was presented in a case study from a complex karst landscape of the Slovak Karst. The applied complex workflow of processing the massive TLS and ALS data resulted in digital models which represent surface and subsurface geomorphological features. The models are directly and effectively accessible in a GIS environment. This integration of models provides a new tool which has not been available before greatly improving our understanding of spatial relations and processes forming the karst landscape.

Acknowledgments This work originated within the research projects APVV-0176-12 supported by the Slovak Research and Development Agency and VEGA 1/0474/16 supported by the Slovak Research Grant Agency VEGA.

References

Bella P (2001) Geomorfologické pomery okolia jaskyne Domica. Aragonit 3:5–11 (in Slovak)

Bella P (2012) Zraniteľnosť, ekostabilizujúce faktory a narušenie jaskynného prostredia. Geografický časopis 64:201–218 (in Slovak)

Bella P, Braucher R, Holec J, Veselský M (2014) Burial age of cemented quartz gravel in the upper level of the Domica Cave, Slovakia. Book of abstracts, 8th scientific conference of the Association of Slovak Geomorphologists at Slovak Academy of Sciences, Snina, Slovakia 6–8 October 2014, pp 10–11

Bishop MP, James LA, Shroder JF Jr, Walsh SJ (2012) Geospatial technologies and digital geomorphological mapping: concepts, issues and research. Geomorphology 137:5–26

CAVERBOB (2015). http://www.caverbob.com/wlong.htm. Dec 2015

Cignoni P, Ranzuglia G (2014) MeshLab, visual computing lab–ISTI–CNR. http://meshlab.sourceforge.net/. June 2015

Droppa A (1970) Príspevok k vývoju jaskyne Domica. Československý kras 22:65–72 (in Slovak)

Fogl M, Moudrý V (2015) Influence of vegetation canopies on solar potential in urban environments. Appl Geogr 66:73–80

Gaál Ľ, Vlček L (2011) Tektonická stavba jaskyne Domica (Slovenský kras). Aragonit 16:3–11 (in Slovak)

Gaálová B, Donauerová A, Seman M, Bujdáková H (2014) Identification and ß-lactam resistance in aquatic isolates of *Enterobacter cloacae* and their status in microbiota of Domica Cave in Slovak Karst (Slovakia). Int J Speleol 43:69–77

Gallay M, Kaňuk J, Hochmuth Z, Meneely JD, Hofierka J, Sedlák V (2015a) Large-scale and high-resolution 3-D cave mapping by terrestrial laser scanning: a case study of the Domica Cave, Slovakia. Int J Speleol 44:277–291

Gallay M, Kaňuk J, Hofierka J, Hochmuth Z, Meneely J (2015b) Mapping and geomorphometric analysis of 3-D cave surfaces: a case study of the Domica Cave, Slovakia. In: Zwoliński Z, Jasiewicz J, Mitasova H, Hengl T (eds) Geomorphometry 2015 conference proceedings, 22–26 June 2015, Poznań, pp 69–73. http://geomorphometry.org/2015

GDAL (2015) Geospatial data abstraction library. http://www.gdal.org/. Nov 2015

Hengl T, Reuter HI (eds) (2008) Geomorphometry: concepts, software, applications. Developments in Soil Science, vol 33, Elsevier, Amsterdam

Hochmuth Z (2004) Rozdiely v intenzite povrchového skrasovatenia na jednotlivých planinách slovenského krasu. Gemorphol Slovaca 4:30–35

Hochmuth Z (2014) História mapovania a vytvorenie spojitého meračského ťahu ako podkladu pre reambulovanie mapy Domice. Slov kras Acta Carsol Slovaca 52(2):173–190 (in Slovak)

Isenburg M (2014) LAStools—efficient LiDAR processing software (version 141017, unlicensed). http://rapidlasso.com/LAStools. Nov 2015

Jakál J (1975) Kras Silickej planiny. Osveta, Martin

Kazhdan MM, Bolitho M, Hoppe H (2006) Poisson surface reconstruction. In: Sheffer A, Polthier K (eds) Symposium on geometry processing, ser. ACM international conference proceeding series, vol 256. Eurographics Association, pp 61–70

Kováč Ľ, Rusek J (2012) Redescription of two troglobiotic species of the genus Pseudosinella Schäffer, 1897 (Collembola, Entomobryidae) from the Western Carpathians. Zootaxa 3341:32–45

Mihailović DT, Krmar M, Mimić G, Nikolić-Đorić E, Smetanová I, Holý K, Zelinka J, Omelka J (2015) A complexity analysis of 222Rn concentration variation: a case study for Domica cave, Slovakia for the period June 2010–June 2011. Radiat Phys Chem 106:88–94

Neteler M, Mitasova H (2008) Open source GIS: a GRASS GIS approach. Springer, New York

Nováková A (2009) Microscopic fungi isolated from the Domica Cave system (Slovak Karst National Park, Slovakia). A review. Int J Speleol 38:71–82

Papáč V, Hudec I, Kováč Ľ, Ľúptáčik P, Mock A (2014) Bezstavovce jaskyne Domica (Invertebrates of Domica) In: Gaál Ľ, Gruber P (eds) Jaskynný systém Domica-Baradla. Jaskyňa, ktorá nás spája. Jósvafő, Hunagry: Aggteleki Nemzeti Park Igazgatóság, pp 267–279; (in Slovak and Hungarian)

Silvestre I, Rodrigues JI, Figueiredo MJG, Veiga-Pires C (2013) Cave chamber data modeling and 3D web visualization. In: 17th international conference on information visualisation (IV), 16–18 July 2013, pp 468–473

Svitavská-Svobodová H, Andreas M, Krištůfek V, Beneš J, Novák J (2015) The thousand-year history of the Slovak Karst inferred from pollen in bat guano inside the Domica Cave (Slovakia). Folia Geobot 50:49–61

van Oosterom P, Martinez-Rubi O, Ivanova M, Horhammer M, Geringer D, Ravada S, Tijssen T, Kodde M, Gonçalves R (2015) Massive point cloud data management: design, implementation and execution of a point cloud benchmark. Comput Graph 49:92–125

Vosselman G, Maas H (eds) (2010) Airborne and terrestrial laser scanning. Taylor & Francis, Routledge

Errors in the Short-Term Forest Resource Information Update

Ville Luoma, Mikko Vastaranta, Kyle Eyvindson, Ville Kankare, Ninni Saarinen, Markus Holopainen and Juha Hyyppä

Abstract Currently the forest sector in Finland is looking towards the next generation's forest resource information systems. Information used in forest planning is currently collected by using an area-based approach (ABA) where airborne laser scanning (ALS) data are used to generalize field-measured inventory attributes over an entire inventory area. Inventories are typically updated at 10-year interval. Thus, one of the key challenges is the age of the inventory information and the cost-benefit trade-off between using the old data and obtaining new data. Prediction of future forest resource information is possible through growth modelling. In this paper, the error sources related to ALS-based forest inventory and the growth models applied in forest planning to update the forest resource information were examined. The error sources included (i) forest inventory, (ii) generation of theoretical stem distribution, and (iii) growth modelling. Error sources (ii) and (iii) stem from the calculations used for forest planning, and were combined in the investigations. Our research area, Evo, is located in southern Finland. In all, 34 forest sample plots (300 m^2) have been measured twice tree-by-tree. First measurements have been carried out in 2007 and the second measurements in 2014 which leads to 7 year updating period. Respectively, ALS-based forest inventory data were available for 2007. The results showed that prediction of theoretical stem distribution and forest growth modelling affected only slightly to the quality of the predicted stem volume in short-term information update when compared to forest inventory error.

V. Luoma (✉) · M. Vastaranta · K. Eyvindson · V. Kankare · N. Saarinen · M. Holopainen
Department of Forest Sciences, University of Helsinki, Latokartanonkaari 7,
00014 Helsinki, Finland
e-mail: ville.luoma@helsinki.fi

J. Hyyppä
National Land Survey, Finnish Geospatial Research Institute,
Geodeetinrinne 2, 04310 Masala, Finland
e-mail: juha.hyyppa@nls.fi

V. Luoma · M. Vastaranta · V. Kankare · N. Saarinen · M. Holopainen · J. Hyyppä
Centre of Excellence in Laser Scanning Research,
Finnish Geospatial Research Institute, 02431 Masala, Finland

I. Ivan et al. (eds.), *The Rise of Big Spatial Data*, Lecture Notes
in Geoinformation and Cartography, DOI 10.1007/978-3-319-45123-7_12

Keywords Growth modelling · Forest planning · GIS · Airborne laser scanning · Forest inventory

1 Introduction

Accuracy of forest resource information has a decisive impact to decision making concerning forest management and wood procurement. Forest inventory information is used in decision support systems that are further used for making computations of the current state of the forest holding as well as future projections. Errors in input data for the execution of long model chains increase in magnitude and cause significant output errors, e.g. when forest management plans are updated (e.g. Ojansuu et al. 2002; Haara and Korhonen 2004; Haara 2005; Vastaranta et al. 2010; Holopainen et al. 2010a, b). The longer the reference period, the larger the output errors; thus, inaccurate input data are especially problematic in forestry yield value determination. In addition, inaccurate input data cause significant losses in forest planning and forest silviculture as the timing of various treatments starts to differ from the optimal timings (e.g. Eid 2000; Eid et al. 2004; Holopainen and Talvitie 2006; Holopainen et al. 2009).

Airborne laser scanning (ALS) has been generalized as a technique used for forest inventory with an aim for collecting information for forest planning. The applied method is known as area-based approach (ABA). In ABA, ALS data are used to generalize field-measured forest inventory attributes over an entire inventory area. ABA has provided accuracies ranging between 10 and 27 % for the mean stem volume at stand- or plot-level (e.g. Næsset et al. 2004; White et al. 2013). In ABA, forest inventory attributes, such as species-specific stem volume (V), basal-area (G), Lorey's height (Hg), basal-area weighted mean diameter (Dg), and stem number (N) are predicted for grid cells covering the entire inventory area. However, models used in forest-planning simulation (including attribute update) require measured or estimated stem diameter distributions that are not directly inventoried. Thus, stem diameter distributions are typically formed by predicting/recovering parameters of some theoretical distribution function such as the probability functions of beta, Weibull or Johnson SB distributions based on the forest inventory attributes (Kilkki et al. 1989; Maltamo and Kangas 1998; Siipilehto 1999; Kangas and Maltamo 2000; Holopainen et al. 2010c). Then, forest growth modelling is done at tree-level by using growth models for basal-area and height (Hynynen et al. 2002). In this way, forest inventory attribute updating systems that are based on tree-level models are subject to (1) inventory errors, (2) errors in the predicted stem diameter distribution, and (3) errors in the growth modelling.

Our objective was to analyse the effects of these error sources on the short-term forest inventory attribute update in boreal managed forest conditions. The analyses of the error sources were partitioned into two parts. The first part dealt with the errors related to the forest inventory using ABA. The second part dealt with the effect of stem distribution prediction error and growth modelling error. The overall aim related to the study is to develop methods for updating grid-level forest inventory attributes for forest management planning purposes.

2 Materials and Methods

2.1 Study Area

The 5 by 5 km study area is located in Evo, southern Finland which belongs to the southern boreal forest zone. It consists of approximately 2000 ha of mainly managed boreal forest having an average stand size slightly less than 1 ha. The elevation of the area varies from 125 to 185 m above sea level. Scots pine (*Pinus sylvestris* L.) and Norway spruce [*Picea abies* (L.) H. Karst.] were the dominant tree species in the study area contributing 49 and 28 % of the total stem volume, respectively. The share of deciduous trees was 23 % of the total stem volume (Fig. 1).

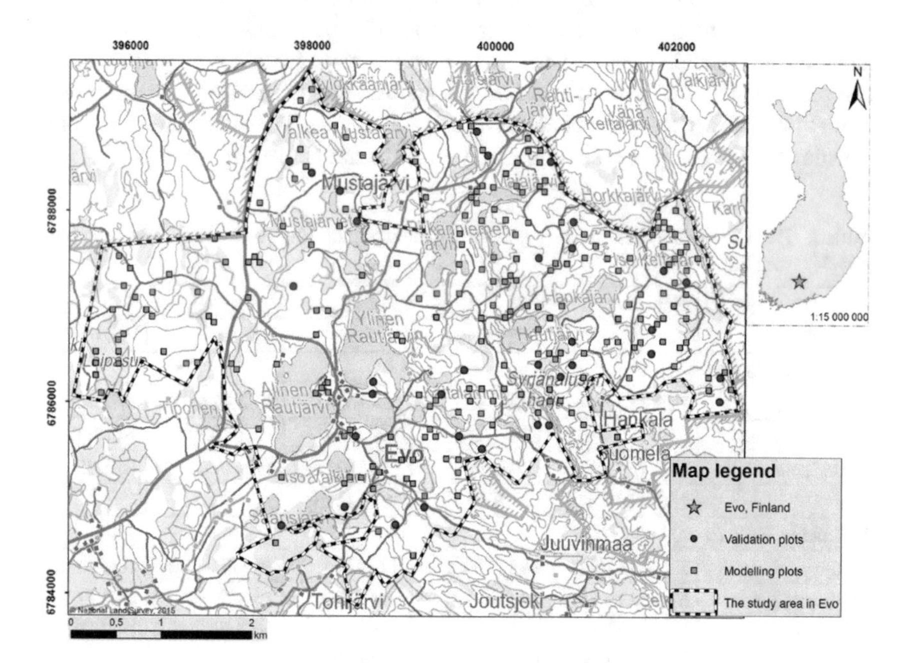

Fig. 1 Map of the study area containing the modelling (n = 246) and validation (n = 34) plots used in the study

2.2 Field Data from 2007 and 2014

Field measurements were undertaken in summer 2007 on 246 circular plots (modelling plots) with 9.77 m radius. The modelling plots were selected based on pre-stratification of existing stand inventory data (Kankare et al. 2013). All trees having a diameter-at-breast-height (DBH) of over 5 cm were tallied and tree height, DBH, and species were recorded. Tree heights were measured using Vertex clinometers as DBH was measured with steel callipers. The stem volumes were calculated with standard Finnish species-specific stem volume models (Laasasenaho 1982). The plot-level data were obtained by summing the tree data. From the 246 modelling plots, a further sample of 34 plots was selected in year 2014 to be used as validation plots in this study. The validation plots were distributed over the study area among the modelling plots to cover all the various site types, stand development classes, and tree species. The unnatural changes to modelling plots, such as clear-cuts or thinnings, limited the number of validation plots available. The descriptive statistics of modelling plots (n = 246) and validation plots (n = 34) are summarized in Table 1. The plot centres were measured with a Trimble's GEOXM 2005 Global Positioning System (GPS) device (Trimble Navigation Ltd., Sunnyvale, CA, USA), and the locations were post-processed with local base station data, resulting in an average error of app. 0.6 m.

The 34 validation plots were re-measured in 2014 with the exactly similar plot set up as year 2007. Again all trees on the plot with DBH over 5 cm were measured and DBH, tree height and species were recorded. The sample plots were located based on the recorded coordinates for the plot centres from 2007 measurements. The plot centres were even marked with signposts during the 2007 measurements so that the exact plot centre could be found for re-measurements. The descriptive statistics for sample plots measured on year 2014 are shown in Table 2.

Table 1 Field inventoried mean attributes of modelling plots (n = 246) and validation plots (n = 34) from 2007

	Field inventory 2007				
	V, m^3/ha	G, m^2/ha	N, /ha	Dg, cm	Hg, m
Modelling plots (n = 246)					
Scots pine	70.6 (86.1)	7.9 (8.5)	282 (358)	23.2 (9.6)	17.7 (6)
Norway spruce	67.9 (95.9)	7.3 (8.6)	361 (344)	17.8 (10.2)	14 (7.4)
Deciduous	48.2 (56)	6 (6.4)	386 (422)	17.8 (9.2)	16.1 (5.5)
All	186.6 (110.4)	21.2 (9.5)	1029 (618)		
Validation plots (n = 34)					
Scots pine	110.8 (116.2)	12 (10.4)	359 (326)	25 (11.1)	18.6 (5.6)
Norway spruce	63.3 (119.6)	6.8 (10.3)	399 (456)	15.3 (10.3)	12.5 (7.7)
Deciduous	56.4 (78.2)	6.7 (8.2)	415 (541)	16.7 (8)	15.9 (5.2)
All	230.4 (125.5)	25.4 (9.5)	1173 (750)		

Standard deviations are provided in the parenthesis. The validation plots were also included in modelling plots

Table 2 Field inventoried mean attributes of validation plots (n = 34) from 2014

	Field inventory 2014				
	V, m^3/ha	G, m^2/ha	N,/ha	Dg, cm	Hg, m
Scots pine	131.1 (115.1)	13.1 (10.5)	340 (317)	26.7 (10.6)	20.6 (4.8)
Norway spruce	75.4 (119.2)	7.9 (10.1)	458 (479)	17.2 (10.4)	14.4 (7.2)
Deciduous	63.4 (88.5)	7.1 (9)	433 (540)	16 (7)	16.5 (4.8)
All	270 (105.8)	28 (8.4)	1231 (814)		

Standard deviations are provided in the parenthesis
n = 34 plots

2.3 *Remote Sensing-Based Forest Inventory from 2007*

The remote sensing data were collected in midsummer 2006. ALS was performed using Optech ALTM3100C-EA system operating with a pulse rate of 100 kHz. Data were acquired at a flight altitude of 1900 m resulting in an average pulse density of 1.3 pulses per square meter in non-overlapping areas and a footprint of 70 cm in diameter. The system was configured to record up to four returns per pulse, i.e. first, last, only, and intermediate. Reported positioning accuracy was 40 cm and 15 cm for horizontal and vertical direction respectively. Same-date aerial photographs were obtained with a digital camera and the photographs were orthorectified, resampled to pixel size of 0.5 m and mosaicked to a single image covering the entire data. Near-infrared (NIR), red (R) and green (G) bands were available.

ALS data were first classified into ground or non-ground points using the TerraScan (TerraSolid, Helsinki, Finland) based on the method explained in Axelsson (2000). A digital terrain model (DTM) was then calculated using classified ground points. Laser heights above ground (normalized height or canopy height) were calculated by subtracting ground elevation from corresponding laser measurements. The expected accuracy of the ALS-derived DTM varies in boreal forest conditions by around 10–50 cm (Hyyppä et al. 2009). Canopy heights closed to zero are the ground returns and those greater than 2 m are considered as vegetation returns. The data between them are considered as returns from ground vegetation or bushes. Only the returns from vegetation were used for feature extraction. Statistical metrics describing canopy structure were extracted for the sample plots (radius 9.77 m) following suggestions by White et al. (2013). Also several statistical and textural features were extracted from the aerial photographs, such as the means and standard deviations of spectral values (Holopainen et al. 2008). The Haralick textural features (Haralick et al. 1973; Haralick 1979) were derived from the spectral values.

Species specific basal area (G), basal area-weighted mean diameter (Dg), Lorey's height (Hg), stem volume (V), and number of stems per hectare (N) were predicted by means of remote sensing metrics using random forest (RF, Breiman 2001) based k nearest-neighbor (NN) approach. Forest inventory attributes measured in the field were used as target observations, and plot-specific metrics derived from remote sensing data sets were used as predictors. The RF approach was applied in the search of nearest neighbors. In the RF method, several regression trees are generated by drawing a replacement from two-thirds of the data for training and one-third for testing for each tree. The samples that are not included in training are called out-of-bag samples, and they can act as a testing set in the approach. The measure of nearness in RF is defined based on the observational probability of ending up in the same terminal node in classification. The R statistical computing environment (R Core Team) and yaImpute library (Crookston and Finley 2008) were applied in the predictions. In the present study, 1200 regression trees were generated, and the square root of the number of predictor variables was picked randomly at the nodes of each regression tree. The number of neighbors was set to one to keep the original variance in the data (see, e.g. Hudak et al. 2008; Franco-Lopez et al. 2001). Prior to the final modeling, RF was used to reduce the number of predictor variables. The aim of the variable reduction was to build up parsimonious models that are capable of accurate prediction. In the variable selection, RF iterated 100 times per model and the best variables based on their importance for each model were selected. Then, only the most important variables based on the results were used for the final imputations. The used predictors were the vegetation ratio from first and last pulses, the heights where 30 and 90 % of first laser returns and 30 % of last returns had been received, mean height in the pixel window, local homogeneity 90° of height, the average NIR and standard deviation of NIR.

To improve the accuracy of the species specific estimates, the sample plots were divided into four strata according to existing stand register information. The first stratum included Scots pine dominated stands, the second stratum Norway spruce dominated stands, the third stratum included stands dominated by deciduous trees and the fourth stratum had stands with approximately equal share of pine and spruce trees with a mixture of deciduous trees. The first stratum comprised 92 sample plots, the second 56, the third 41 and the fourth 57 sample plots respectively. The final imputations were carried out for each stratum separately.

2.4 Simulation of Forest Growth

The forest attribute update calculations from 2007 to 2014 were carried out using SIMO software (SIMO simulation framework, Rasinmäki et al. 2009). SIMO is a common platform for various stand simulators including Finnish tree- and stand-level simulators. The simulation logic is described in XML documents (eXtensible Markup Language) and lends itself to be easily adapted for various types of calculations. The non-spatial tree-level growth models found in SIMO are, for the most part, the same as those found in the MELA2002 and MOTTI simulators (Hynynen et al. 2002; Salminen et al. 2005). They include growth models for all sites and tree species in Finland, including separate models for peatlands. The tree-level simulator can be used to simulate the growth of either sample trees

Table 3 Principles used in the error analyses

Alternative	Source of error	Input data	Reference data
1	Inventory error	Species-specific stem volumes derived from ABA	Species-specific stem volumes derived from field measurements from 2007
2	Prediction of theoretical stem diameter distribution and growth modelling error	Species-specific stem volumes derived from field measurements from 2007	Species-specific stem volumes derived from field measurements from 2014
3	Combined errors	Species-specific stem volumes derived from ABA	Species-specific stem volumes derived from field measurements from 2014

measured in the field or descriptive trees generated on the basis of a theoretical diameter/height distribution. The simulation is performed at the single-tree level. The statistics for the strata and stands are derived as the sums and means of the simulated tree properties.

2.5 *Evaluation of the Errors*

The accuracy of the ABA and updated stem volumes estimates was evaluated by calculating bias and root-mean-square error (RMSE) for three different alternatives (Table 3):

$$BIAS = \frac{\sum_{i=1}^{n}(y_i - \hat{y}_i)}{n} \quad BIAS\% = 100 * \frac{BIAS}{\bar{y}}$$

$$RMSE = \sqrt{\frac{\sum_{i=1}^{n}(y_i - \hat{y}_i)^2}{n}} \quad RMSE\% = 100 * \frac{RMSE}{\bar{y}}$$

where n is the number of plots, y_i is the observed value (by tree-wise measurements from 2014) for plot i, $\hat{y}_i$ is updated attribute for plot i and $\bar{y}_i$ is the observed mean of the species-specific—or total stem volume.

3 Results and Discussion

The results from the remote sensing data based prediction of forest inventory attributes in year 2007 are presented in Table 4. For the validation plots (n = 34) the empirical 95 % interval of total stem volume was between 42.4 and

Table 4 Mean values for predicted forest inventory attributes using ABA

	ABA predicted forest inventory attributes 2007				
	V, m^3/ha	G, m^2/ha	N, /ha	Dg, cm	Hg, m
Scots pine	104.4 (85)	11.5 (8.5)	367 (292)	22.5 (8.9)	17.8 (6.4)
Norway spruce	49.8 (76.9)	5.7 (7.3)	382 (361)	14.4 (8.4)	11.6 (6.4)
Deciduous	56.6 (67.5)	6.7 (6.9)	483 (512)	15.7 (6.5)	15.2 (5.8)
Total	210.9 (108.1)	23.8 (8.8)	1231 (685)		

Standard deviations are provided in the parenthesis

431.2 m^3/ha. The species specific empirical 95 % intervals for stem volume were for pine from 0 to 266.4 m^3/ha, for spruce from 0 to 239.0 m^3/ha and for deciduous trees from 0.5 to 225.7 m^3/ha, respectively.

The RMSE of forest inventory for total stem volume was 25.2 % as the bias was 8.5 % (Table 5). Species-specific RMSEs and biases varied from 80.0 to 134.3 % and from −0.5 to 21.3 %, respectively. At the sample plot-level the range in inventory error (difference) was from −83.6 to 167.4 m^3/ha (Fig. 2). Based on Hudak et al. (2008) and Franco-Lopez et al. (2001) increasing the number of neighbors would improve the prediction accuracy. However, inventory RMSEs are in line with the previous studies in the same study area (Holopainen et al. 2010a; Yu et al. 2010; Vastaranta et al. 2011, 2012, 2013; Kankare et al. 2015). Controversially, ABA inventory in this study included bias. Bias can be resulted

Table 5 Effect of inventory error on predicted species-specific stem volumes as well as on the total stem volume (V) on the validation plots

	Error source			
	V	V_pine	V_spruce	V_dec
Forest inventory error				
RMSE, m^3/ha	58.0	88.7	80.8	75.7
RMSE-%	25.2 %	80.0 %	127.6 %	134.3 %
Bias, m^3/ha	19.6	6.4	13.5	−0.3
Bias-%	8.5 %	5.7 %	21.3 %	−0.5 %
Growth modelling and prediction of theoretical stem diameter distribution error				
RMSE, m^3/ha	50.7	35.4	17.4	41.8
RMSE-%	18.8 %	27.0 %	23.1 %	65.9 %
Bias, m^3/ha	18.1	7.8	6.8	3.5
Bias-%	6.7 %	5.9 %	9.0 %	5.5 %
Combined error of forest inventory, prediction of theoretical stem diameter distribution and forest growth modelling				
RMSE, m^3/ha	66.3	86.3	82.4	67.4
RMSE-%	24.6 %	65.8 %	109.2 %	106.3 %
Bias, m^3/ha	35.5	13.0	20.1	2.3
Bias-%	13.1 %	9.9 %	26.7 %	3.7 %

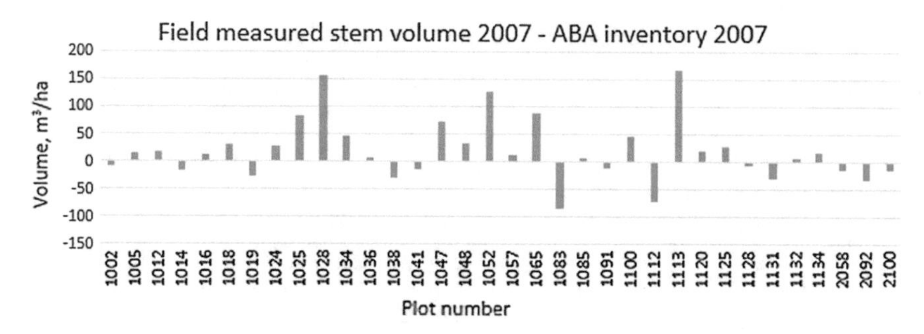

Fig. 2 Field measured stem volume (m^3/ha) (2007) compared to stem volume estimate based on ABA (2007)

from the rather limited number of validation plots (n = 34) as well as from slight differences in forest inventory attributes measured from modelling plots used in ABA compared to validation plots (see Table 1). For example, the mean stem volume was 230.4 m^3/ha in the validation plots ranging from 54.7 to 575.4 m^3/ha as the respective numbers from modelling plots were 186.6 m^3/ha (mean), 0 m^3/ha (min) and 575.4 m^3/ha (max). To avoid more bias number of nearest neighbors was chosen to be 1.

Prediction of stem diameter distribution and growth modelling errors caused 6.7 % bias and 18.8 % RMSE to the updated stem volume. Species-specific RMSEs and biases varied from 23.1 to 65.9 % and from 5.5 to 9.0 %, respectively. The RMSEs are lower than the ones for the ABA forest inventory of year 2007. Based on the previous studies (Vastaranta et al. 2010; Holopainen et al. 2010c) it can be assumed that the majority of this error is caused by the growth modelling and only a minor component from the generated stem distribution. Although, error of predicting stem diameter distribution cannot be separated from the growth modelling error in this study, it has been shown that its effect is marginal in this kind of study design (e.g. Holopainen et al. 2010c). At the sample plot-level the range in error of prediction of stem distribution and growth modelling error (difference) was from −134.7 to 93.7 m^3/ha (Fig. 3).

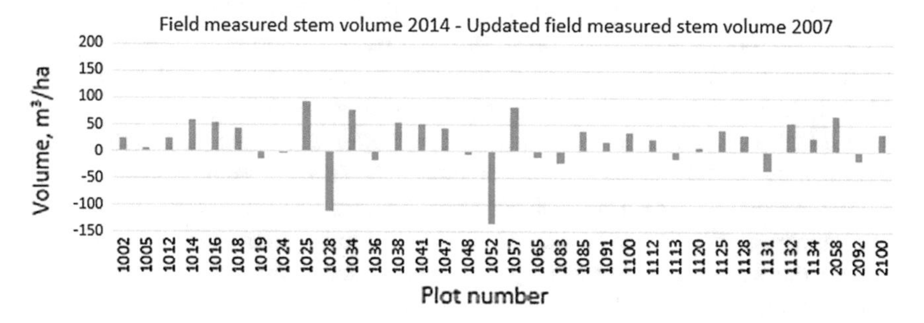

Fig. 3 Field measured stem volume (m^3/ha) (2014) compared to field measured stem volume from 2007 updated to year 2014. The update was done by utilizing growth models

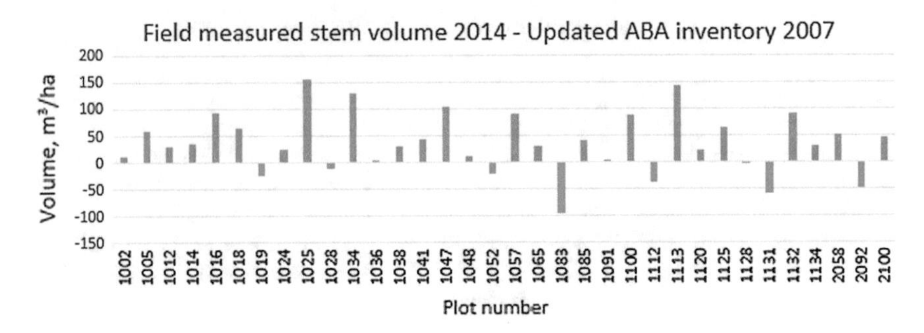

Fig. 4 Field measured stem volume (m^3/ha) (2014) compared to stem volume estimate based on ABA from 2007 updated to year 2014. The update was done by utilizing growth models

Combined error of forest inventory, prediction of theoretical stem distribution and forest growth modelling caused 13.1 % bias and 24.6 % RMSE to the updated stem volume. Species-specific RMSEs and biases varied from 65.8 to 109.2 % and from 3.7 to 26.7 %, respectively. At the sample plot-level the range in combined errors was from −95.3 to 156.8 m^3/ha (Fig. 4).

Compared to attribute update from error free data (errors of prediction of stem distribution and growth modelling), it can be seen that biases are 5–15 % points larger for total stem volume as well as for species specific stem volumes when all the error sources are combined. Similarly, RMSE for total stem volume is roughly 10 % points larger. Species-specific errors increase more. Accuracy of the species-specific stem volumes is ranging from 80.0 to 134.3 % with ABA (inventory error) and thus it can be expected that these errors shift to outputs of the update process.

4 Conclusion

The objective here was to analyse the effects of error sources on the short-term forest inventory attribute update in boreal managed forest conditions. The analyses of the error sources were partitioned into two parts. The first part dealt with the errors related to the forest inventory using ABA. The second part dealt with the effect of stem distribution prediction error and growth modelling error. The results showed that prediction of theoretical stem distribution and forest growth modelling affected only slightly to the quality of the predicted stem volume in short-term information update. The results of our study confirm that the quality of the input data is the most effective error source in short-term forest information update. Thus, further studies are required especially for obtaining species-specific forest inventory information more accurately.

Acknowledgments Our study was made possible by financial aid from the Finnish Academy project Centre of Excellence in Laser Scanning Research (CoE-LaSR, decision number 272195). We also wish to thank M.Sc. Risto Viitala at the HAMK University of Applied Sciences for organizing part of the field data collections.

References

Axelsson P (2000) DEM generation from laser scanner data using adaptive TIN models. Int Arch Photogramm Remote Sens Amst 33(B4):110–117

Breiman L (2001) Random forests. Mach Learn 45(1):5–32

Crookston NL, Finley AO (2008) yaImpute: an R package for kNN imputation. J Stat Softw 23 (10):1–16

Eid T (2000) Use of uncertain inventory data in forestry scenario models and consequential incorrect harvest decisions. Silva Fenn 34:89–100

Eid T, Gobakken T, Næsset E (2004) Comparing stand inventories for large areas based on photo-interpretation and laser scanning by means of cost-plus-loss analyses. Scand J For Res 19:512–523

Franco-Lopez H, Ek AR, Bauer ME (2001) Estimation and mapping of forest stand density, volume, and cover type using k-nearest neighbors method. Remote Sens Environ 77(3):251–274

Haara A (2005) The uncertainty of forest management planning data in Finnish non-industrial private forestry. Doctoral thesis, Dissertationes Forestales 8, 34 p

Haara A, Korhonen K (2004) Kuvioittaisen arvioinnin luotettavuus. Metsätieteen aikakauskirja 4:489–508 (in Finnish)

Haralick R, Shanmugan K, Dinstein I (1973) Textural features for image classification. IEEE Trans Syst Man Cybern 3(6):610–621

Haralick R (1979) Statistical and structural approaches to texture. Proc IEEE 67(5):786–804

Holopainen M, Talvitie T (2006) Effects of data acquisition accuracy on timing of stand harvests and expected net present value. Silva Fenn 40(3):531–543

Holopainen M, Haapanen R, Tuominen S, Viitala R (2008) Performance of airborne laser scanning- and aerial photograph-based statistical and textural features in forest variable estimation. In: Hill R, Rossette J, Suárez J. Silvilaser 2008 proceedings, pp 105–112

Holopainen M, Vastaranta M, Mäkinen A, Rasinmäki J, Hyyppä J, Hyyppä H, Kaartinen H (2009) The use of tree level ALS data in forest management planning simulations. Photogramm J Finl 21(2):13–24

Holopainen M, Mäkinen A, Rasinmäki J, Hyyppä J, Hyyppä H, Kaartinen H, Kangas A (2010a) Effect of tree-level airborne laser-scanning measurement accuracy on the timing and expected value of harvest decisions. Eur J For Res 129(5):899–907

Holopainen M, Mäkinen A, Rasinmäki J, Hyytiäinen K, Bayazidi S, Vastaranta M, Pietilä I (2010b) Uncertainty in forest net present value estimations. Forests 1(3):177–193

Holopainen M, Mäkinen A, Rasinmäki J, Hyytiäinen K, Bayazidi S, Pietilä I (2010c) Comparison of various sources of uncertainty in stand-level net present value estimates. For Policy Econ 12 (5):377–386

Hudak AT, Crookston NL, Evans JS, Hall DE, Falkowski MJ (2008) Nearest neighbor imputation of species-level, plot-scale forest structure attributes from LiDAR data. Remote Sens Environ 112:2232–2245

Hynynen J, Ojansuu R, Hökkä H, Siipilehto J, Salminen H, Haapala P (2002) Models for predicting stand development in MELA system. Finn For Res Inst Res Pap 835

Hyyppä J, Hyyppä H, Yu X, Kaartinen H, Kukko A, Holopainen M (2009) Forest inventory using small-footprint airborne LiDAR. In: Shan J, Toth CK (eds) Topographic laser ranging and scanning: principles and processing. CRC Press/Taylor & Francis Group, Boca Raton, pp 335–370

Kangas A, Maltamo M (2000) Performance of percentile based diameter distribution prediction and Weibull method in independent data sets. Silva Fenn 34:381–398

Kankare V, Vastaranta M, Holopainen M, Räty M, Yu X, Hyyppä J, Hyyppä H, Alho P, Viitala R (2013) Retrieval of forest aboveground biomass and volume with airborne scanning LiDAR. Remote Sens 5(5):2257–2274

Kankare V, Vauhkonen J, Holopainen M, Vastaranta M, Hyyppä J, Hyyppä H, Alho P (2015) Sparse density, leaf-off airborne laser scanning data in aboveground biomass component prediction. Forests 6:1839–1857

Kilkki P, Maltamo M, Mykkänen R, Päivinen R (1989) Use of the Weibull function in estimating the basal-area diameter distribution. Silva Fenn 23:311–318

Laasasenaho J (1982) Taper curve and volume functions for pine, spruce and birch. Communicationes. Institute Forestalis Fenniae 108:74 p

Maltamo M, Kangas A (1998) Methods based on k-nearest neighbour regression in the prediction of basal area diameter distribution. Can J For Res 28:1107–1115

Næsset E, Gobakken T, Holmgren J, Hyyppä H, Hyyppä J, Maltamo M, Nilsson M, Olsson H, Persson Å, Söderman U (2004) Laser scanning of forest resources: the Nordic experience. Scand J For Res 18(19):482–499

Ojansuu R, Halinen M, Härkönen K (2002) Metsätalouden suunnittelujärjestelmän virhelähteet männyn esiharvennuskypsyyden määrittämisessä. Metsätieteen aikakauskirja 3(2002):441–457 (in Finnish)

Rasinmäki J, Kalliovirta J, Mäkinen A (2009) SIMO: an adaptable simulation framework for multiscale forest resource data. Comput Electron Agric 66:76–84

Siipilehto J (1999) Improving the accuracy of predicted basal-area diameter distribution in advanced stands by determining stem number. Silva Fenn 34:331–349

Salminen H, Lehtonen M, Hynynen J (2005) Reusing legacy FORTRAN in the MOTTI growth and yield simulator. Comput Electron Agric 49:103–113

Vastaranta M, Ojansuu R, Holopainen M (2010) Puustotunnusten laskennallisen ajantasaistuksen luotettavuus–tapaustutkimus Pohjois-Savossa. Metsätieteen aikakauskirja 4:367–381 (in Finnish)

Vastaranta M, Holopainen M, Yu X, Haapanen R, Melkas T, Hyyppä J, Hyyppä H (2011) Individual tree detection and area-based approach in retrieval of forest inventory characteristics from low-pulse airborne laser scanning data. Photogramm J Finl 22(2):1–13

Vastaranta M, Kankare V, Holopainen M, Yu X, Hyyppä J, Hyyppä H (2012) Combination of individual tree detection and area-based approach in imputation of forest variables using airborne laser data. ISPRS J Photogramm Remote Sens 67:73–79

Vastaranta M, Wulder MA, White JC, Pekkarinen A, Tuominen S, Ginzler C, Kankare V, Holopainen M, Hyyppä J, Hyyppä H (2013) Airborne laser scanning and digital stereo imagery measures of forest structure: comparative results and implications to forest mapping and inventory update. Can J Remote Sens 39(5):382–395

White JC, Wulder MA, Varhola A, Vastaranta M, Coops NC, Cook BD, Pitt D, Woods M (2013) A best practices guide for generating forest inventory attributes from airborne laser scanning data using the area-based approach. Information report FI-X-10. Natural Resources Canada, Canadian Forest Service, Canadian Wood Fibre Centre, Pacific Forestry Centre, Victoria, BC, 50 p

Yu X, Hyyppä J, Holopainen M, Vastaranta M (2010) Comparison of area-based and individual tree-based methods for predicting plot-level forest attributes. Remote Sens 2:1481–1495

Accuracy of High-Altitude Photogrammetric Point Clouds in Mapping

Topi Tanhuanpää, Ninni Saarinen, Ville Kankare, Kimmo Nurminen, Mikko Vastaranta, Eija Honkavaara, Mika Karjalainen, Xiaowei Yu, Markus Holopainen and Juha Hyyppä

Abstract During the past decade, airborne laser scanning (ALS) has established its status as the state-of-the-art method for detailed forest mapping and monitoring. Current operational forest inventory widely utilizes ALS-based methods. Recent advances in sensor technology and image processing have enabled the extraction of dense point clouds from digital stereo imagery (DSI). Compared with ALS data, the DSI-based data are cheap and the point cloud densities can easily reach that of ALS. In terms of point density, even the high-altitude DSI-based point clouds can be sufficient for detecting individual tree crowns. However, there are significant differences in the characteristics of ALS and DSI point clouds that likely affect the accuracy of tree detection. In this study, the performance of high-altitude DSI point clouds was compared with low-density ALS in detecting individual trees. The trees were extracted from DSI- and ALS-based canopy height models (CHM) using watershed segmentation. The use of both smoothed and unsmoothed CHMs was tested. The results show that, even though the spatial resolution of the DSI-based CHM was better, in terms of detecting the trees and the accuracy of height estimates, the low-density ALS performed better. However, utilizing DSI with shorter ground sample distance (GSD) and more suitable image matching algorithms would likely enhance the accuracy of DSI-based approach.

T. Tanhuanpää (✉) · N. Saarinen · V. Kankare · M. Vastaranta · M. Holopainen
Department of Forest Sciences, University of Helsinki, Latokartanonkaari 7, 00014 Helsinki, Finland
e-mail: topi.tanhuanpaa@helsinki.fi

K. Nurminen · E. Honkavaara · M. Karjalainen · X. Yu · J. Hyyppä
National Land Survey, Finnish Geospatial Research Institute, Geodeetinrinne 2, 04310 Masala, Finland
e-mail: kimmo.nurminen@nls.fi

T. Tanhuanpää · N. Saarinen · V. Kankare · M. Vastaranta · M. Karjalainen · X. Yu · M. Holopainen · J. Hyyppä
Centre of Excellence in Laser Scanning Research, Finnish Geospatial Research Institute, 02431 Masala, Finland

I. Ivan et al. (eds.), *The Rise of Big Spatial Data*, Lecture Notes in Geoinformation and Cartography, DOI 10.1007/978-3-319-45123-7_13

Keywords Image-based point clouds · LiDAR · Height models · Tree detection · Forest

1 Introduction: DSM for Solar Radiation Calculation

Detailed and spatially accurate datasets form the core of operative forest management. This data allows for the selection of various management actions based on precise knowledge on forest structure. Airborne laser scanning (ALS) has been the latest significant breakthrough in operational forest resource assessment. ALS technology, combines accurate distance measurements with laser light, positioning with global navigation satellite system (GNSS) and precise orientation of the measurement unit with an inertial measurement unit (IMU) (Wehr and Lohr 1999). The methodology enables forming spatially accurate three dimensional (3D) point clouds that represent the shape and structure of the scanned object. In a forested environment, the point clouds have proven to be an accurate means for assessing the vertical structure of the forest canopy (e.g., Lim et al. 2003). Two distinct methodologies can be distinguished for assessing forest attributes from ALS data. In the so-called area-based approach (ABA), various statistical features describing height distribution of the ALS point cloud are calculated in cell-wise manner for a forest area. The features are utilized in developing e.g., regression models (Næsset 2002) together with information from field plots for predicting forest inventory attributes for unmeasured grid cells. Typically, low-density ALS data (<1 point/m^2) are utilized in ABA. An alternative approach utilizing more dense point clouds is referred as individual tree detection (ITD). The methodology is based on identifying the points from the point clouds representing each individual tree crown, calculating tree-level ALS features, and generating tree attributes for the detected trees (Hyyppä and Inkinen 1999). The partitioning of the point clouds can be done utilizing, for instance the pouring algorithm (Koch et al. 2006) which utilizes canopy height models (CHM) derived from the point clouds. The method seeks the local maxima from the CHM and treats them as seed pixels, i.e., tree tops. The crown area is formed by adding neighboring pixels to the seed as far as their height value is lower than the last added pixel. The CHM is often smoothed prior to the segmentation process in order to decrease the amount of noise in the data. ITD generally requires denser ALS data than ABA (>5 points/m^2). However, several studies have proven that the structure of the forest and the method used for delineation affect the ITD accuracy more than the point density (Kaartinen et al. 2012; Vauhkonen et al. 2012). The price of the ALS datasets increases as the point density gets higher.

Aerial images and photogrammetry have a long tradition in the assessment of forest resources. Image-based 3D measurement of forests have been studied from 1940s (see Korpela 2004). With means of stereophotogrammetry, detecting a common point (e.g., a tree top) from at least two images enables defining its 3D coordinates (XYZ). During recent years, aerial imagery has been widely utilized especially in tree species interpretation, due to its sufficient spatial and spectral resolution (e.g., Yu et al. 1999;

Held et al. 2003). However, advances in sensor technology and image processing have enabled generating dense digital stereo imagery (DSI)-based point clouds that offer the same levels of point density as ALS. The DSI-based point clouds are generated with automatic image matching algorithms, e.g., semi-global matching (Hirschmüller 2008). The density of the final point cloud depends on the image resolution and the matching algorithm used (White et al. 2013). In the case of forestry, DSI-based point clouds can be utilized in, e.g., predicting forest attributes in the same manner as ALS-based point clouds (Nurminen et al. 2013).

In terms of forest resource assessment, both ALS and DSI-based data collection have their strengths and weaknesses. Although highly dependent on the density of the ALS data and the ground sample distance (GSD) of the DSI, the cost of DSI data is approximately from one third to half of the cost of ALS data (Holopainen et al. 2014; White et al. 2013). Considering the spatial resolution (XY), a given point cloud density can be achieved from higher flying altitude when using DSI, which greatly affects the cost of the data. Also, DSI have much higher spectral resolution which can be utilized together with the extracted point cloud, e.g., for use in species classification. However, the DSI-based point clouds can only cover the upper parts of the canopy that are visible in the images, whereas ALS penetrates the canopy to some extent, reaching the suppressed tree crowns and the ground (Ackermann 1999). In order to achieve reliable estimates for tree height, the ground elevation has to be known accurately. The current operational forest inventory systems commonly use ALS data for generating the digital terrain model (DTM) and the three dimensional (3D) structure of the trees, whereas DSI are often used only for species interpretation. However, the recent developments, especially in image processing, have made it possible to achieve sufficient point clouds for ABA, using only DSI and existing DTMs (Järnstedt et al. 2012). Also, considering only the point densities of the image-based point clouds, even ITD should be possible using solely materials derived from DSI. Recent studies have shown the differences between the two types of point clouds (Vastaranta et al. 2013; White et al. 2015), but the DSI-based data have rarely been utilized at the level of single trees, i.e., in ITD.

The aim of this study was to investigate the capability of high-altitude DSI-based point clouds in detecting single tree crowns in mature boreal forests. The performance of DSI-based was evaluated by the number of correctly matched field-measured reference trees, the accuracy of individual height measurements, and the correctness of the overall plot-level height distributions. For comparison, the same procedures were applied to CHMs derived from low-density ALS data.

2 Materials and Methods

2.1 Study Area

The study area of 5 × 5-km is located in Evo, southern Finland (61.19°N, 25.11°E; Fig. 1). It belongs to the southern Boreal Forest Zone and contains approximately

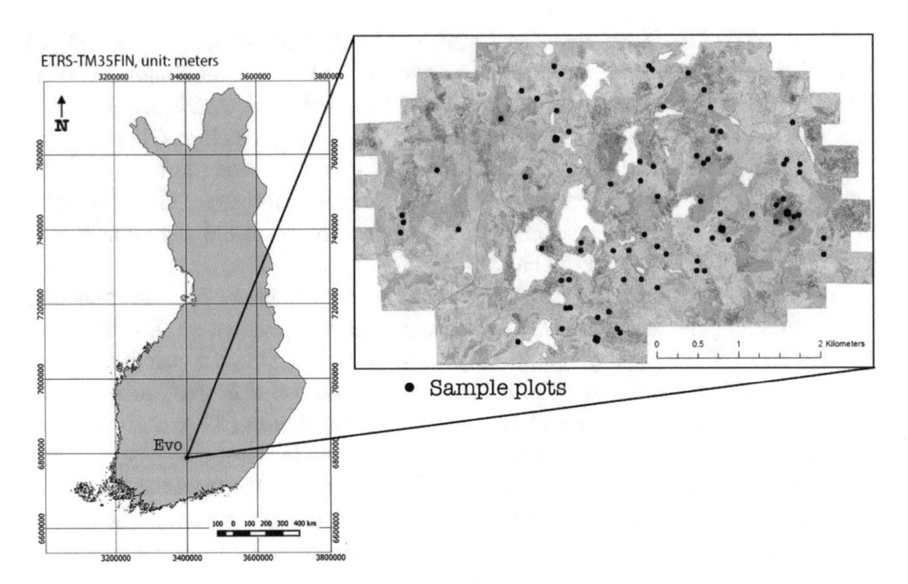

Fig. 1 The study area: *black dots in the right-hand side* indicate the location of the field plots

2000 ha of managed boreal forest. The average stand size is slightly less than 1 ha. The area consists of a mixture of forest stands, varying from natural to intensively managed forests. The elevation of the area varies from 125 to 185 m above sea level. The dominant tree species are Scots pine (*Pinus sylvestris* L.) and Norway spruce (*Picea abies* (L.) H. Karst) contributing 40 and 35 % of the total volume, respectively. The share of all deciduous trees together is 24 % of the total volume.

2.2 Field Data

The field data, consisting of 91 plots (32 m × 32 m), was collected in summer of 2014. A tree map was generated for all plots using a terrestrial laser scanner. The tree maps were validated in the field: the trees missing from the maps were added to it and incorrect (i.e., non-existing) trees were removed from the map. After the validation, all trees with diameter-at-breast-height (DBH) over 5 cm were measured manually. DBH was measured with steel caliper and the tree heights with an electronic hypsometer.

The locations of the individual trees were brought into a common coordinate system by determining the position of each sample plot. The locations were calculated using the geographic coordinates of the plot centers and four corners. Plot center positions were measured using differential GPS (Leica GPS 120, Switzerland) with sub-meter accuracy. A total station (Leica TS02, Switzerland) was used to measure the exact distance to each plot center. The plot position was

Table 1 Definitions of study subgroups

Sub-group	Group description	n
Pine	Plots with pine representing over 70 % of the basal area	12
Spruce	Plots with spruce representing over 70 % of the basal area	15
Mixed	Mixed plots with none of the species representing over 70 % of the basal area	12
Total	All plots	39

further adjusted manually using ALS data. In this procedure, the locations of all trees on the tree map were used to help find the true location of the plot in the ALS point cloud. The plot was shifted and rotated so that the tree locations within the map aligned properly with the point cloud.

For this study, only the mature plots (i.e., the plots with basal area-weighted DBH over 26 cm) were included in this study. This subset of 39 plots was further divided into sub-groups by the dominant tree species, contributing over 70 % of the basal area. The classes were defined as described in Table 1. From the 39 mature plots, only the trees with commercial value were taken into consideration. Hence the trees with height under 14 m were excluded from the data. Altogether, there were 1684 trees in the final field reference. The descriptive statistics of all sub-groups are presented in Table 2.

2.3 *Aerial Images*

The aerial images were acquired on the 22nd of May 2014 with a Z/I Imaging DMC (Digital Mapping Camera) photogrammetric aerial camera. The image block consisted of two flying strips of 12 images. The forward and side overlap of the pictures were 80 and 64 %, respectively. The images were acquired from altitude of approximately 5000 m above the mean ground level which leads to 50 cm ground sample distance (GSD). The width of an image strip was 6.9 km, and the distance between adjacent flight lines was 2.5 km. The image orientation was done in BAE Systems Socet Set software (San Diego, California, USA) on the basis of exterior orientation values. The final orientation was based on automatic tie points and 40 interactively measured ground control points. The ground control points were derived from elevation model and orthophotos provided by Finnish National Land Survey (NLS). Three radial distortion parameters were solved with on-the-job calibration. The root mean square error (RMSE) values of the adjustment were 0.266 m (X), 0.400 m (Y) and 1.187 m (Z).

The calculation of stereo models was carried out with NGATE (Next Generation Automatic Terrain Extraction) module of the Socet Set software bundle using an altered strategy file for forestry applications. Extraction of the digital surface model (DSM) was done for each stereo model of consecutive images in the same strip, but inter-strip stereo models were not used in the calculation. National DTM was

Table 2 The descriptive statistics of 39 sample plots of 32 × 32 m used in this study

	Minimum	Maximum	Mean	SD
Pine				
Mean height (m)	21.4	32.1	25.8	3.6
Mean DBH (cm)	26.3	46.4	30.8	6.0
Basal area (m^2/ha)	17.3	40.3	26.7	7.1
Volume (m^3/ha)	164.5	518.4	300.0	109.4
Plot density (trees/ha)	391	1035	565	201
Spruce				
Mean height (m)	25.4	33.4	29.2	2.6
Mean DBH (cm)	26.0	42.1	33.9	5.4
Basal area (m^2/ha)	22.1	38.9	32.8	5.1
Volume (m^3/ha)	242.6	484.9	390.8	75.0
Plot density (trees/ha)	342	879	585	159
Mixed				
Mean height (m)	23.1	31.6	27.4	2.7
Mean DBH (cm)	26.6	41.6	33.4	5.0
Basal area (m^2/ha)	15.2	43.2	33.1	8.0
Volume (m^3/ha)	177.7	508.2	349.1	96.4
Plot density (trees/ha)	342	2217	909	482
Total				
Mean height (m)	21.4	33.4	27.6	3.2
Mean DBH (cm)	26.0	46.4	32.8	5.5
Basal area (m^2/ha)	15.2	43.2	31.0	7.2
Volume (m^3/ha)	164.5	518.4	350.0	98.4
Plot density (trees/ha)	342	2217	678	335

applied to normalize DSM into CHM. The national DTM was derived from ALS data and had resolution of 2 m. The image-based CHM is referred to as CHM_{image}.

2.4 ALS Data

ALS data were acquired in late May 2014, using a Leica ALS70-HA SN7202 system (Leica Geosystems AG, Heerbrugg, Switzerland) operating at a pulse rate of 105 kHz. Flying altitude was 2500 m above sea level and flying speed 150 knots. On average, the measurement density was 0.7 pulses/m^2. The system was configured to record up to five echoes per pulse, i.e. first or only, last and 1–3 intermediates. Hence, the point density of the ALS point clouds was approximately 1–4 points/m^2 depending on the vegetation structure.

The ALS data were processed with Terra Scan software (Terrasolid, Helsinki, Finland) as follows. First, the point cloud density and pattern were unified by minimizing the flightline overlap. After this, all water surfaces were masked out from the data using water boundaries. Also, all points under ground level were removed from the data with classification tools of TerraScan. The ground points were classified with tools based on methods by Axelsson (2000). The digital surface model (DSM) was created from all vegetation points with resolution of 1 m. The CHM was obtained by subtracting the national DTM from the created DSM. The ALS-based CHM is referred to as CHM_{als}.

2.5 *Tree Delineation*

Individual crown segments were delineated from the CHMs with watershed segmentation process (see e.g., Pitkänen et al. 2004; Koch et al. 2006). Both smoothed and unsmoothed CHMs were tested for tree extraction. A simple 3 × 3 pixel moving average filter was used for the smoothing process. The unsmoothed CHMs derived from aerial images and ALS data were denoted CHM_{image} and CHM_{als}. The smoothed CHMs were denoted CHM_{image_smooth} and CHM_{als_smooth}, respectively.

The height and location of the crown segments were adopted from the maximum value of the unsmoothed CHM (i.e., CHM_{image} and CHM_{als}) inside each segment. If the CHM maximum was a "plateau" consisting of several pixels, the location was determined as the mean location of the plateau pixels. The final tree candidates were chosen among the crown segments in terms of tree height. Focusing on mature trees, only crown segments with CHM value (i.e., tree height) of at least 14 m were considered to represent actual trees.

2.6 *Tree Matching*

For assessing the tree-level accuracy of the tree detection procedure, the tree candidates were matched to field measured reference trees with the method described in Kaartinen et al. (2012). The procedure matches location points of reference trees and tree candidates that are most alike in terms of location. Tree candidates that were more than 5 m from reference trees were automatically treated as undetected. The matching started by determining a distance matrix that consisted of distances between every reference tree and tree candidate. The first reference tree to be matched was the one with the nearest tree candidate (shortest distance in the distance matrix). After the match was made, both the reference tree and the matched tree candidate were removed from the distance matrix. The procedure was repeated until there were no tree candidates left within 5 m of the remaining reference trees.

2.7 Accuracy Assessment

The accuracy resulted from using either CHM_{image}, CHM_{als}, CHM_{image_smooth}, or CHM_{als_smooth}, was estimated through six different measures: the percentage of matched trees (Eq. 1.), relative error of omission and commission (Eqs. 2 and 3.), root mean square error (RMSE) and bias of the predicted height of matched trees (Eqs. 4 and 5). In addition, the correctness of the plot-level height distributions was evaluated using the Reynolds error index (Reynolds et al. 1988) (EI, Eq. 6.) and the relative error index (Packalén and Maltamo 2008) (EIrel, Eq. 7.). EI shows the absolute difference between distributions whereas EIrel shows the difference in relation to the total number of observations. The height distributions were created from the CHM-derived heights using 1 m bin size. All measures were calculated separately for all plots, pine and spruce dominated plots, and mixed plots.

$$Matched\ trees_{rel} = \frac{N_{matched}}{N_{ref}}, \tag{1}$$

$$Error\ of\ omission_{rel} = \frac{N_{missed}}{N_{ref}}, \tag{2}$$

$$Error\ of\ comission_{rel} = \frac{N_{extra}}{N_{ref}} \tag{3}$$

where $N_{matched}$ is the number of successfully matched trees, N_{missed} the number of the undetected trees, N_{extra} the number of unmatched tree candidates and N_{rel} the number of reference trees.

$$RMSE = \sqrt{\frac{\sum_{i=1}^{n}\left(x_{obs,i} - x_{model,i}\right)^2}{n}}, \tag{4}$$

$$bias = \frac{\sum_{i=1}^{n}\left(x_{obs,i} - x_{model,i}\right)}{n}, \tag{5}$$

where n is the number of trees, x_{obs} the field measured tree height i, and x_{model} the CHM derived tree height i.

$$EI = \sum_{i=1}^{k} w_i \left|f_i - \hat{f}_i\right|, \tag{6}$$

$$EI_{rel} = \sum_{i=1}^{k} 0.5\left|\frac{f_i}{N} - \frac{\hat{f}_i}{\hat{N}}\right|, \tag{7}$$

where k is the number of height classes, w_i is the weight of class i, f_i is the true number of trees in height class i, $\hat{f}_i$ is the predicted number of trees in height class i, N is the true number, and $\hat{N}$ the predicted number of trees on the plot.

To find the optimal procedure for detecting and identifying individual trees within this study, the approaches (CHM_{image}, CHM_{als}, CHM_{image_smooth} or CHM_{als_smooth}) were ranked according to their performance in terms of the six measures. The ranking was made on the basis of the relative values. For each classification, the approach which was the best received a rank of 1, and the second best rank 2, etc. Thus, the approach with the smallest overall sum of ranks could be argued as the best within the study. Because the number of found trees and the error of commission are actually two sides of the same phenomenon, only the latter was taken into consideration when summing up the ranks.

3 Results and Discussion

When considering the results, the difference in the costs of the DSI- and ALS-based data has to be taken into consideration. Even though the cell size of the image-based CHMs was significantly lower (1 m for ALS and 0.5 m for DSI), the acquisition of DSI materials was significantly more efficient. For DSI, the whole study area was covered from 5000 m with two flight lines. Similarly, covering the area by means of ALS from 2500 m required 4 main flight lines and two lines across the main lines. On the other hand, comparing with ALS data, generating surface models from DSI requires heavy pre-processing (White et al. 2013), which also affects the overall efficiency of the method.

The accuracies concerning the detection and matching of individual trees are presented in Table 3. The results are reported separately for each approach (i.e., different height models) and sub-group. Columns denoted abs show the absolute number of trees, whereas the relative amounts are reported in rel columns. Column rank shows the rank of each approach within sub-groups.

Depending on the sub-group and the CHM used for detecting the trees, the segmentation process identified 39–93 % of the trees. The detection rate was the best when using CHM_{image} (unfiltered image-based CHM), whereas CHM_{als_smooth} resulted in the lowest detection rate. Logically, the error of omission follows the same pattern. However, when examining the error of commission, the performance of the approaches changes drastically. Altogether, 4–46 % of the tree candidates produced could not be matched to any reference tree. Here, CHM_{als_smooth} gave the most accurate results whereas CHM_{image} produced the most extra trees in every sub-group. In terms of the total number of the produced tree candidates (i.e., the number of found trees and the error of commission together), CHM_{image} and CHM_{als_smooth} represent the study extremes. The former produces the largest number of tree candidates (2355) whereas the latter results in the smallest number of tree candidates (774). The mean ranks of detection accuracy for CHM_{image},

Table 3 The detection accuracy of watershed segmentation when applied to four different height models

	Found trees			Error of omission			Error of commission		
	abs	rel	Rank	abs	rel	Rank	abs	rel	Rank
CHM_{image}									
Pine	434	0.93	1	31	0.07	1	362	0.46	4
Spruce	564	0.85	1	101	0.15	1	342	0.38	4
Mixed	427	0.77	1	127	0.23	1	226	0.35	4
Total	1425	0.85	1	259	0.15	1	930	0.40	4
CHM_{image_smooth}									
Pine	341	0.73	3	124	0.27	3	87	0.20	2
Spruce	348	0.52	3	317	0.48	3	80	0.19	3
Mixed	278	0.50	3	276	0.50	3	67	0.19	3
Total	967	0.57	3	717	0.43	3	234	0.20	3
CHM_{als}									
Pine	400	0.86	2	65	0.14	2	158	0.28	3
Spruce	455	0.68	2	210	0.32	2	80	0.15	2
Mixed	344	0.62	2	210	0.38	2	60	0.15	2
Total	1199	0.71	2	485	0.29	2	298	0.20	2
CHM_{als_smooth}									
Pine	233	0.50	4	232	0.50	4	15	0.06	1
Spruce	288	0.43	4	377	0.57	4	12	0.04	1
Mixed	214	0.39	4	340	0.61	4	12	0.05	1
Total	735	0.44	4	949	0.56	4	39	0.05	1

The results are given separately for pine and spruce dominated plots, plots with mixed vegetation, and all plots together

CHM_{image_smooth}, CHM_{als}, and CHM_{als_smooth} were 2.5, 2.9, 2.1, and 2.5 respectively.

According to the presented mean ranks and within the methods tested, utilizing CHM_{als} seems to perform the best in detecting individual tree crowns. Although the resolution of the DSI-based CHMs was higher, the ALS-based CHMs captured the variation of the canopy better. When the noise (i.e., the number of false peaks in the surface) was reduced the number of resulting tree candidates reduced to a significant underestimate. The strong decline in the number of detected trees between CHM_{image} and CHM_{image_smooth} can result from using too heavy filtering. Using the mean filter seems to smoothen the surface too much as the number of segmentation seed points (i.e., the local maxima) decreases rapidly. The filtering also affects the ALS-based CHMs. Even though the vertical canopy structure is captured better in CHM_{als}, the filtering seems to be too heavy for the coarse resolution which leads to poor results with CHM_{als_smooth}. Because of the difference in CHM resolutions (i.e., 1 m for ALS-based CHMs and 0.5 m for DSI-based CHMs) the changes are not directly comparable. Utilizing a different kind of filter for both ALS and

DSI-based data might decrease the number of tree candidates but still help to preserve more of the height variation caused from actual tree tops. For example, Hyyppä et al. (2001) utilized a filter weighing the center of the filter window. Still, when compared with previous studies on ALS-based tree detection, also the performance of CHM_{image} seems reasonable. In an international comparison of ALS-based detection of individual trees by Kaartinen et al. (2012), the methods utilizing low-density ALS detected 25–90 % of the trees whereas the error of commission was between 0 and 34.7 %.

The results concerning the accuracy of the height measurements on the matched trees and the correctness of height distributions are presented in Table 4. Again, column rank shows the goodness within all CHMs used. In terms of height measurement accuracy, CHM_{als} performs the best in nearly all subgroups. For CHM_{als}, the RMSE and bias of estimated tree heights were between 0.11–0.16 and 0.1–0.5 m, respectively. The mean ranks for the accuracy of tree height for CHM_{image}, CHM_{image_smooth}, CHM_{als}, and CHM_{als_smooth} were 2.6, 2.9, 1.1, and 3.4 respectively.

Table 4 The accuracy of height estimates for the detected trees

	RMSE of height			Bias of height			Error indices		
	abs (m)	rel (%)	Rank	abs (m)	rel (%)	Rank	EI	EI_{rel}	Rank
CHM_{image}									
Pine	4.17	0.17	2	1.12	0.05	3	61	0.52	4
Spruce	4.32	0.17	3	0.93	0.04	2	52	0.52	2
Mixed	5.36	0.23	3	−0.71	−0.03	2	50	0.48	2
Total	4.61	0.23	3	0.50	−0.03	2	54	0.49	3
CHM_{image_smooth}									
Pine	4.34	0.18	3	0.73	0.03	1	38	0.49	2
Spruce	4.65	0.18	4	0.98	0.04	3	36	0.52	3
Mixed	5.64	0.24	4	−0.82	−0.03	3	37	0.49	3
Total	4.85	0.24	4	0.37	−0.03	3	37	0.49	2
CHM_{als}									
Pine	2.82	0.12	1	1.10	0.05	2	38	0.45	1
Spruce	2.78	0.11	1	0.48	0.02	1	26	0.43	1
Mixed	3.82	0.16	1	0.17	0.01	1	34	0.42	1
Total	3.13	0.16	1	0.60	0.01	1	32	0.40	1
CHM_{als_smooth}									
Pine	4.40	0.18	4	2.36	0.10	4	34	0.52	3
Spruce	4.42	0.16	2	2.47	0.09	4	34	0.55	4
Mixed	4.69	0.18	2	2.17	0.09	4	37	0.52	4
Total	4.49	0.18	2	2.35	0.09	4	35	0.49	4

The results are given separately for each height model used and for pine and spruce dominated plots, plots with mixed vegetation, and all plots together. Columns denoted abs show the absolute values, whereas the relative values are reported in *rel* columns. Column *EI* shows the Reynolds error index and EI_{rel} the relative error index

The tree height was underestimated for nearly all sub-groups with all four methods, which is typical for both DSI- and ALS-based approaches (e.g., St-Onge et al. 2004; Gaveau and Hill 2003). The sub-group consisting of mixed plots makes an exception. With both DSI-based CHMs, the height of trees on mixed sub-group was overestimated. This suggests, that the DSI-based CHMs were not able to reach the crowns of the suppressed trees in more heterogeneous surroundings. For example, in Vastaranta et al. (2013) it was found that the lower percentiles of DSI-based height distributions were higher than those based on ALS data. When considering the RMSE of the height estimates, the ALS data perform better than DSI in all sub groups. Earlier studies on the relative accuracies of ALS- and DSI-based height measurements in complex terrain are rare. However, individual accuracies of both methods have been widely tested. For example, Korpela (2004) measured individual tree heights from DSI with a standard deviation of 0.3–1.0 m. Similarly, Persson et al. (2002) measured tree heights from ALS data with a RMSE of 0.63 m. In this study, the accuracy of height measurements was considerably lower for both DSI- and ALS-based methods. This is likely resulting from the coarser resolution of both ALS and DSI datasets used in this study.

CHM_{als} performs the best also with this respect to EI_{rel}, that is actually a combination of the number of tree candidates and the accuracy of their height estimates. However, as EI_{rel} varies between 0.40 and 0.55, approximately half of the tree candidates are placed in erroneous height bins. Both missed and extra trees (i.e., error of omission and commission) are included in the figure.

According to the summarizing score in Table 5, CHM_{als} seems to perform better than the two image-based CHMs in mapping individual tree crowns, whereas CHM_{als_smooth} performed the worst from the four CHMs tested. Even though the spatial resolution of the image-based CHMs was higher than in ALS-based CHMs, they fall behind low-density ALS in the ability of capturing crown-level variation and small openings in the canopy.

The relatively high spatial resolution leads to over-segmentation if the image-based CHM is not smoothed, however if a mean filter is used, too much of the variation within the CHM is lost which leads to under-segmentation. This also reflects in the matching results. If the canopy is heavily over-segmented, most of the reference trees can be linked to a tree candidate within the constraints of the algorithm. In the case of under-segmentation the number of matched trees decreases

Table 5 Total rank score sums (referred as abs) and mean ranks for all four height models

	Total rank score	
	abs	Mean
CHM_{image}	51	2.6
CHM_{image_smooth}	58	2.9
CHM_{als}	30	1.5
CHM_{als_smooth}	61	3.1

because there are not enough tree candidates to match. The problem could be approached through utilizing a more detailed algorithm when matching the tree candidates with the field measured trees. For example, Yu et al. (2006) utilized the principle of Hausdorff distance when matching the trees between two subsequent scannings according to their locations, whereas Olofsson et al. (2008), included both tree location and height in the matching procedure. However, in this study we settled for using a very simple matching algorithm when comparing the different approaches.

Considering the ranking on grounds of the five criteria (i.e., the errors of omission and commission, RMSE and bias of height, and EI_{rel}), the method utilizing CHMals seem to perform the best, whereas the second best overall rank was achieved using CHM_{image}. For both ALS- and DSI-based CHMs, applying a simple 3 × 3 pixel average filter resulted in worse overall results than using the using the unsmoothed CHM. The results indicate that, despite the lower spatial resolution, CHM_{als} is more capable in detecting small-scale variation in the canopy than $CHM_{image.}$ However, this does not necessarily mean that ALS-based CHMs would be superior to the DSI-based CHMs. With the cost of the low-density ALS, a DSI dataset with much shorter GSD could be acquired. Also, a more suitable image matching algorithm could improve the accuracy of DSI-based height models, even with the GSD used in this study.

4 Conclusion

In terms of the cost of data acquisition, the high-altitude DSI point clouds seem to offer an efficient means for creating fine-scale CHMs for mapping individual tree crowns. However, the fine resolution does not necessarily provide the same level of accuracy in describing the forest canopy as the ALS-based CHMs of the same resolution. Further research is needed for assessing the effect of different image matching algorithms in forest conditions as well as for utilizing DSI with shorter GSD.

Acknowledgments This study has been conducted with funding from the European Community's Seventh Framework Programme [FP7/2007-2013] under Grant Agreement Number 606971, Finnish Cultural Foundation under grant 00150939, and from the Academy of Finland in the form of the Centre of Excellence in Laser Scanning Research (Project Number 272195).

References

Ackermann F (1999) Airborne laser scanning—present status and future expectations. ISPRS J Photogramm Remote Sens 54(2):64–67

Axelsson P (2000) DEM generation from laser scanner data using adaptive TIN models. Int Arch Photogramm Remote Sens 33(B4/1; PART 4):111–118

Gaveau DL, Hill RA (2003) Quantifying canopy height underestimation by laser pulse penetration in small-footprint airborne laser scanning data. Can J Remote Sens 29(5):650–657

Held A, Ticehurst C, Lymburner L, Williams N (2003) High resolution mapping of tropical mangrove ecosystems using hyperspectral and radar remote sensing. Int J Remote Sens 24 (13):2739–2759

Hirschmüller H (2008) Stereo processing by semiglobal matching and mutual information. IEEE Trans Pattern Anal Mach Intell 30(2):328–341

Holopainen M, Vastaranta M, Hyyppä J (2014) Outlook for the next generation's precision forestry in Finland. Forests 5(7):1682–1694

Hyyppä J, Inkinen M (1999) Detecting and estimating attributes for single trees using laser scanner. Photogramm J Finl 16(2):27–42

Hyyppä J, Kelle O, Lehikoinen M, Inkinen M (2001) A segmentation-based method to retrieve stem volume estimates from 3-D tree height models produced by laser scanners. IEEE Trans Geosci Remote Sens 39(5):969–975

Järnstedt J, Pekkarinen A, Tuominen S, Ginzler C, Holopainen M, Viitala R (2012) Forest variable estimation using a high-resolution digital surface model. ISPRS J Photogramm Remote Sens 74:78–84

Kaartinen H, Hyyppä J, Yu X, Vastaranta M, Hyyppä H, Kukko A, Holopainen M, Heipke C, Hirschmugl M, Morsdorf F, Næsset E (2012) An international comparison of individual tree detection and extraction using airborne laser scanning. Remote Sens 4(4):950–974

Koch B, Heyder U, Weinacker H (2006) Detection of individual tree crowns in airborne lidar data. Photogramm Eng Remote Sens 72(4):357–363

Korpela I (2004) Individual tree measurements by means of digital aerial photogrammetry, vol 3. Finnish Society of Forest Science Helsinki, Finland

Lim K, Treitz P, Wulder M, St-Onge B, Flood M (2003) LiDAR remote sensing of forest structure. Prog Phys Geogr 27(1):88–106

Næsset E (2002) Predicting forest stand characteristics with airborne scanning laser using a practical two-stage procedure and field data. Remote Sens Environ 80(1):88–99

Nurminen K, Karjalainen M, Yu X, Hyyppä J, Honkavaara E (2013) Performance of dense digital surface models based on image matching in the estimation of plot-level forest variables. ISPRS J Photogramm Remote Sens 83:104–115

Olofsson K, Lindberg E, Holmgren J (2008) A method for linking field-surveyed and aerial-detected single trees using cross correlation of position images and the optimization of weighted tree list graphs. In: Proceedings of SilviLaser 2008: 8th international conference on LiDAR applications in forest assessment and inventory, Edinburgh, UK, 17–19 September 2008, pp 95–104

Packalén P, Maltamo M (2008) Estimation of species-specific diameter distributions using airborne laser scanning and aerial photographs. Can J For Res 38(7):1750–1760

Persson A, Holmgren J, Söderman U (2002) Detecting and measuring individual trees using an airborne laser scanner. Photogramm Eng Remote Sens 68(9):925–932

Pitkänen J, Maltamo M, Hyyppä J, Yu X (2004) Adaptive methods for individual tree detection on airborne laser based canopy height model. Int Arch Photogramm Remote Sens Spatial Inf Sci 36(8):187–191

Reynolds MR, Burk TE, Huang W-C (1988) Goodness-of-fit tests and model selection procedures for diameter distribution models. For Sci 34(2):373–399

St-Onge B, Jumelet J, Cobello M, Véga C (2004) Measuring individual tree height using a combination of stereophotogrammetry and lidar. Can J For Res 34(10):2122–2130

Vastaranta M, Wulder MA, White JC, Pekkarinen A, Tuominen S, Ginzler C, Kankare V, Holopainen M, Hyyppä J, Hyyppä H (2013) Airborne laser scanning and digital stereo imagery measures of forest structure: comparative results and implications to forest mapping and inventory update. Can J Remote Sens 39(5):382–395

Vauhkonen J, Ene L, Gupta S, Heinzel J, Holmgren J, Pitkänen J, Solberg S, Wang Y, Weinacker H, Hauglin KM, Lien V (2012) Comparative testing of single-tree detection algorithms under different types of forest. Forestry 85(1):27–40

Wehr A, Lohr U (1999) Airborne laser scanning—an introduction and overview. ISPRS J Photogramm Remote Sens 54(2):68–82

White JC, Stepper C, Tompalski P, Coops NC, Wulder MA (2015) Comparing ALS and image-based point cloud metrics and modelled forest inventory attributes in a complex coastal forest environment. Forests 6(10):3704–3732

White JC, Wulder MA, Vastaranta M, Coops NC, Pitt D, Woods M (2013) The utility of image-based point clouds for forest inventory: a comparison with airborne laser scanning. Forests 4(3):518–536

Yu B, Gong P, Pu R (1999) Penalized discriminant analysis of in situ hyperspectral data for conifer species recognition. IEEE Trans Geosci Remote Sens 37(5):2569–2577

Yu X, Hyyppä J, Kukko A, Maltamo M, Kaartinen H (2006) Change detection techniques for canopy height growth measurements using airborne laser scanner data. Photogramm Eng Remote Sens 72(12):1339–1348

Outlook for the Single-Tree-Level Forest Inventory in Nordic Countries

Ville Kankare, Markus Holopainen, Mikko Vastaranta, Xinlian Liang, Xiaowei Yu, Harri Kaartinen, Antero Kukko and Juha Hyyppä

Abstract In Nordic countries, the forest resource information systems have advanced to a state where substand-level information can be utilized. The demand of high detail up-to-date forest resource information has become a prerequisite for many of the operators working with the data, but the information on the forest attributes such as species-specific timber assortments and tree quality cannot be obtained accurately enough from the current inventory systems. Therefore, the forest organizations are actively looking forward and started the development of the next generation forest inventory platforms. The vision is, a radical leap in the cost effectiveness of forestry and wood supply will be gained through new digital services and Big Data applications. The most prominent solution for the increased demand on the information detail is single-tree-level forest inventory through various laser scanning technologies (airborne-, terrestrial- and mobile laser scanning, (ALS, TLS and MLS, respectively)) but it has not yet been adapted into operational forestry mainly due to the higher costs and challenges in data processing. Many studies have already concluded that single-tree-level information will play an important role in the next generation's forest mapping systems that will be based on multisource approach. The challenges in multisource approach have been the data acquisition, automatic tree attribute measurements and the optimal data combinations. MLS and harvester data are of high interest technologies in the

V. Kankare (✉) · M. Holopainen · M. Vastaranta
Department of Forest Sciences, University of Helsinki, Latokartanonkaari 7, 00014 Helsinki, Finland
e-mail: ville.kankare@helsinki.fi

X. Liang · X. Yu · H. Kaartinen · A. Kukko · J. Hyyppä
National Land Survey, Finnish Geospatial Research Institute, Geodeetinrinne 2, 04310 Masala, Finland

V. Kankare · M. Holopainen · M. Vastaranta · X. Liang · X. Yu · H. Kaartinen · A. Kukko · J. Hyyppä
Centre of Excellence in Laser Scanning Research, Finnish Geospatial Research Institute, 02431 Masala, Finland

I. Ivan et al. (eds.), *The Rise of Big Spatial Data*, Lecture Notes in Geoinformation and Cartography, DOI 10.1007/978-3-319-45123-7_14

reference data acquisition but have not yet been implemented into practical applications. The goal of the present outlook was to evaluate and discuss the potential and challenges of the laser scanning technologies (especially ALS and TLS) in single-tree-level forest inventory as a part of multisource approach which, when implemented, will create a scenario for vast forest big data.

Keywords Remote sensing · Laser scanning · Single-tree-level · Forest inventory · GIS Ostrava 2016

1 Introduction

In Finland, intensive small-scale forestry is practiced mainly in privately owned forests. There are over 600,000 forest owners, and the size of the average forest holding is only 25 ha with an average stand size less than 2 ha. Forest resource mapping, management planning and decision-making relies on the precise knowledge of forest structural attributes, especially in the small-scale forestry that is practiced e.g. in Nordic countries. With detailed and unbiased forest resource information it is possible to e.g. optimize the forest management operations and the flow of raw material from the forest to the designated factories. Previously, a typical approach in the forest mapping has been based on generalizing field sample plot measurements using coarse- or medium-resolution remote sensing (RS) data. The accuracy of forest attributes estimation based on coarse resolution of the RS data is not sufficient for small scale (forest stands or small areas) forest management planning and therefore preharvest field visits are commonly used in operational forest management and timber sales. Also, current forest-planning applications are highly affected by the quality of the required input data (Holopainen et al. 2010, 2011; Vastaranta et al. 2010) due to the long model chains. In the last few years, the forest resource information systems have advanced to a state where substand-level information can be used. However, decision-making, planning and optimization of the forest management operations are still mainly based on stand-level mean attributes (mean diameter, basal area, height or age), in which a great deal of information concerning forest structure variability is lost. Thus, the most significant attributes describing the timber quality and log yield are stem form and diameter distribution. Therefore, the demand for more detailed forest resource information especially at single-tree-level is steadily growing, particularly from the more mature stands where the added value for the high detail data is the highest. Traditional field measurements at single-tree-level are considered as costly to obtain, especially if attributes such as tree quality, biomass or timber assortments are considered. Due to the more labour intensive nature of these particular attributes, they been estimated using existing models and easily measurable tree attributes (tree species, DBH and height); e.g. stem form has been modelled using taper curve models (Laasasenaho 1982) with knowledge of tree species, diameters [diameter at breast height (DBH, 1.3 m) and diameter at 6 m] and height. The downside to this modelling approach is the locality of the models or models might not even exist in the specific target area.

During the last two decades, significant technological and methodological leaps in three-dimensional (3D) forest resource mapping have revolutionized the field of study, with the introduction of airborne -, terrestrial -, mobile - and personal laser scanning (ALS, TLS, MLS and PLS, respectively) systems. Currently the forest inventory information in Finland is acquired using an area-based approach (ABA) and low density ALS data (0.5–1 pulses/m^2). This approach has succeeded in replacing traditional stand-wise field inventories (SWFIs). Compared to SWFI, ABA has generally provided more precise estimation of the inventory attributes, as well as cost savings. The forest resource information of the privately owned forests is managed by the Finnish Forestry Centre (FFC) that has been one of the pioneers in the utilization of ALS ABA forest inventory method in operational forestry. FFC has just launched open access, web-based application for forest owners (metsään.fi). This is the first step towards wide utilization of forest big data in Finland.

Despite the operator (forest owner, industry or policy maker) the detailed and up-to-date forest information is mandatory but the information on the forest resource attributes such as species-specific timber assortments, diameter distribution, tree quality and biomass distribution cannot be obtained accurately enough from the current inventory system. Therefore, the forest organizations are actively looking forward to next generation's forest inventory techniques e.g. to improve the current wood procurement practices. Vision of forest industry is that radical leap in the cost effectiveness of forestry and wood supply will be gained through new digital services and Big Data applications. Added value could be produced in efficient forest management, customer-oriented wood harvesting and improved wood-based products value chains. In 2014–2016 Finnish forest industry has led Forest Big Data (FBD) research and development project. In FBD one main task is to develop next generation forest information acquisition systems. Starting point is that in the future, the data solution for detailed forest management and wood procurement will use multi-source and -sensor 3D precision forestry information (Holopainen et al. 2014). One prominent solution for the increased demand on the information detail is including single-tree-level forest inventory information to the systems but it has not yet been adapted into operational forestry mainly due to the higher cost and challenges in data processing. Holopainen et al. (2014) concluded that the base layer of the next generation's forest inventory system is up-to-date forest attribute maps with resolution of 10–20 m and single-tree-level information can be obtained and included in the most valuable stands where the added value of the information detail is the highest. Therefore, the goal of the present outlook was to evaluate and discuss the potential and challenges of the laser scanning technologies (especially ALS and TLS) in single-tree-level forest inventory as a part of multisource approach.

2 Single-Tree-Level Forest Inventory

2.1 Airborne Laser Scanning

ALS is increasingly used in forest mapping and monitoring tasks due to its capability to collect spatially accurate data efficiently over large areas with resolution from which individual trees can be detected. A widespread of methods have been developed for this purpose (individual tree detection, ITD) during the last two decades (see summaries from Kaartinen et al. 2012; Vauhkonen et al. 2012; Eysn et al. 2015). The most influential factors affecting successful ITD have been: (1) the used tree detection algorithm and (2) forest structure (Kaartinen et al. 2012; Vauhkonen et al. 2012). ALS ITD has been capable to detect from 25 to 100 % (e.g. Hyyppä and Inkinen 1999; Popescu et al. 2003; Yu et al. 2011; Kaartinen et al. 2012; Vauhkonen et al. 2012) of the total number of trees. ITD algorithms detect most of the tree from the dominant canopy layer even with lower pulse densities (see e.g. Yu et al. 2011) but the non-dominant trees that are rarely visible from the data creates a challenge that is not yet fully solved. In forest management point-of-view this might not be a significant problem especially in latter part of the management chain (older forests) because previous studies (see e.g. Persson et al. 2002; Vastaranta et al. 2011) have showed that 60.2–99.9 % of the detected number of the trees still contributed 75.9–100 % of the total volume in the specific study areas.

Thus, ALS provides important 3D information about the area of interest, it is capable to measure only tree height related metrics, such as maximum height, height distribution within area of interest (e.g. inside single-tree segment) and density (e.g. penetration). All other single-tree-level attributes have to be estimated using either allometric models or ABA approach. The challenges in tree attribute modelling using ALS data is that the existing tree-level allometric models are tree species specific, developed mainly for field measurements and the models are highly dependent on the measured tree height accuracy. Uncertainties are likely to increase when longer modelling chains are required (see Fig. 1). Previous studies (see e.g. Maltamo et al. 2004; Rönnholm et al. 2004; Kankare et al. 2013b) have showed that ALS tends to underestimate tree height by 0.4–1.5 m due to pulse

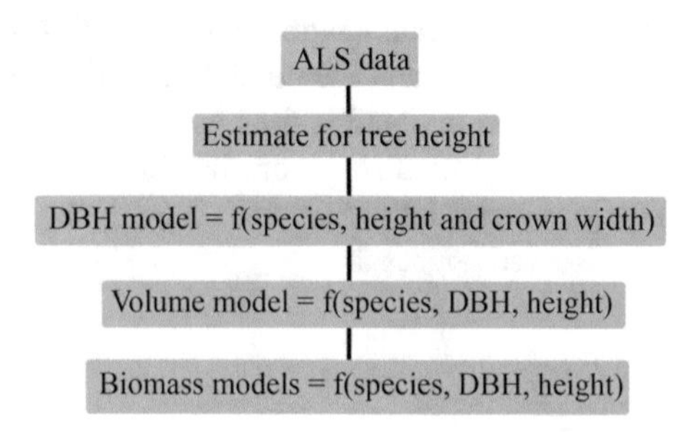

Fig. 1 Model-based estimation of tree attributes

interactions within the canopy where pulse returns are rarely recorded from exactly the highest point of the canopy (low density scanning with small foot print laser).

As shown in Fig. 1, all other tree attributes are estimated based on existing models which have been developed mainly for traditional field measurements. This creates a need for model calibrations or the creation of completely new models that should be based on characteristics measurable using laser scanning. This approach was shown to have significant effect e.g. in biomass component modelling accuracies in Kankare et al. (2013b), where models were developed based on ALS metrics and compared to the existing allometric models used with ALS derived tree height and DBH. Developed models increased e.g. the accuracy of total biomass by 39.8 %. Tree DBH have been estimated based on the allometric relationship between height and DBH (see e.g. Kalliovirta and Tokola 2005) or ABA approach. Especially the modelled height-DBH relationship is prone to bias due to its variability and dependence on site fertility and structure causing similar height trees to have quite varying values of DBH. DBH accuracies have varied between 1.3 cm and 5.1 cm depending on the used estimation approach (see e.g. Popescu et al. 2007; Maltamo et al. 2009; Vauhkonen et al. 2010; Yu et al. 2011, Kankare et al. 2013b, 2014b). Similar approach has been used when either tree volume (e.g. Laasasenaho 1982) or biomass (e.g. Repola 2008, 2009) are modelled but these models utilize the estimated DBH, which can increase uncertainties in the volume or biomass estimates. The existing models also rely on precise knowledge of the tree species. Tree species recognition has not yet been fully solved using laser scanning techniques (Vauhkonen et al. 2014). With ALS data, tree species classification from ITD is based on different types of metrics (e.g. structural, intensity or waveform) which are then used as a basis for the classification (see Vauhkonen et al. 2014). Tree species classification accuracies for Scandinavian boreal forests, where the number of commercially important tree species is rather low, have varied between 60 and 93 % (see e.g. Holmgren and Persson 2004; Holmgren et al. 2008; Korpela et al. 2010; Vauhkonen et al. 2009, 2010; Hovi et al. 2016).

2.2 *Terrestrial Laser Scanning*

TLS offers new opportunities for the high detail 3D mapping of smaller areas such as sample plots or individual trees. TLS is seen as an efficient and objective option for acquiring the required field data to be used, e.g. as a reference for forest mapping over large areas (Liang et al. 2011; Lindberg et al. 2012) using ABA or when developing new tree attribute models, such as biomass components (see e.g. Kankare et al. 2013a; Hauglin et al. 2013). TLS is capable to detect also the non-dominant layer trees that causes challenges for ALS given that the data acquisition is executed with multiple scans and taking the varying forest structures into consideration in data acquisition. Like ALS, the quality and accuracy of TLS based ITD is highly dependent on forest structure (see e.g. Kankare et al. 2015). The shadowing effect (occlusion; visibility from the scan location) caused by the forest structure (tree stem, branches and other understorey vegetation) has the most significant effect on tree detection accuracy and it should be taken into account when collecting the data in the field. ITD accuracy with multiple scans has varied between 91.7 and 100 %, whereas with single-scan, the corresponding accuracy has varied between 55.3 and 90 %, depending on forest structure (e.g. tree density) of the study areas in question (Maas et al. 2008; Liang et al. 2012; Liang and Hyyppä 2013).

TLS, compared to ALS, is capable to determine tree height and diameters (stem curve) directly from the point cloud with the presumption of encompassing data. Therefore, it is possible to reduce the uncertainty caused by the modelling required in ALS e.g. for estimating DBH. Shadowing effect was shown to have significant effect on the manual stem curve measurements in Kankare et al. (2013a), where diameters were successfully measure up to 86.9 and 48.2 % of the tree height for Scots pine (*Pinus sylvestris* L.) and Norway spruce (*Picea abies* (L.) H. Karst.), respectively. The automatic data processing of TLS data is commonly based on the stem point recognition and fitting of 3D cylinders (see e.g. Liang et al. 2014). Tree volume and diameters can be deduced from the fitted cylinders but the downside is that the stem form is assumed to be circular, which will cause uncertainty in the diameter estimates especially in more diverse forest conditions (e.g. in old forests or city forests). Stem curve measurement accuracies have been approximately 1 cm to 1.1 cm for the visible part of the stem (see e.g. Maas et al. 2008; Liang et al. 2014). The accuracies of the DBH measurements from TLS data have varied between 0.74–3.25 cm depending on the data acquisition strategy (one or multiple scans) (e.g. Liang and Hyyppä 2013; Maas et al. 2008; Yao et al. 2011). Tree height accuracy plays also an important role in automated data processing due to the modelling of the highest part of the stem (stem points occluded by the canopy). Previous studies have shown that tree height is underestimated due to occluded visibility of the highest tip of the tree canopy from the limited number of view points. Tree height accuracy compared to field measurements have varied between 1.36 and 6.53 m in previous studies (e.g. Liang and Hyyppä 2013; Maas et al. 2008). Tree species recognition automatically from TLS point cloud has been left for smaller attention although it is critical in the point of the data would be utilized in operational forestry. As far as the authors know, Puttonen et al. (2010) and Othmani et al. (2013) have published the only results in tree species recognition using either a combination of TLS and a hyperspectral sensor or solely TLS data.

TLS enables new possibilities not only in gained added value through additional tree attributes that are costly to measure with traditional means, such as diameter distribution, timber assortments and external quality, but also in single-tree-level modelling and model development. Figure 2 demonstrates the different detail levels available with comprehensive TLS data. Adaptation of these attributes into the next generation's forest inventory system would benefit all parties utilizing the forest inventory data. Traditionally e.g. timber assortments have been validated in area-level (e.g. stand) and there exists only few studies where distribution of the

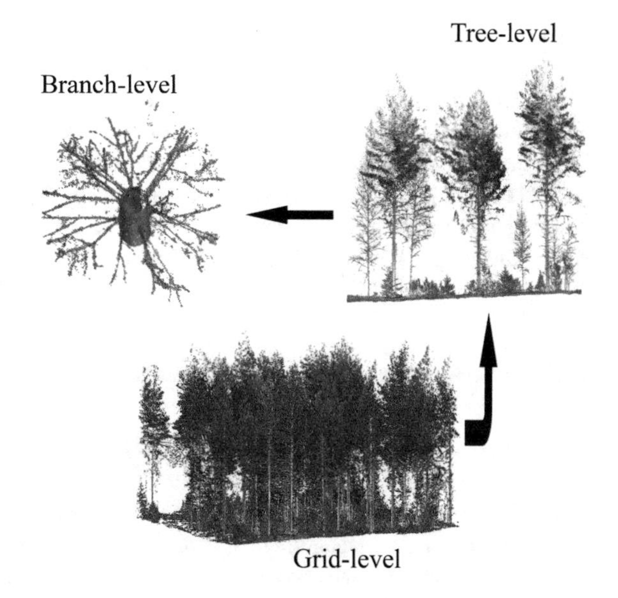

Fig. 2 Small area coverage of TLS point cloud from grid-level (16 m × 16 m) to branch-level

stem volume into various timber assortments have been evaluated at tree-level using either ALS or TLS data (see Maltamo et al. 2009; Kankare et al. 2014b; Vastaranta et al. 2014). Kankare et al. (2014b) showed that TLS is capable of accurately estimating timber assortment, especially saw wood volume. External quality attributes were also found to have significant effect on timber assortment estimation especially for Scots pine (see e.g. Kankare et al. 2014a, b). The millimeter level-of-detail in TLS data enables stem external quality related analysis, such as trunk and branch-level information, where from ALS data only different canopy or single-tree crown properties, such as canopy width or living crown density or height, can be deduced. TLS has already been used to measure branch size distribution (Raumonen et al. 2013), stem form (taper, sweep and lean) (Pfeifer and Winterhalder 2004; Thies et al. 2004; Liang et al. 2014) and bark characteristics (see Kretschmer et al. 2013), although accurate reference information has not been always available for example when branch size distributions have been measured. Stängle et al. (2014) also demonstrated that external bark quality attributes can be linked with the internal quality of logs, as determined with strong correlation by X-ray computer tomography which could be utilized in planning the raw wood supply chain.

TLS data can also be used as a basis for new tree attribute model development. TLS enables non-destructive sampling of stem curve and canopy metrics, which can be utilized e.g. in biomass component model development (see e.g. Yu et al. 2013; Kankare et al. 2013a; Hauglin et al. 2013). Yu et al. (2013) demonstrated that stem reconstruction and correspondingly derived stem volume can be used as explanatory variables for accurately modelling stem biomass. Kankare et al.

(2013a) and Hauglin et al. (2013) demonstrated that more accurate biomass component (stem, branches and canopy) estimation can be achieved using TLS than with existing allometric models that rarely utilize any canopy related metric. Hauglin et al. (2014) studied the use of TLS measurements as a reference for ALS-based branch biomass model development but they found only a small increase (3 %) in the branch biomass accuracy if TLS is used in ALS model training compared to when branch biomass is estimated based on ALS-derived DBH and height with existing allometric models.

2.3 Multisource Single-Tree-Inventory

The multisource single-tree-inventory (MS-STI) relies on highly accurate and detailed tree maps that are derived from TLS, MLS or harvester data. Holopainen et al. (2014) concluded that the utilization of accurate tree maps would be beneficial for planning forest management operations, as input data for the development of next generation's growth models and also adding diameter distribution and tree quality information to the system. The concept of MS-STI was introduced in Vastaranta et al. (2014) where they estimated logging recoveries of mature Scots pine stand using ALS data combined with existing information of the tree species and positions. The benefit of the method is that the two major bottlenecks, (1) tree species recognition and (2) detection of non-dominant layer trees, of state-of-the-art ITD techniques used with ALS can be minimized. Tree map including the location and required tree attributes, such as tree species, diameters, height and quality, is used as auxiliary information and combined with ITD segments from lower resolution RS data, e.g. ALS, using location information. Saarinen et al. (2014) tested the MS-STI approach further in urban environment with encouraging results. Kankare et al. (2015) evaluated the accuracy of TLS based tree maps as a basis of MS-STI approach in varying forest conditions in Evo, Finland. The accuracy of the used tree maps was found to be highly affected by the forest structure due to the accuracy level decrease in tree detection and tree attribute estimation. The results in Kankare et al. (2015) support the vision presented in Holopainen et al. (2014) where the MS-STI and tree map information is suggested to be used in more mature forest where also the added value of the more detailed data is the highest.

The concept of MS-STI can also be utilised in tree-level model development. In this approach the tree map data is combined similarly to lower resolution RS data. The approach utilizes detailed stem measurements from TLS and combines it to height measurement from ALS. Kankare et al. (2014b) utilized this approach when evaluating tree-level timber assortment estimation. The results in this particular study showed that with this combinations, the accuracies for total volume and timber assortments can be improved. However, this approach warrants further research to evaluate its performance in different modelling scenarios.

3 Discussion and Conclusion

Small scale forestry, practiced e.g. in Finland, creates a demanding surroundings for RS techniques. The demand of high detail up-to-date forest resource information has become a prerequisite for many of the operators working with the data, but the information on the forest attributes such as species-specific timber assortments and tree quality cannot be obtained accurately enough from the current inventory system. Therefore, the forest organizations are actively looking forward and started the development of the next generation forest inventory platforms. The vision of the forest industry is that radical leap in the cost effectiveness of forestry and wood supply will be gained through new digital services and Big Data applications. Current forest inventory platforms have already advanced to a state were the implementation of substand-level (grid-level, 16 m × 16 m) information has been started and the next step in the future will be to include single-tree-level information to the system. Visions of the new platforms are based on multisource and—sensor idea, in which the used data source can vary on demand (see Holopainen et al. 2014). For example, the ALS data could be acquired with 10-year cycle whereas cheaper RS data (e.g. aerial images) could be acquired with higher temporal resolution.

The introduction of high detail 3D RS techniques has revolutionized the forest mapping and monitoring applications in the past 15 years (see in depth comparison study Yu et al. 2015). ALS is the most prominent technique at the moment, which has already been adapted into operational use. Technique's capability to collect spatially accurate data efficiently over large areas with resolution from which individual trees can be detected, is one of the main reasons of its success. However, high density ALS data has not yet been adapted into operational forestry mainly due to the higher cost. Currently utilized ALS data has a pulse density of 0.5–1/m^2, in Nordic countries, but we believe with the pace of the technical and methodological development, the pulse density will inevitably increase in the near future. Also the introduction of new platforms (TLS and MLS) will provide the much needed leaps in high detail data acquisition of smaller areas. TLS could be used in static plot- or grid-level data acquisition where MLS (see e.g. Kukko 2013) is more operational platform and it could be mounted into harvesters in future. Harvester data is seen as one of the most prominent solution to provide the reference data for updating the forest resource information. The added value of MLS technique in harvester comes also from the new innovations in driver support systems that could monitor the thinning density and provide additional support for decision making of the driver.

The addition of single-tree-level information to the current forest resource platforms, especially from more mature forests, will create vast opportunities for new innovations and options e.g. for optimization of wood supply chain. The development of MS-STI methodologies can provide the tools for measuring the required single-tree-level information but also for predicting the attributes for larger areas. Method requires spatially accurate tree maps that could be produced with harvester data. The challenge in harvester data is the location accuracy which is currently not sufficient for single-tree-level data matching. The location of the harvester is recorded under forest canopy approximately 3 m (Rasinmäki and Melkas 2005) and the location is measured for the harvester not for the crane. Kaartinen et al. (2015) conducted a comparison study of different position measurement techniques under forest canopy and concluded that the required positional accuracy level (under 1 m) cannot be achieved with stand-alone global navigation satellite system (GNSS) devices. Results in this particular study showed that with differential GNSS and inertial measurement unit (IMU) the required positional accuracy can be achieved but it warrants further testing.

In the near future, TLS data could also be utilized in tree attribute model development and calibration. TLS offers the possibilities to acquire accurate measurements of tree stem and branch dimensions non-destructively. There has been increasing discussion and need for new methods to update or calibrate existing tree attribute models due to their locality or non-existence. Few studies have already demonstrated the potential of TLS data in stem volume and biomass modelling (see e.g. Hauglin et al. 2013; Kankare et al. 2013a; Liang et al. 2014; Yu et al. 2013), but the use of TLS as a method to collect reference data warrants further research. Main questions are, how to collect optimal data and process it automatically.

To conclude, the use of single-tree-level information and MS-STI warrants further technical development and cost reductions. With the current pricing of devices e.g. for laser scanners, it is not possible to mount them into harvester machines. We strongly believe, that increased use of laser scanning based techniques in various industry fields (built environment, cars, etc.) will drive the development towards smaller, cheaper and also more consumer based devices. There is also a demand on methodological development and evaluation, in which the following subjects should and will be addressed in the near future:

1. MLS data and harvester head stem diameter measurements should be evaluated as a reference data,
2. further development of TLS and MLS automatic processing algorithms, and
3. to evaluate algorithm performance in more diverse and challenging forest conditions in international comparison studies [EuroSDR led by Finnish Geospatial Research Institute (FGI)].

Acknowledgments Our study was made possible by financial aid from the Finnish Academy project Centre of Excellence in Laser Scanning Research (CoE-LaSR, decision number 272195). We would like to acknowledge the partners in Forest Big Data (FBD)/Data to Intelligence (D2I) Digile/Tekes Program.

References

Eysn L et al (2015) A benchmark of lidar-based single tree detection methods using heterogeneous forest data from the alpine space. Forests 6:1721–1747

Hauglin M, Astrup R, Gobakken T, Næsset E (2013) Estimating single-tree branch biomass of Norway spruce with terrestrial laser scanning using voxel-based and crown dimension features. Scand J For Res 28:456–469

Hauglin M, Gobakken T, Astrup R, Ene L, Næsset E (2014) Estimating single-tree crown biomass of Norway spruce by airborne laser scanning: a comparison of methods with and without the use of terrestrial laser scanning to obtain the ground reference data. Forests 5:384–403
Holmgren J, Persson Å (2004) Identifying species of individual trees using airborne laser scanner. Remote Sens Environ 90:415–423
Holmgren J, Persson Å, Söderman U (2008) Species identification of individual trees by combining high resolution LiDAR data with multi-spectral images. Int J Remote Sens 29:1537–1552
Holopainen M, Mäkinen A, Rasinmäki J, Hyyppä J, Hyyppä H, Kaartinen H, Viitala R, Vastaranta M, Kangas A (2010) Effect of tree level airborne laser scanning accuracy on the timing and expected value of harvest decisions. Eur J For Res 129:899–910
Holopainen M, Vastaranta M, Rasinmäki J, Kalliovirta J, Mäkinen A, Haapanen R, Melkas T, Yu X, Hyyppä J (2011) Uncertainty in timber assortment estimates predicted from forest inventory data. Eur J For Res 129:1131–1142
Holopainen M, Vastaranta M, Hyyppä J (2014) Outlook for the next generation's precision forestry in Finland. Forests 5:1682–1694
Hovi A, Korhonen L, Vauhkonen J, Korpela I (2016) LiDAR waveform feature for tree species classification and their sensitivity to tree- and acquisition related parameters. Remote Sens. Environ. 173:224–237
Hyyppä J, Inkinen M (1999) Detecting and estimating attributes for single trees using laser scanner. Photogramm J Finl 16:27–42
Kaartinen H, Hyyppä J, Yu X, Vastaranta M, Hyyppä H, Kukko A, Holopainen M, Heipke C, Hirschmugl M, Morsdorf F, Næsset E, Pitkänen J, Popescu S, Solberg S, Wolf BM, Wu J-C (2012) An international comparison of individual tree detection and extraction using airborne laser scanning. Remote Sens 4:950–974
Kaartinen H, Hyyppä J, Vastaranta M, Kukko A, Jaakkola A, Yu X, Pyörälä J, Liang X, Liu J, Wang Y, Kaijaluoto R, Melkas T, Holopainen M, Hyyppä H (2015) Accuracy of kinematic positioning using global satellite navigation systems under forest canopies. Forests 6:3218–3236
Kalliovirta J, Tokola T (2005) Functions for estimating stem diameter and tree age using tree height, crown width and existing stand database information. Silva Fenn 39:227–248
Kankare V, Holopainen M, Vastaranta M, Puttonen E, Yu X, Hyyppä J, Vaaja M, Hyyppä H, Alho P (2013a) Individual tree biomass estimation using terrestrial laser scanning. ISPRS J Photogram Remote Sens 75:64–75
Kankare V et al (2013b) Single tree biomass modelling using airborne laser scanning. ISPRS J Photogramm Remote Sens 85:66–73
Kankare V et al (2014a) Estimation of timber quality of Scots pine with terrestrial laser scanning. Forests 5:1879–1895
Kankare V et al (2014b) Accuracy in estimation of timber assortments and stem distribution—a comparison of airborne and terrestrial laser scanning techniques. ISPRS J Photogramm Remote Sens 97:89–97
Kankare V, Liang X, Vastaranta M, Yu X, Holopainen M, Hyyppä J (2015) Diameter distribution estimation with laser scanning based multisource single tree inventory. ISPRS J Photogramm Remote Sens 108:161–171
Korpela I, Ørka HO, Maltamo M, Tokola T (2010) Tree species classification using airborne LiDAR—effects of stand and tree parameters, downsizing of training set, intensity normalization, and sensor type. Silva Fenn 44:319–339
Kretschmer U, Kirchner N, Morhart C, Spiecker H (2013) A new approach to assessing tree stem quality characteristics using terrestrial laser scans. Silva Fenn 47:1–14
Kukko A (2013) Mobile laser scanning—system development, performance and applications, vol 153. Finnish Geodetic Institute, Kirkkonummi
Laasasenaho J (1982) Taper curve and volume functions for pine, spruce and birch, vol 108. Communicationes Institute Forestalis Fenniae

Liang X, Litkey P, Hyyppä J, Kaartinen H, Kukko A, Holopainen M (2011) Automatic plot-wise tree location mapping using single-scan terrestrial laser scanning. Photogramm J Finl 22:37–48

Liang X, Litkey P, Hyyppa J, Kaartinen H, Vastaranta M, Holopainen M (2012) Automatic stem mapping using single-scan terrestrial laser scanning. IEEE Trans Geosci Remote Sens 50:661–670

Liang X, Hyyppä J (2013) Automatic stem mapping by merging several terrestrial laser scans at the feature and decision levels. Sensors 13:1614–1634

Liang X, Kankare V, Yu X, Hyyppä J, Holopainen M (2014) Automated stem curve measurement using terrestrial laser scanning. IEEE Trans Geosci Remote Sens 52:1739–1748

Lindberg E, Holmgren J, Olofsson K, Olsson H (2012) Estimation of stem attributes using a combination of terrestrial and airborne laser scanning. Eur J For Res 131:1917–1931

Maas H-G, Bienert A, Scheller S, Keane E (2008) Automatic forest inventory parameter determination from terrestrial laser scanner data. Int J Remote Sens 29:1579–1593

Maltamo M, Mustonen K, Hyyppä J, Pitkänen J, Yu X (2004) The accuracy of estimating individual tree variables with airborne laser scanning in a boreal nature reserve. Can J For Res 34:1791–1801

Maltamo M, Peuhkurinen J, Malinen J, Vauhkonen J, Tokola T (2009) Predicting tree attributes and quality characteristics of scots pine using airborne laser scanning data. Silva Fenn 43:507–521

Othmani A, Lew Yan Voon LFC, Stolz C, Piboule A (2013) Single tree species classification from terrestrial laser scanning data for forest inventory. Pattern Recognit Lett 34(16):2144–2150

Persson A, Holmgren J, Söderman U (2002) Detecting and measuring individual trees using an airborne laser scanner. Photogramm Eng Remote Sens 68:925–932

Pfeifer N, Winterhalder D (2004) Modelling of tree cross sections from terrestrial laser scanning data with free-form curves. Int Arch Photogramm Remote Sens Spat Inf Sci 36(8/W2):76–81

Popescu SC, Wynne RH, Nelson RF (2003) Measuring individual tree crown diameter with lidar and assessing its influence on estimating forest volume and biomass. Can J Remote Sens 29:564–577

Popescu SC (2007) Estimating biomass of individual pine trees using airborne lidar. Biomass Bioenergy 31:646–655

Puttonen E, Suomalainen J, Hakala T, Räikkönen E, Kaartinen H, Kaasalainen S, Litkey P (2010) Tree species classification from fused active hyperspectral reflectance and LiDAR measurements. For Ecol Manag 260:1843–1852

Rasinmäki J, Melkas T (2005) A method for estimating tree composition and volume using harvester data. Scan J For Res 20:85–95

Raumonen P, Kaasalainen M, Åkerblom M, Kaasalainen S, Kaartinen H, Vastaranta M, Holopainen M, Disney M, Lewis P (2013) Fast automatic precision tree models from terrestrial laser scanner data. Remote Sens 5:491–520

Repola J (2008) Biomass equations for birch in Finland. Silva Fenn 42:605–624

Repola J (2009) Biomass equations for Scots pine and Norway spruce. Silva Fenn 43:625–647

Rönnholm P, Hyyppä J, Hyyppä H, Haggrén H, Yu X, Kaartinen H (2004) Calibration of laser-derived tree height estimates by means of photogrammetric techniques. Scan J For Res 19:524–528

Saarinen N, Vastaranta M, Kankare V, Tanhuanpää T, Holopainen M, Hyyppä J, Hyyppä H (2014) Urban-tree-attribute update using multisource single-tree inventory. Forests 5:1032–1052

Stängle SM, Brüchert F, Kretschmer U, Spiecker H, Sauter UH (2014) Clear wood content in standing trees predicted from branch scar measurements with terrestrial LiDAR and verified with x-ray computed tomography. Can J For Res 44:145–153

Thies M, Pfeifer N, Winterhalder D, Gorte BGH (2004) Three-dimensional reconstruction of stems for assessment of taper, sweep and lean based on laser scanning of standing trees. Scand J For Res 19:571–581

Vastaranta M, Ojansuu R, Holopainen M (2010) Puustotietojen ajantasaistuksen luotettavuus. Metsätieteen aikakauskirja 4:367–381

Vastaranta M, Holopainen M, Yu X, Hyyppä J, Mäkinen A, Rasinmäki J, Melkas T, Kaartinen H, Hyyppä H (2011) Effects of individual tree detection error sources on forest management planning calculations. Remote Sens 3:1614–1626

Vastaranta M, Saarinen N, Kankare V, Holopainen M, Kaartinen H, Hyyppä J, Hyyppä H (2014) Multisource single-tree inventory in the prediction of tree quality variables and logging recoveries. Remote Sens 6:3475–3491

Vauhkonen J, Tokola T, Packalén P, Maltamo M (2009) Identification of Scandinavian commercial species of individual trees from airborne laser scanning data using alpha shape metrics. For Sci 55:37–47

Vauhkonen J, Korpela I, Maltamo M, Tokola T (2010) Imputation of single-tree attributes using airborne laser scanning-based height, intensity, and alpha shape metrics. Remote Sens Environ 114:1263–1276

Vauhkonen J et al (2012) Comparative testing of single-tree detection algorithms under different types of forest. Forestry 85:27–40

Vauhkonen J, Ørka HO, Holmgren J, Dalponte M, Heinzel J, Koch B (2014) Tree species recognition based on airborne laser scanning and complementary data sources. In: Maltamo M, Næsset E, Vauhkonen J (eds) Forestry application of airborne laser scanning: concepts and case studies, managing forest ecosystems, vol 27. Springer, Berlin, pp 135–156

Yao T, Yang X, Zhao F, Wang Z, Zhang Q, Jupp D, Lovell J, Culvenor D, Newnham G, Ni-Meister W, Schaaf C, Woodcock C, Wang J, Li X, Strahler A (2011) Measuring forest structure and biomass in New England forest stands using Echidna ground-based lidar. Remote Sens Environ 11(15):2965–2974

Yu X, Hyyppä J, Vastaranta M, Holopainen M, Viitala R (2011) Predicting individual tree attributes from airborne laser point clouds based on the random forests technique. ISPRS J Photogramm Remote Sens 66:28–37

Yu X, Liang X, Hyyppä J, Kankare V, Vastaranta M, Holopainen M (2013) Accurate stem biomass estimation based on stem reconstruction from terrestrial laser scanning point clouds. Remote Sens Lett 4(4):344–353

Yu X et al (2015) Comparison of laser and stereo optical, SAR and inSAR point clouds from air- and space-borne sources in the retrieval of forest inventory attributes. Remote Sens 7:15933–15954

Proximity-Driven Motives in the Evolution of an Online Social Network

Ákos Jakobi

Abstract Although early theoretical works dealing with the effects of information communication technologies on people's relation to spatiality claim that distance is no longer important in the information age, there is a growing number of empirical results stressing on the contrary the importance of geographical factors. In the era of big data now we have the chance to give more insights on the geography of the internet-related social processes, since there are unprecedentedly large enough samples to analyse social behaviour as well as to understand the changing role of geography. Accordingly, the following paper is focusing on the geographical analysis of a nowadays very popular topic: the online social networks. Examples of iWiW, the largest Hungarian social media site, are applied to show that such networks are evolving and are structured not independently of spatial constraints. The paper attempts to present many proximity-driven characteristics of this network starting from distance-based examples of space-time evolution and with examples of proximity-focused statistical analysis of the spatial structure. The calculations highlighted that—although it was changing in time—proximity-driven processes have been predominant in city-to-city diffusion, especially when dealing with intracity spreading, but also at cases of short distance neighbourhood diffusion. Calculations by comparing factors like the average strength of connectivity, population size and average relative distance rates of cities also confirmed that proximity had an influence on the network structure.

Keywords Online social networks · Proximity · Distance · Network geography · Hungary

Á. Jakobi (✉)
Department of Regional Science, Faculty of Sciences, Eötvös Loránd University, Pázmány Péter Sétány 1/c, Budapest 1117, Hungary
e-mail: jakobi@caesar.elte.hu

I. Ivan et al. (eds.), *The Rise of Big Spatial Data*, Lecture Notes in Geoinformation and Cartography, DOI 10.1007/978-3-319-45123-7_15

1 Introduction

In the internet era, online social networks (OSN) are one of the major platforms of communication (see Lazer et al. 2009), supporting place-independent social life; however, recent findings suggest that geographical location of users strongly affect network topology (Takhteyev et al. 2012). Although on the one hand cyberspace is clearly present in the vanishing distance dependent costs of online telecommunication, leading to the claim of the "Death of Distance" thesis (Cairncross 1997), on the other hand the role of geographical location and distance is not clear at all regarding online communication and online involvement itself, because internet seems to stimulate local offline communication (Storper and Venables 2004) and users mostly interact with their strongly connected cliques but are also able to extend their interactions to more distant places than ever before (Wellman 2002). It seems that physical place and distance has a determining power on online communities (Liben-Nowell et al. 2005), and internet infrastructure (Tranos and Nijkamp 2012).

The above mentioned debate raised the question that to what extent an online social network is spatially bounded. Does proximity matters or OSNs are realized forms of absolute spatial independency? To see it clearer we should note that social network sites are supplemental forms of communication between people who have known each other primarily in real life (Ellison et al. 2006) and OSNs are "biased versions of real-life networks" (Ugander et al. 2011). We claim that virtual space and physical world are strongly interrelated, since it is assumed that flesh and blood users document their offline friendships in the online environment. According to this statement online social networks should be geographically determined, but it is still unclear what spatial motives are decisive in the formation of an OSN. Therefore, the following paper has the aim to give evidences how geography influences OSNs by analysing one of the most important spatial factors of network evolution: the proximity.

2 The Dateset

The following analysis was made by the application of data of the once largest Hungarian online social network site, named iWiW (International Who Is Who). The iWiW was launched in April 2002 and the service became highly popular and reached a few hundred thousands of people by 2005. By the introduction of new functions in 2005–2006 the number of registered users grew rapidly from 1.5 to more than 4 million until December 2008. Later, the website could not meet the challenges of competing with market-leading OSNs (namely Facebook) and, after a long declining period, the service was shut down on June 30, 2014.

The examinations were based on a data collection for January 2013 (and provided for research purposes by the data owner company). Location of users was defined by profile information, which is occasionally considered to be problematic in papers focusing on OSN user and social media content localization (Hecht et al. 2011). In iWiW, however, it was compulsory to choose a town of residence from a scroll-down menu when registering as user; thus, location is documented in every profile. This place could be easily changed afterwards and certainly there was no eligibility check. One might consider our location indicator based on user profiles a biased and occasionally updated census-type data. The geolocated individual user data have been summed up to the level of cities, and because user profile information also contained friendship data (data of people a user is in connection with), we could draw the connectivity network of the cities as well. This was registered in the database in forms of settlement pairs.

Altogether 2562 cities had active user data with a sum of 4,058,505 users. The users have established 785,841,313 friendship ties in the website, out of which 369,789,373 ties remained within settlement borders (considered as intra-city loops) and 415,653,749 ties were established between users from two distinct settlements. Concerning the city-level aggregated data, the network database covered 1,369,978 settlement-to-settlement pairs.

Additionally, data were appropriate to trace the evolution of the network in time. Until the very late periods of iWiW life-cycle, new users could register a profile only after an invitation had been sent from a member. The ID of the inviter was involved in each user's profile. Therefore we were able to trace the diffusion of iWiW across time and space because we know not only the location of each new user, but also the location from where the invitation was sent to each new user, and the timestamp of the acceptance of the invitation. In that way, we could investigate more than 2.7 million geo-located invitations between April 2002 and June 2012. Since our data was collected in 2013, inviter ID was missing in cases when the profile of the inviter was already deleted or the profile was registered after June 2012 when invitation wasn't needed anymore for registration, therefore the following analysis was performed for data between 2002 and 2012. Concerning the number of invitations from 2002 until the end of 2005 only 1–2 thousand invitations were sent by members per month. Then, the number of invitations jumped to more than 50,000 monthly and increased to a peak about 90,000 invitations per month until the middle of 2007. After that period invitations per month started to decrease rapidly. Although free registration was also introduced in 2012, this previously only mode of diffusion remained the major means of spreading.

3 Space-Time Evolution of the Network: The Importance of Proximity

According to the main findings of the international literature (Oh et al. 2008; Lan et al. 2011; Takhteyev et al. 2012) we assume that the evolution of our OSN has also followed certain geographical characteristics. Our general assumption is that the formation of new connections between old and new members is largely depending on distance between them. Although it is possible to get in connection with anyone in networks of the cyberspace, we still believe that the majority of new friendships are evolving among closely located people. Naturally, there are exceptions, but the share of random or not proximity-driven new connections is expected to be small.

On the other hand it is also presumed that the importance of proximity-driven formation of new connections could be different in certain time periods in the life of the OSN. We expect that at the beginning, when only few people are involved, the importance of close friendship is supposed to be larger, than at later periods, when the number of users is much higher and the chance of getting in connection with new distant acquaintances is also larger. All in all, we nevertheless assume the dominance of proximity-driven invitation processes throughout the whole examined period.

In our examination we followed the main concepts of spatial innovation diffusion theories (Hägerstrand 1967; Gould 1975), which highlight that many of the new things are spreading in space not randomly, but often as determined by geographical or other proximity factors (Boschma 2010). In that sense, new things appear first close to its origin, then in the next phase it appears at the closest neighbouring areas, while at a later period also distant places adopt the new thing. In other words, neighbourhood diffusion refers to the spreading when an innovation will likely be adopted first close to its source and later at greater distances following a distance-decay pattern. Concepts of this approach are widely applied in the literature (Cliff 1968; Johnston and Pattie 2011) confirming that there are many spatial evolutionary processes, which follow proximity rules.

To test our assumption we analysed the geo-located invitation data for each months. Since the location of both the inviter and the invitee were known in the dataset, it was easy to determine the geographical distance between the two people, or at least their two cities. The calculation of the distance between the city of the sender and the city of the receiver has been done for all invitation cases. Based on that, at first the average distance of invitations was calculated for each month (Fig. 1). Consequently, the smaller the average distances of invitations the higher the probability that neighbourhood diffusion has a large proportion within all diffusion cases.

Concerning the average distance of the invitations we found evidence of the assumption that proximity factors were changing in time. According to the results a continuous increase of distance values were observable until 2006 and a decrease afterwards. At the beginning of the examined period the average distance between the city of the inviter and the city of the invitee was slightly larger than 15 km. It reached as high as approximately 45 km in April 2006 and started to decrease and to stabilise later at around 20–25 km. We suppose that the noteworthy break of the trend after 2006 was in relation with the increase of the total number of new invitations.

In order to test the importance of proximity-driven diffusion forms, then we classified the invitation cases into proximity based (or short distance) and

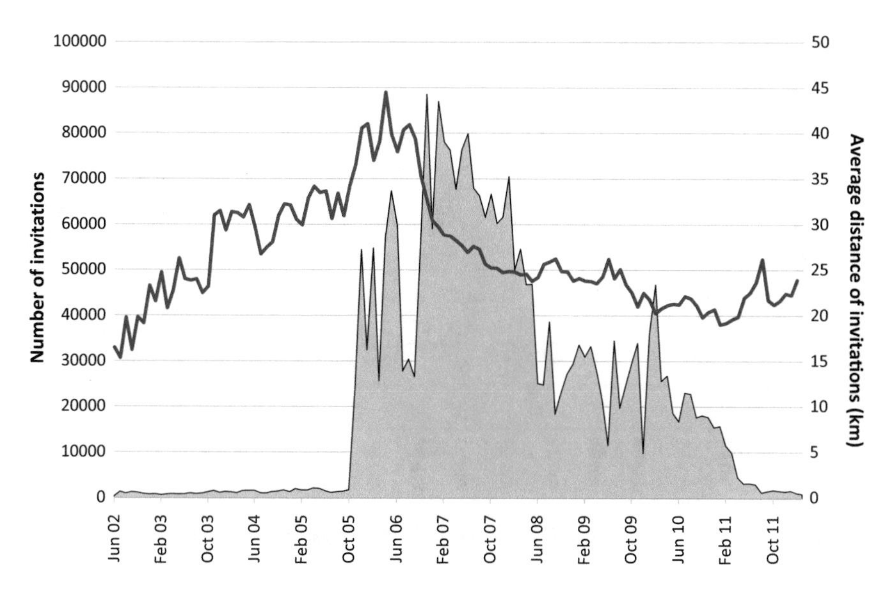

Fig. 1 The total number of iWiW invitations (*grey area*) and the average distance of iWiW invitations (*line*) by months (June 2002–March 2012)

non-proximity based (or long distance) categories. Since iWiW data were only possible to be geo-located on the level of cities (and not on the level of addresses or streets, etc.), zero distances were registered for invitations, where the inviter and the invitee were located in the same city. We consider these cases as intracity diffusion (or loop diffusion), which are in fact also proximity-based, since distances between inviter and invitee within a settlement are surely not zero. We know that there could be a distance-decay also within a city, however, our dataset was not able to detect intracity location differences. By the way, we still reckon loop diffusion as spreading to the closest distance, while the rest remained as cases of simple short distance or long distance diffusion.

It can be declared that intracity loop invitations happened to be the most prevalent category every time during the observed years. Invitations within the same city had always a dominant role, since the share of loops within all connections was permanently above 50 % (Fig. 2). Proportional values started from as high as 80 % and decreased to 52 % until 2006, when turned to rise again roughly until the end of the examined period. The curve more or less inversely followed the change of the total number of invitations.

Concerning the rest of the invitation data, we distinguished the group of short distance (but not loop) diffusion to the neighbouring zones and the group of long distance (not proximity-driven) spreading cases. In order to find short distance (neighbourhood) diffusion cases we had chosen threshold distance metrics instead of topological adjacency of cities, since it was assumed that virtual space connections took administrative topology less into account, distance on the other hand

Fig. 2 The share of invitations sent within the same city (June 2002–March 2012)

seemed to play notable role. The threshold value was set to 15 km in line with results of calculations, where the probability of links as a function of distance had been calculated and the values returned a slight break of probability at around 15 km (Lengyel et al. 2015). This threshold possibly well separated neighbourhood diffusion zones from other spreading areas.

The share of invitations that were sent to a maximum of 15 km covered approximately 2–5 % of all cases in the early years (Fig. 3). Then, after 2006, the proportion of short distance invitations was doubled and stayed relatively high almost until the end. Consequently, the share of neighbourhood diffusion cases has

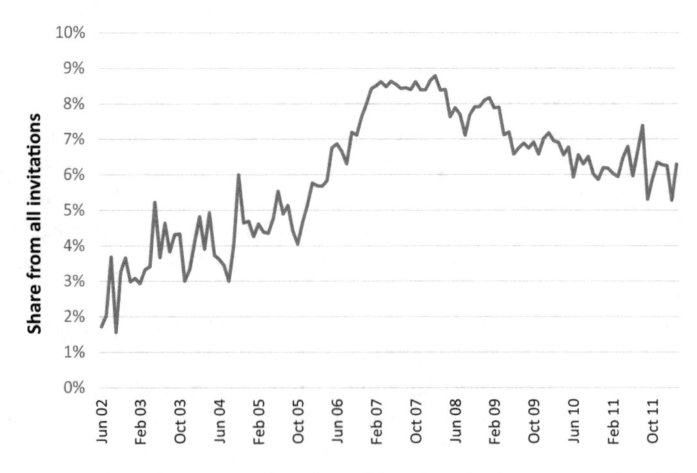

Fig. 3 The share of invitations sent to less than 15 km, intracity loops are excluded (June 2002–March 2012)

been considerably increased in the second half of the time. Apparently the proximity dimension was strengthened and became more important after 2006 resulting some decrease in the average distance of the invitations.

After taking all proximity-driven (loop and neighbourhood) diffusion cases away, the remaining city-to-city invitations were possible to be declared as independent of geographical distance. The overall share of such cases were approximately 15 % at the beginning, then increase to a top about 33 % around the year of 2006, while continuously decreased thereafter to a level of 15 % until the end of the examined period. Accordingly, the share of distance-independent cases has never been predominant during the years, and it seems that it could reach somewhat larger proportion only around 2006, when the annual number of new invitations was extraordinary high.

4 Proximity-Driven Motives in the Spatial Network Structure

Previously we have seen that proximity played an important role in the development of our online social network. It is a question, on the other hand, whether proximity has also a notable effect on user and connectivity patterns of the network structure. We assume that the locations, where iWiW was adapted earlier, would have higher rates of user penetration compared to those applying OSN services later. And because iWiW was firstly and primarily used by people in Budapest, we presumed higher user rates in the capital city (claimed as the city of origin) and smaller rates in farther distances from the city. In order to reveal such proximity-driven motives, we compared the rate of iWiW users and the distance from Budapest for each settlements.

According to the results a negative relationship could be found between the rate of users among the local population and the distance from Budapest, in which the departure from the experienced maximum level is, in fact, growing in negative terms (Fig. 4). Although the fitted linear regression model had not so large R-square results, the outcomes still reflect that proximity matters in OSN presence: as the distance increases, the probability of a lower user rate increases.

Although data of the rate of users revealed that this online social network can not be considered as aspatial, naturally, there are other geographical motives possible to be explored in a network structure, especially in connection with network ties. Since this OSN dataset is a typical example of big data, we may assume that the entities or nodes (in our case the localities, namely the cities) have the chance to get network connections with almost all the others. It is also known, however, that some of the cities have large number of connections with high variety, while others are rather connected to only few cities. The possible determining factor behind is

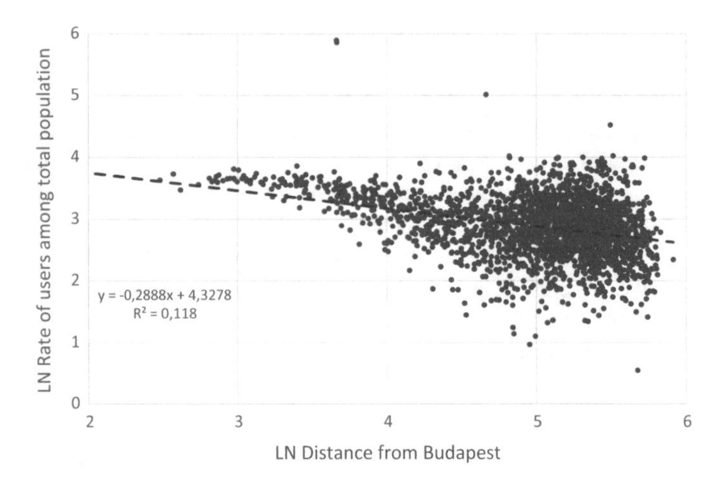

Fig. 4 The connection between the rate of iWiW users and the distance from Budapest. All variables are transformed to natural logarithm values

the size of the city, since large cities could have more connections and more divers network structure than small villages with only a couple of users, who are connected with users from only a few other places. On the other hand it is assumed, that cities having connections with not many other cities are possibly tied to others stronger than those having connections with large pool of cities.

In order to deal with different strength of network connections between cities we compared the ratio of the observed and randomly expected city-to-city connection weights for each pair of cities. The observed or raw weights have been calculated as the sum of connections between users of the two cities (Eq. 1):

$$w_{ij} = c_{ij} + c_{ji} \tag{1}$$

where w_{ij} is the observed (or raw) weight of connections between city i and j, c_{ij} is the number of connections between users, who are located in city i and have friends in city j, while c_{ji} is the number of connections between users, who are located in city j and have friends in city i.

The expected city-to-city connection weights have been calculated as follows (Eq. 2):

$$e_{ij} = \frac{s_i s_j}{\sum_{i=1,j=1}^{n} w_{ij}} \tag{2}$$

Here $s_i = \sum_j^n w_{ij}$ is the strength of node i, namely the total number of connections in the city, and e_{ij} is the expected number of links between cities i and j based purely on the total number of links at those cities assuming random tie formation.

Finally we calculated the log likelihood ratios of the above detailed observed (or raw) and randomly expected components (Eq. 3):

$$LLR_{ij} = Log\left(\frac{w_{ij}}{e_{ij}}\right) = Log\left(w_{ij}/\frac{s_i s_j}{\sum_{i=1,j=1}^{n} w_{ij}}\right) \quad (3)$$

in which LLR_{ij} refers to the log-likelihood ratio between settlement i and j. Note that LLR_{ij} can be negative or positive depending on the ratio of the measured weight and the expected one. The higher positive LLR refers to strong city-to-city ties, while negative LLR represents weak intercity connections.

As mentioned above, we assume that geography plays a role in connectivity strength between settlements. When looking at the scatter plot of average strength of connectivity (average LLR score by settlements) against the size of the cities (natural logarithm of population), it could be definitely noticed that the smallest settlements have on average the strongest connections with others (top left on Fig. 5). Such cities are typically connected to only few number of other cities but with strong relations.

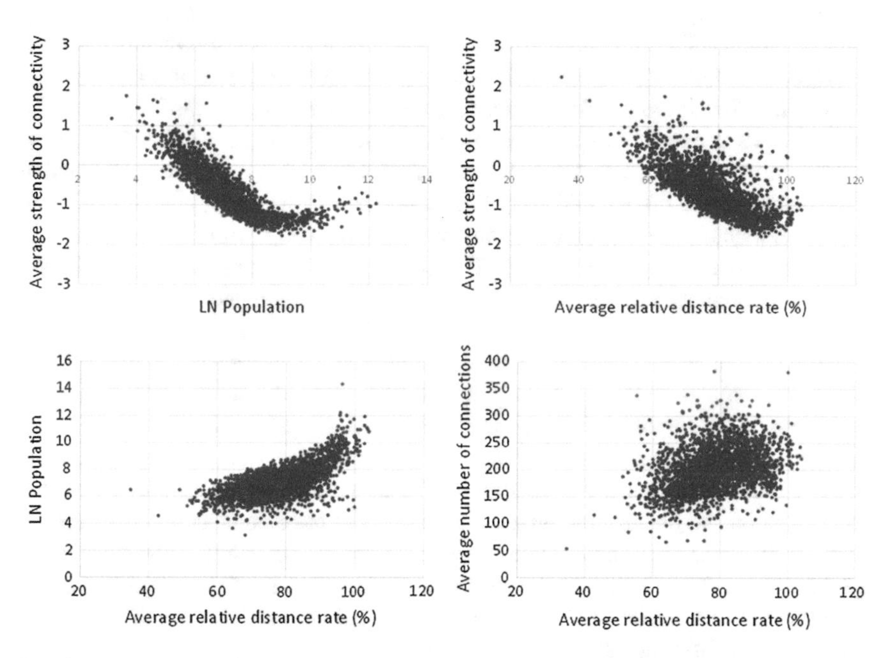

Fig. 5 *Top left* connection between the average strength of connectivity (average LLR score) and size (natural logarithm of population). *Top right* connection between the average strength of connectivity (average LLR score) and average relative distance rate. *Bottom left* connection between size (natural logarithm of population) and average relative distance rate. *Bottom right* connection between average number of connections and average relative distance rate

For ourselves it is nevertheless more interesting that also distance has a possible influence on the average strength values of cities. But we would get a biased picture if absolute distances between connected cities would have been applied, since absolute distance of connections largely depends on central or peripheral geoposition of a settlement, instead we suggest to use relative distance rates for non-biased network proximity. Average relative distance rates were calculated by the comparison of the observed and expected distance averages (Eq. 4):

$$ARD_i = \left(\frac{\sum_{i=k}^{k} d_{ij}}{k} \Bigg/ \frac{\sum_{i=1}^{n} d_{ij}}{n} \right) \times 100 \tag{4}$$

in which ARD_i refers to the average relative distance rate of settlement i, d_{ij} is the distance between settlement i and j, k is the observed number of connected settlements and n is the total (or expected) number of settlements.

By the comparison of average strength of connectivity (average LLR score) and average relative distance rates of cities (ARD), we should declare that the closer associates a city generally owns, the stronger the connections it has on average (top right on Fig. 5). The fitted linear regression model resulted a significant R-square above 0.5. Consequently it seems that the tightest and strongest network connections do not stretch too far. Additionally, it is also observable that the larger the city, the farther its connections are reaching on average (bottom left on Fig. 5). Finally, by the comparison of the average number of connections and average relative distance rates (ARD) we could more or less notice that the cities, which have more distant connections in general, are having users typically with larger number of friendships (bottom right on Fig. 5). This relationship is, however, less significant than the previously detailed ones, since the scatter plot reflects evidently larger standard deviation and also linear R-square results happened to be small (0.1).

Examples of the analysis of individual cities also confirmed that tighter connections in virtual space are falling in line with short distances in real physical geography, even though it is in principle the same simple to access any points in cyberspace. The example of Herend, a middle-sized city, well reflects that cyberspace is not independent from constraints of real physical space, although it is also observable on the picture that there is the chance to have connections with distant cities as well (Fig. 6). This city has basically strong connections with close cities (generally less than 40 km), but some strong connections are from larger distances. Based on that the relationship between distance and connectivity weight (strength) is not deterministic rather stochastic.

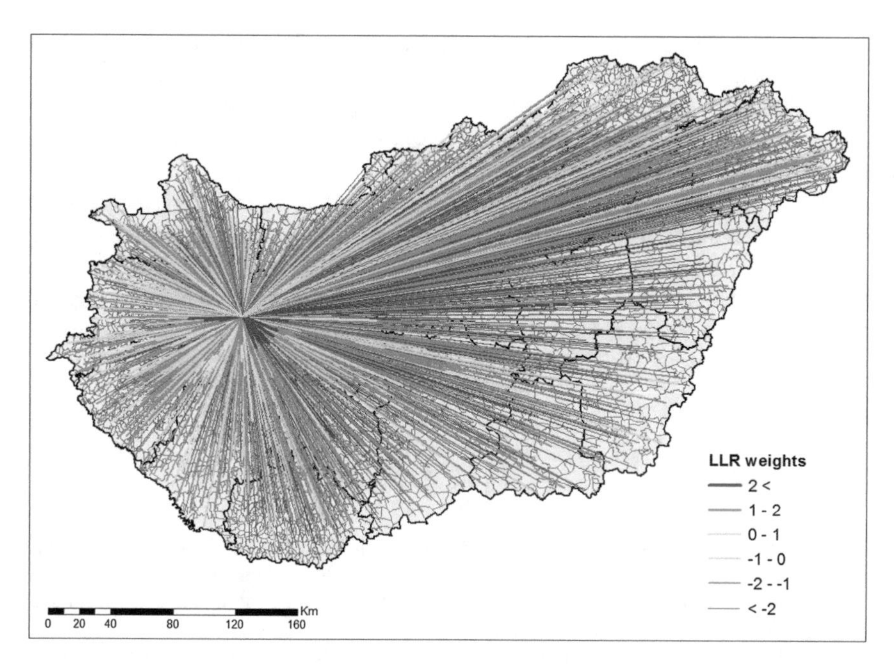

Fig. 6 The connectivity network or Herend according to the strong and weak connections (based on LLR scores)

5 Summary and Conclusions

OSNs are large-scale networks in which users are the nodes and their connections with other users are the edges. They are also defined as web-based services that 'enable users to articulate and make visible their social networks' (Boyd and Ellison 2007, p. 212.). The definition claims that OSNs are supplemental forms of communication between people who have known one another primarily in real life. In other words, major OSNs are not used to meet new people but rather to articulate relationships with people in their existing offline network. Furthermore, the degree distribution of online social networks, like Facebook, is very close to the degree distribution of real-life social networks (Ugander et al. 2011), in other words, OSNs clearly differ from other web-based networks, such as internet infrastructure. The latter are led by power law tie-distribution: a small share of webpages accounts for an outstandingly high number of links (Barabási and Albert 1999). In our understanding, OSNs are showing strong geography-related network characteristics and not a typical power law pattern.

The paper demonstrated that OSNs are definitely place-dependent, because many aspects of network connectivity happened to show geographical relatedness as well. Despite the fact that online social networks are virtual creations it was found that diffusion processes related to network evolution are not independent of spatiality. It was pointed out that proximity-driven diffusion processes are predominant, especially when dealing with intracity spreading, but also at cases of short distance neighbourhood diffusion. Although cyberspace allows creating connections independently of distance, the majority of new registrations arose in the geographic vicinity of earlier ones.

Also by determining the strongest and most important city-to-city connections or the average relative distance rates it turned out that distance does significantly matter in network formation. Many of the network structure characteristics were happened to be proximity-driven. As a combined result we should claim that big datasets of online social networks now has a good chance to give evidence on why geography matters (de Blij 2012) in the information age.

References

Barabási AL, Albert R (1999) Emergence of scaling in random networks. Science 286:509–512

Boschma R, Frenken K (2010) The spatial evolution of innovation networks. A proximity perspective. In: Boschma R, Martin R (eds) The handbook of evolutionary economic geography. Edward Elgar, Cheltenham, pp 120–135

Boyd D, Ellison NB (2007) Social network sites: definition, history, and scholarship. J Comput Mediat Commun 13:210–230

Cairncross F (1997) The death of distance. How the communication revolution will change our lives. Harvard Business School Press, Boston

Cliff AD (1968) The neighbourhood effect in the diffusion of innovations. Trans Inst Br Geogr 44:75–84

de Blij H (2012) Why geography matters: more than ever, 2nd edn. Oxford, New York

Ellison N, Steinfeld C, Lampe C (2006) Spatially bounded online social networks and social capital: the role of Facebook. Paper presented at the annual conference of the international communication association, Dresden, 19–23 June. http://www.ucalgary.ca/files/stas341/Facebook_ICA_2006.pdf. Accessed 30 Nov 2015

Gould PR (1975) Spatial diffusion: the spread of ideas and innovations in geographic space. Learning Resources in International Studies, New York

Hägerstand T (1967) Innovation diffusion as a spatial process. University of Chicago Press, Chicago

Hecht B, Hong L, Suh B, Chi EH (2011) Tweets from Justin Bieber's heart: the dynamics of the "location" field in user profiles. In Proceedings of the ACM conference on human factors in computing systems. ACM Press, New York, pp 237–246

Johnston RJ, Pattie C (2011) Social networks, geography and neighbourhood effects. In: Scott J, Carrington P (eds) The SAGE handbook of social network analysis. SAGE Publications Ltd, London, pp 301–311

Lan T, Lan C, Tserendondog O (2011) Analysis of social network sites diffusion in Mongolia. Afr J Bus Manag 5(23):9889–9895

Lazer D, Pentland A, Adamic L, Aral S, Barabasi AL, Brewer D, Christakis N, Contractor N, Fowler J, Gutmann M, Jebara T, King G, Macy M, Roy D, Van Alstyne M (2009) Computational social science. Science 323(5915):721–723. doi:10.1126/science.1167742

Lengyel B, Varga A, Ságvári B, Jakobi Á, Kertész J (2015) Geographies of an online social network. PLoS One 10(9):e0137248. doi:10.1371/journal.pone.0137248

Liben-Nowell D, Novak J, Kumar R, Raghavan P, Tomkins A (2005) Geographic routing in social networks. Proc Natl Acad Sci USA 102:11623–11628

Oh J, Susarla A, Tan Y (2008) Examining the diffusion of user-generated content in online social networks. SSRN eLibrary. doi:10.2139/ssrn.1182631. Accessed 27 Nov 2015

Storper M, Venables A (2004) Buzz: face-to-face contact and the urban economy. J Econ Geogr 4:351–370

Takhteyev Y, Gruzd A, Wellman B (2012) Geography of Twitter networks. Soc Netw 34:73–81

Tranos E, Nijkamp P (2012) The death of distance revisited: cyberplace, physical and relational proximities. Working Paper Tinbergen Institute, TI 2012-066/3. http://papers.ssrn.com/sol3/papers.cfm?abstract_id=2103024. Accessed 20 Nov 2015

Ugander J, Karrer B, Backstrom L, Marlow C (2011) The anatomy of the Facebook social graph. http://arxiv.org/abs/1111.4503. Accessed 30 Nov 2013

Wellmann B (2002) Little boxes, glocalization, and networked individualism. In: Tanabe M, van den Besselaar P, Ishida T (eds) Digital cities II: computational and sociological approaches. Springer, Berlin, pp 10–25

Mapping Emotions: Spatial Distribution of Safety Perception in the City of Olomouc

Jiří Pánek, Vít Pászto and Lukáš Marek

Abstract Places are strongly linked with emotions and can be considered safe or unsafe, pleasant or ugly, favourite or boring among other emotions. Subjective perceptions of a city can be valuable sources of information for city planners and a local administration. Among the above-mentioned emotions that have an effect on the quality of life for people in a city, the perception of safety has a prominent position. Safety and fear of criminality affect our interaction with public spaces the most. But criminality does not have to be the only reason people feel uncomfortable in a city, they may also be afraid of the darkness or the friendlessness of a place. The paper describes the mapping of unsafe places in the city of Olomouc via a paper-based questionnaire and a web-based crowdsourcing tool PocitoveMapy.cz. In total, the authors collected answers from 661 respondents; 144 used the online tool and 517 used the paper-based version. The final dataset comprises 1516 places (453 online/1063 questionnaire). The data were gathered over the period between 1st October and 2nd December 2015. The authors collected data that are gender specific as well and time of day specific, therefore it was possible to analyse the differences between daytime and night-time fearful places in the city as well as places that are perceived unsafe by women and men. The spatial density analysis, local correlations and hexagonal aggregation revealed hot-spots that are felt by the citizens of Olomouc to be unsafe. The strongest agreement in votes can mainly be found in the three localities with the densest localisation of votes. In these localities,

J. Pánek (✉)
Department of Development Studies, Faculty of Science, Palacký University Olomouc, 17. Listopadu 12, 771 46 Olomouc, Czech Republic
e-mail: Jiri.Panek@upol.cz

V. Pászto
Department of Geoinformatics, Faculty of Science, Palacký University Olomouc, 17. Listopadu 50, 771 46 Olomouc, Czech Republic
e-mail: Vit.Paszto@upol.cz

L. Marek
Department of Geography, University of Canterbury, Private Bag 4800, 8041 Christchurch, New Zealand
e-mail: Lukas.Marek@canterbury.ac.nz

I. Ivan et al. (eds.), *The Rise of Big Spatial Data*, Lecture Notes in Geoinformation and Cartography, DOI 10.1007/978-3-319-45123-7_16

a strong correlation exists also between the perception of fear during the daytime and the night-time. The results of the case study can be used by the local police department or administration authorities in the future development of safety strategies for the city.

Keywords Emotional mapping · Subjective data · Unsafe places · Olomouc · Spatial correlation · Geovisualisation

1 Introduction

Personal security plays an important role in the quality of life and well-being of citizens. Perceived safety and fear of crime have an influence on people's behaviour (Curtis 2012). Participatory mapping and qualitative GIS (sometimes also called GeoParticipation) allow city planners and decision makers to deploy new tools and methods to collect both qualitative and quantitative data about cities, their dynamics and the people living there (Kloeckl et al. 2011). Most geospatial applications rely on objective data only, although there can be discussion concerning the extent to which GIS data are objective, as there is always a level of generalisation, uncertainty and author's bias (Pickles 1995). The call for a more humanised and participatory approach to geospatial information and technologies has been heard since the publication of the *Ground Truth* (Pickles 1995).

The authors collected and analysed subjective data linked to perceived safety in the day-time and in the night-time in the city of Olomouc, Czech Republic. The data were collected in the form of an emotional map and the outcomes can be seen as a version of a Gould-style mental map (Gould 1986). Emotions and space are connected because every location can evoke an emotion (Mody et al. 2009) and places can be felt to be attractive, boring, dangerous or scary, among other emotions (Korpela 2002). In the past 10 years several projects have dealt with georeferenced emotions and the methods used to gather emotional data can be divided into three groups: (1) biometric measurements (Bergner et al. 2011; Nold 2009), (2) extraction from user generated content such as Twitter, Flickr, Facebook, etc. (Biever 2010; Bollen et al. 2011; Mislove et al. 2010), and (3) surveys (Huang et al. 2014; MacKerron and Mourato 2010; Mody et al. 2009). The author's approach can be considered to be a survey.

Griffin and Mcquoid (2012) distinguished between three categories when talking about maps and emotions. These categories are (1) maps of emotions, (2) using maps to collect emotional data, and (3) emotions while using maps. The case study described in this paper is a combination of the first two categories. Maps were used to collect the emotional information, and also to visualise the emotional data.

Emotions are one of the defining characteristics of every human being and yet their presence in maps and spatial data is uncommon (Griffin and Mcquoid 2012). Several authors (Barrett 2006; Reeve 2014; Russell 1980) described emotions as a two-dimensional structure with the axis being pleasant-unpleasant and high-arousal or low-arousal. Geographers, on the other hand, described emotions as subjective relational flows between places and people (Smith et al. 2012) and therefore their definition is not consistent with a two dimensional understanding of emotions.

The analysis between crime and place started with the *Chicago School*, and later in the 1980s when the "new" Chicago School implemented GIS into the ecological studies of crime patterns and theories of environmental criminology (Anselin et al. 2000). Studies proved that perceptions of safety in an urban environment are influenced by environmental characteristics; mainly by what is visible (Fisher and May 2009; Lipscomb 2014). Although historically cartography was mainly focused on treating that which is visible or can be mapped (including air temperature and wind speed) (Wilson 2011), critical and feminist cartographers always advocated mapping a space as people experience it, with emotions included (Pearce 2008). Hauthal and Burghardt (2014) argued that *…mappers of georeferenced emotions are almost exclusively researchers…* using emotional maps in various fields such as tourism (Mody et al. 2009), navigation (Gartner 2012; Huang et al. 2014) and city planning (Raslan et al. 2014).

Mapping of safety and fear of criminality has become a modern trend in various fields such as cartography, GIScience, environmental ecology, behavioural geography, urban planning and psychology. In the Czech/Slovak context only a few authors dealt with the mapping of safety (Jíchová and Temelová 2012; Sessar and Sirotek 2001; Stasíková 2011), and none of them used emotional or mental maps to collect or visualise the data. Němečková (2014) used subjective data to analyse safety in the city, but the spatial element of the data was not even explored in her research. Some researchers (Brown and Polk 1996; Clemente and Kleiman 1977; Oc and Tiesdell 1997) argued that fear of criminality could be as dangerous as criminality itself, therefore the application of GIScience became trendy in analysing spatial patterns of criminality and how it affects people′s behaviour within an urban environment (Chainey and Ratcliffe 2013; Doran and Burgess 2011; Leitner 2013; Santos 2012).

The aim of the paper is to analyse subjective data about the perception of safety in the city of Olomouc. The authors are looking for patterns and hot-spots within the city and if there are any relevant linkages between gender and visibility (day-time vs. night-time) and the spatial distribution of places marked as unsafe. The uniqueness of the presented case-study is its combination of online-crowdsourced data and information collected on the street through paper-based questionnaires.

2 Background

2.1 Location of the Case-Study

Olomouc is a historical city in the Czech Republic with a population of 99.809 (Czech Statistical Office 2015), it is also the seat of The Regional Authority of the Olomouc Region and the sixth largest city in the country. The city is sometimes called the "city of parks" due to the several parks that surround the historical city-centre. The parks, although perceived as a pride of the city, are often a source of danger and criminality, mainly in the evenings and nights. The perception of safety in the city and its temporal variability (day/night) as well as spatial distribution was the main concern of the authors. The Olomouc Region stated, in its Strategy of Criminality Prevention 2013–2016 (Olomouc Region 2015) that one of its four visions is to increase the sense of safety in the region. Based on this document the authors believe that, in the area of criminal prevention, emotional mapping and citizens perception of safety is in alignment with the long-term strategy of the region.

Emotional maps are often a neglected part of cartography, yet they contain relevant information, especially for town and regional planning. There has been limited research into emotional maps within the Czech Republic and almost no practical implementation of such maps within local government administrations. A few examples of research in the area of safety perception were mentioned in the introduction section. Nevertheless, the authors acknowledge that this field deserves much more attention that it has been given up to now.

2.2 Data

The collection of data was managed in two separate ways: an analog, paper-based questionnaire and a crowdsourcing web-based application. All paper-based questionnaires were later digitised into a GIS environment (specifically ArcGIS for Desktop 10.3). A total of 661 respondents completed the questionnaire; 144 used the online tool and 517 used the paper-based version. The final dataset comprises of 1516 places (453 via online tool and 1063 via questionnaire). The data were gathered in the period between 1st October and 2nd December 2015.

Both the analog as well the digital collection procedure had two main questions for the responders:

(a) mark places where you feel unsafe in Olomouc at night (when it is dark),
(b) mark places where you feel unsafe in Olomouc during the day (when there is a light).

For the analog collection of data, trained interviewers (students of geoinformatics at Palacký University Olomouc) approached each respondent at random with

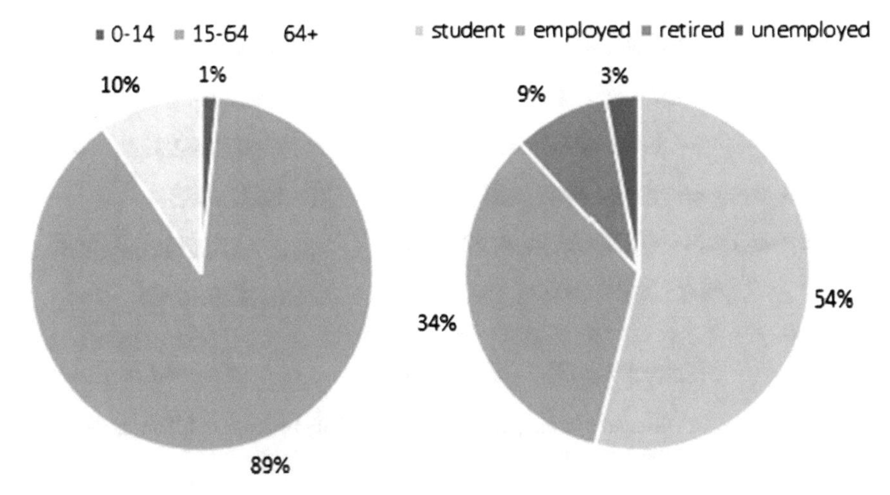

Fig. 1 Paper-based questionnaire respondents (from *left* age groups, occupation status)

the main task being to establish contact with him/her and to convince them to participate in the case study. In total, 236 men (46 %) and 281 women (54 %) participated in the survey. Other basic characteristics about the participants are shown in Fig. 1. The questionnaire items (excluding the main questions above) were as follows:

- gender,
- age group,
- occupation status,
- place of residence (aggregated to city districts),
- frequency of visits in marked unsafe places (Fig. 2),
- date and place of the interview,
- reasons for feeling unsafe (optional).

The second option for collecting the data, as opposed to classical paper-based questionnaires, was crowdsourcing. The authors used their own tool PocitoveMapy.cz to collect the emotional data. Since the main research aim was to identify places associated with unsafe feelings, some of the above mentioned items were omitted from the online tool. The tool itself is designed as a web-application based on Leaflet Library. Similar to other web-based tools for crowdsourced mapping, it allows users to collect spatial data on a slippery map background. Unlike Ushahidi, Umap, ArcGIS Online and many others, PocitoveMapy.cz does not require the registration or installation of any specific software, plug-in or virtual server. The simplicity of the tool helps to engage various target groups, while it is still rich in information gathered. The tool has been used previously in mapping bikers safety in Reykjavík (Iceland), in mapping links between environment and public health in the Czech Republic and for neighbourhood development activities in small town in Bohemia.

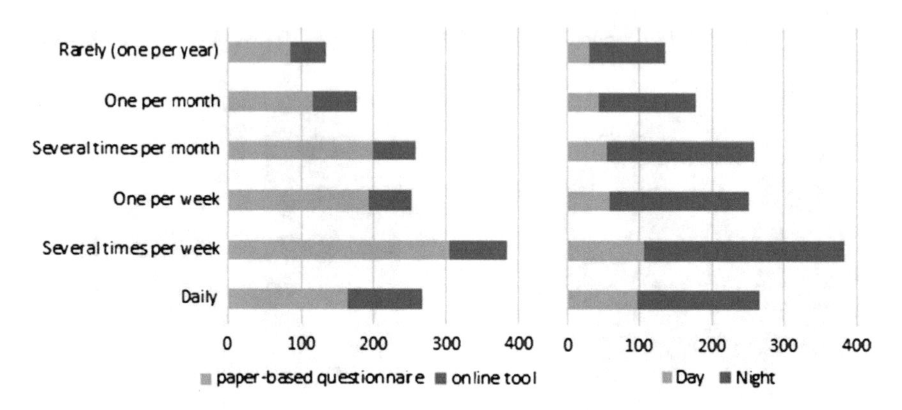

Fig. 2 Frequency of attendance in the places marked as unsafe by the acquisition method (*left*) and by time of day (*right*)

3 Methods

The original data consist of points, which express people's perceptions of fear in individual places. As the data points were dense in some locations they were aggregated into a hexagonal grid with a cell area of 1 ha (Burian et al. 2014). Firstly, the basic GIS tools were used (e.g. frequency analysis, spatial joining, density analysis) for the purpose of visual analytics and data interpretation. The resulting unsafe places were displayed on maps according to various attributes, in order to reveal their spatial patterns and distribution. To emphasize certain localities with a high point density, i.e. places multiply marked as unsafe, density analysis was performed. The significance of such localities was evaluated using 18 neighbouring cells (two-layer hexagons around each particular cell), which allowed highlighting of the most unsafe ones.

Secondly, aggregated grids were entered into the evaluation by the local Spearman's ρ. The neighbourhood of individual grid cells was defined as 1 % of all grid cells, i.e. each hexagon is connected to approximately 100 of the nearest hexagons.

The computation of the correlation coefficient is the most common method to explore and enumerate the association between two (or more) characteristics. However, correlation does not determine the mutual causality of the investigated factors. Pearson's correlation, Spearman's rank correlation and Kendall's rank correlation are widely used methods for the correlation calculation (Reimann et al. 2008). The correlation per se usually provides a global overview of the described association, which may appear to be an inconvenient method in the case of local studies.

Non-parametric Spearman's rank correlation (Spearman's ρ) was undertaken to analyse the associations between people's cognition of fear/lack of safety in the urban environment. Spearman's correlation can explore the non-linear relations of characteristics, and this was the main reason for using it, because the character of the analysed data does not allow us to ensure normal probability distribution. Measured values are substituted for their rank in the calculation of Spearman's correlation.

$$\rho = 1 - \frac{6\sum_{i=1}^{n}(p_i - q_i)^2}{n(n^2 - 1)}$$

Non-parametric Spearman's rank correlation, described above, where the expression $(p_i - q_i)$ means differences in the rank of values corresponding to the measured characteristics and n is a number of pairs. In the study, the authors applied the local version of Spearman's ρ, which calculated the correlation between observations neighbouring in geographical space. The local Spearman's ρ was computed using the R language, utilizing the adapted function based on the package lctools (Kalogirou 2011, 2015).

4 Results

Aggregated grids were entered into the evaluation through the local Spearman's ρ. The neighbourhood of individual grid cells was defined as 1 % of all grid cells, i.e. each hexagon is connected to approximately 100 of the nearest hexagons.

The visualization of marked unsafe places together with density analysis surface is depicted in Fig. 3. The most unsafe places are located at the main train station (eastern part of the city), around the city centre (mainly in surrounding parks and small streets), at localities in suburban residential areas (southern and south-eastern parts of the city) and at places with underpasses.

A similar pattern can be identified from the map in Fig. 4, which uses hexagons with an aggregated points count. It is possible to detect unsafe localities more precisely than using density surface and it is still easier to recognise spatial patterns in comparison with individual points visualization.

The following maps (Fig. 5) display the distribution and count of unsafe places according to the time of day. Respondents tended to mark places connected with fearful emotions during the night (or darkness; 1109 records) approximately three-times more than during the day (407 records). On the one hand this is obvious but on the other hand it means there are still localities demanding higher personal security attention during the daylight. These are, again—the main train station, parks (especially Bezručovy Sady), underpasses and residential areas which are partly treated as socially excluded localities (Nový Svět, Černá cesta). In optional commentaries respondents' argued that these are places with "weird people, the homeless, pickpockets, paupers, drunk people, molestation, heavy traffic, underpasses and Roma people" among other comments.

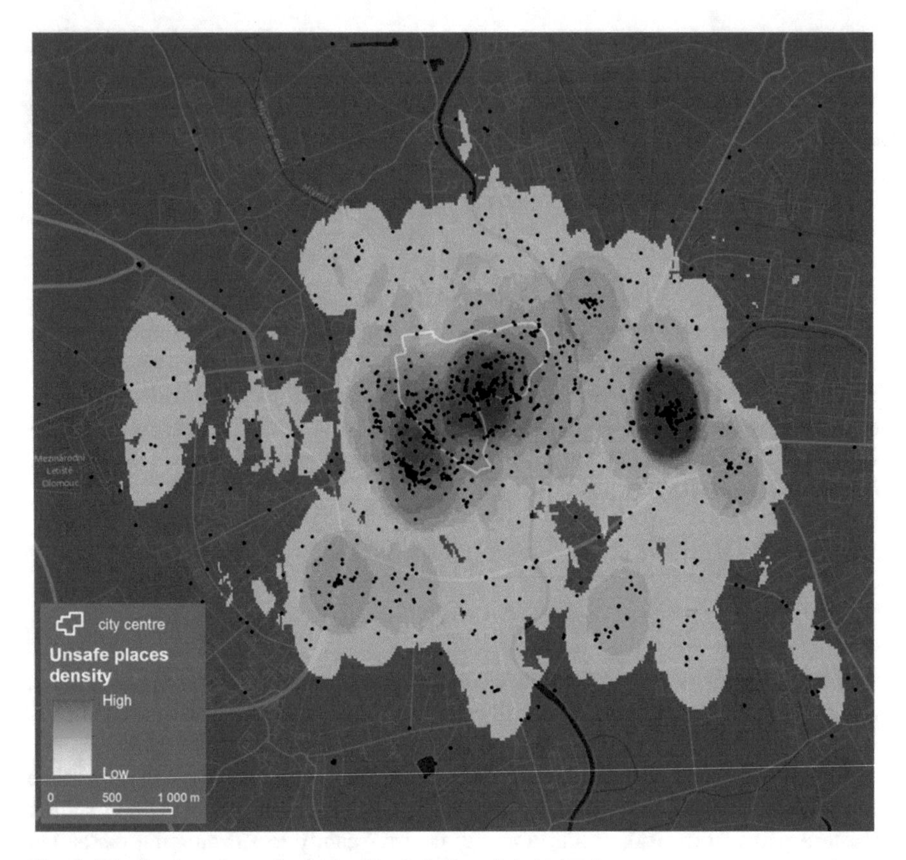

Fig. 3 Distribution of unsafe places (*black dots*) and their density rate

Additionally, unsafe places (marked at least once) cover almost all of the city centre during the night (or periods of darkness). Also localities within large blocks of flats in suburban residential areas are highlighted as unsafe during the night and in darkness. The main reasons respondents feel unsafe are "not enough street lights, dark streets, the homeless, drunk people, bad feelings, drug users, weird people, underpass, weird people".

The main difference between men's and women's unsafe marked places is that men are not as afraid in the city centre as women (Fig. 6). Also parks and underpasses are localities in which men feel less frightened. On the other hand, women did not mark the above-mentioned socially excluded localities as unsafe very often (it is maybe because they rarely visit those localities—according to the questionnaires), whereas men visited them more often. Nevertheless, there were also comments about positive aspects of the city, stating that Olomouc is perceived as much safer than other cities of a comparable size in the Czech Republic.

The visualisation of Spearman's ρ for all pairs of feelings is depicted in Fig. 7. Only positive associations were found between people's evaluations of places, as

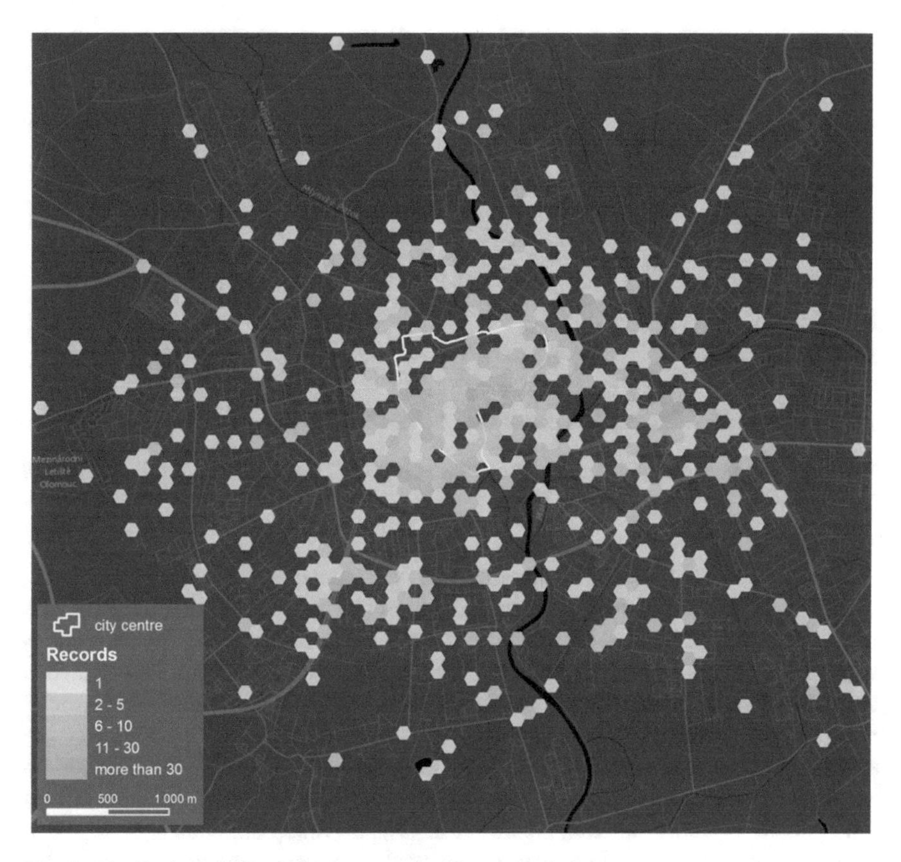

Fig. 4 Distribution of all unsafe places marked by respondents

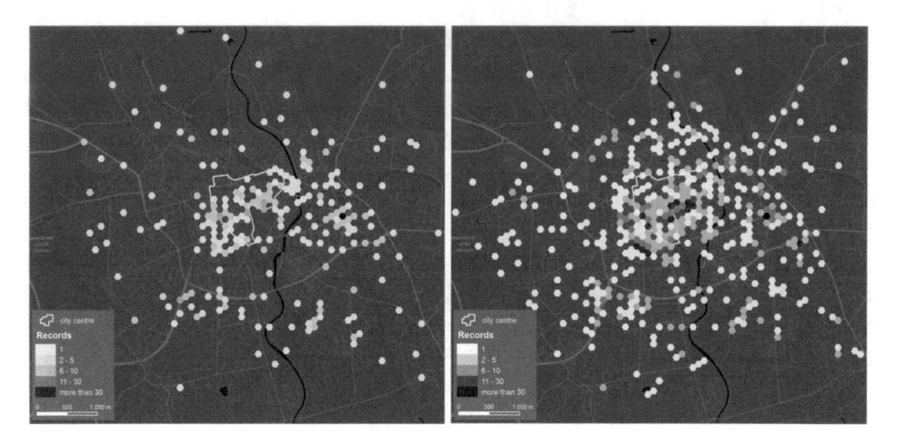

Fig. 5 Distribution of unsafe places by time of day (day on *left*, night on *right*)

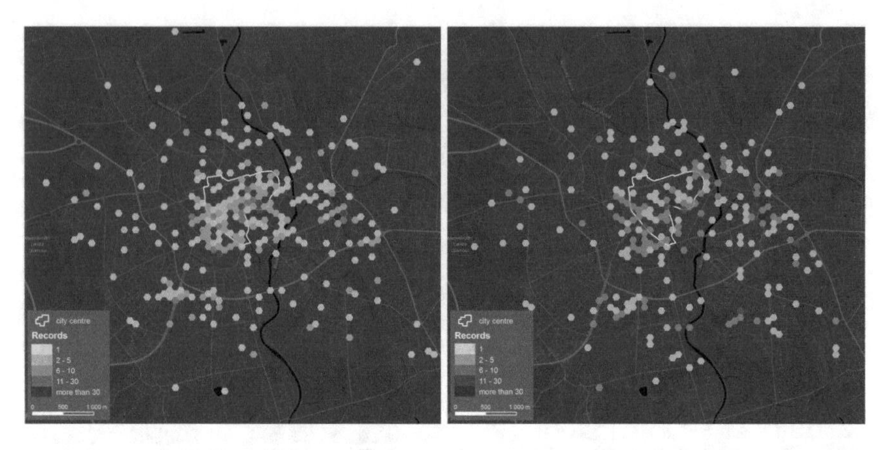

Fig. 6 Distribution of unsafe places by gender (women on *left*, men on *right*). *Note* data from web tool is not included due to technical issues

shown in red. It is important to note that positive associations not only occur among places people voted for more often in both topics (represented by more records), but also among places that were mentioned only rarely. Meaning that correlations express the (local) topical agreement among places. The correlation also depicts a mutual/two-sided relationship. The lighter colour of the area, the weaker associations between feelings exist in the area. Contrarily, the darker the colour of the place, the stronger association is present in the neighbourhood. The areas with the value of the correlation around zero present the places where the associations among characteristics are weak. Individual points (votes) are shown in Fig. 7 in order to provide a more complex overview of the evaluated situation in the results. The information about the frequency of fearful feelings in individual places is depicted by the size of points in Fig. 7.

According to local Spearman's correlation, the strongest association between men's and women's perception of fear and safety can be found in the periphery of the city, and this is caused by the sparse number of votes in these places. The strongest agreements in votes can mainly be found in the three localities (corresponding with previous findings) with the densest localisation of votes; (A) the main train station, (B) city parks, and (C) the old town near the church of St. Michael. In these localities, a strong correlation exists also between the perception of fear during the day and during the night. Moreover, other places were identified as places to be fearful of during the daytime and the night-time—(D) neighbourhood Černá cesta and (E) tř. Svobody.

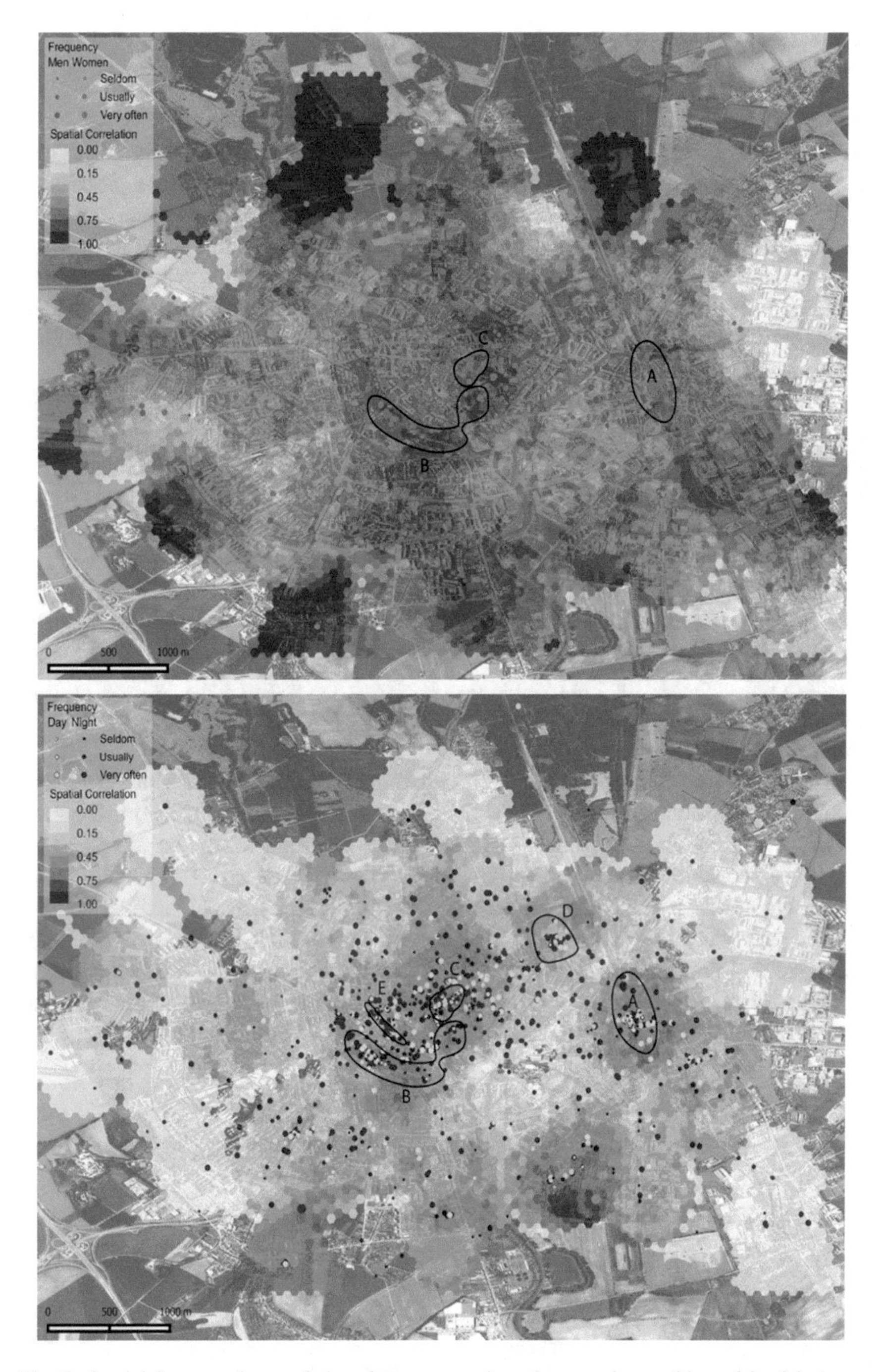

Fig. 7 Spatial Spearman's correlations between men's and women's cognition of fearful places (*top*); and correlations between citizens' perception of fear during the day and during the night (*bottom*)

5 Discussion and Conclusions

Emotional mapping plays an important role in the processes of community planning, participatory development and the creation of knowledge about common spaces and the environment. Emotional maps allow users to get involved in the process of collecting information related to their emotional links with their environment. The idea is grounded in the GeoParticipation approach—using spatial tools in order to involve citizens in community participation and decision-making processes related to public spaces. The outcome of this case-study combines data from analog paper-based, traditional and top-down questionnaires with data collected by digital crowdsourcing and the participatory web-application PocitoveMapy.cz.

The authors analysed the spatial distribution of locations marked as unsafe in Olomouc by gender groups of respondents (women/men), as well as by time of day respondents feeling unsafe in a specific location (daytime/night-time). There were locations that show spatial correlation among these categories. There were also locations that created spatial hot-spots of unsafety. The weakest point in the local correlation analysis is the evaluation of places where votes are seldom placed. The detached places might influence the visual interpretation of spatial associations. Therefore, the visualisation of the results of the analysis and the original votes is suitable, in order to provide a complex overview of the phenomenon. Using quantitative methods for the evaluation of people's sense of the city may help to identify places that should be of interest to local governments and city councils in order to help them manage public spaces in an effective way.

The future outlook of the research can include spatial analysis based on the different age groups of respondents, or based on their relationship with the city. Citizens who live in the city may have different perceptions of the unsafe places to citizens who only commute to Olomouc due to work or study issues. Nevertheless, the first analysis proved that the inclusion of qualitative and subjective data can produce relevant spatial as well as temporal data, which can be used in local planning and community development. The combination of subjective layer with "official" data, for example from the police department, can enhance the information richness of the dataset and reveal new relationships among various datasets.

References

Anselin L, Cohen J, Cook D, Gorr W, Tita G (2000) Spatial analyses of crime. Crim Justice 4 (2):213–262

Barrett LF (2006) Solving the emotion paradox: categorization and the experience of emotion. Personal Soc Psychol Rev 10(1):20–46

Bergner B, Zeile P, Papastefanou G (2011) Emotional barrier-GIS—a new approach to integrate barrier-free planning in urban planning processes. In: Proceedings REAL CORP, pp 247–257. Retrieved from http://realcorp.at/archive/CORP2011_27.pdf

Biever C (2010) Twitter mood maps reveal emotional states of America. New Sci 207(2771):14

Bollen J, Mao H, Zeng X (2011) Twitter mood predicts the stock market. J Comput Sci 2(1):1–8

Brown M, Polk K (1996) Taking fear of crime seriously: the Tasmanian approach to community crime prevention. Crime Delinq 42(3):398–420. doi:10.1177/0011128796042003004

Burian J, Pászto V, Langrová B (2014) Possibilities of the definition of city boundaries in GIS—the case study of a medium-sized city. In: 14th SGEM GeoConference on informatics, geoinformatics and remote sensing, vol 3, pp 777–784. doi:10.5593/SGEM2014/B23/S11.099

Chainey S, Ratcliffe J (2013) GIS and crime mapping. Wiley, New York

Clemente F, Kleiman MB (1977) Fear of crime in the United States: a multivariate analysis. Soc Forces 56(2):519–531. doi:10.1093/sf/56.2.519

Curtis JW (2012) Integrating sketch maps with GIS to explore fear of crime in the urban environment: a review of the past and prospects for the future. Cartogr Geogr Inf Sci 39 (4):175–186. doi:10.1559/15230406394175

Czech Statistical Office (2015) Population of municipalities of the Czech republic, 1 Jan 2015. Retrieved 25 Sept 2015 from, https://www.czso.cz/documents/10180/20556287/1300721503.pdf/33e4d70e-e75f-4596-930c-63406c9068d0?version=1.1

Doran BJ, Burgess MB (2011) Putting fear of crime on the map: investigating perceptions of crime using geographic information systems. Springer, Berlin

Fisher BS, May D (2009) College students' crime-related fears on campus: are fear-provoking cues gendered? J Contemp Crim Justice 25(3):300–321. doi:10.1177/1043986209335013

Gartner G (2012) Putting emotions in maps—the wayfinding example. Mountaincartography.org, pp 61–65. Retrieved from http://www.mountaincartography.org/publications/papers/papers_taurewa_12/papers/mcw2012_sec3_ch08_p061-065_gartner.pdf

Gould P (1986) Mental maps. Taylor & Francis, London

Griffin AL, Mcquoid J (2012) At the intersection of maps and emotion: the challenge of spatially representing experience. Kartographisch Nachrichten 6:291–299

Hauthal E, Burghardt D (2014) Mapping space-related emotions out of user-generated photo metadata considering grammatical issues. Cartogr J 53:78–90

Huang H, Gartner G, Klettner S, Schmidt M (2014) Considering affective responses towards environments for enhancing location based services. ISPRS Int Arch Photogramm Remote Sens Spat Inf Sci 1:93–96

Jíchová J, Temelová J (2012) Kriminalita a její percepce ve vnitřním městě: případová studie pražského Žižkova a Jarova. Geografie 3(117):329–348

Kalogirou S (2011) Testing local versions of correlation coefficients. Jahrbuch Für Regionalwissenschaft 32(1):45–61. doi:10.1007/s10037-011-0061-y

Kalogirou S (2015) lctools: local correlation, spatial inequalities and other tools. Retrieved from http://cran.r-project.org/package=lctools

Kloeckl K, Senn O, Di Lorenzo G, Ratti C (2011) Live singapore!—an urban platform for real-time data to program the city. In: Computers in urban planning and urban management (CUPUM), vol 4, Lake Louise, Alberta, Canada, 5–8 July 2011

Korpela K (2002) Children's environment. In: Robert B. Bechtel, Arza Churchman (eds) Handbook of environmental psychology, Wiley, New York, USA, pp 363–373

Leitner M (2013) Crime modeling and mapping using geospatial technologies, vol 8. Springer, Berlin

Lipscomb S (2014) Visualizing perceived safety in a campus environment. Retrieved 1 Dec 2015, from http://www.imagin.org/awards/sppc/2014/papers/sam_lipscomb.pdf

MacKerron G, Mourato S (2010). Mappiness. Retrieved 12 Sept 2015, from http://www.mappiness.org.uk/

Mislove A, Lehmann S, Ahn Y-Y, Onnela J-P, Rosenquist JN (2010) Pulse of the nation: U.S. mood throughout the day inferred from Twitter. Retrieved 12 Sept 2015, from http://www.ccs.neu.edu/home/amislove/twittermood/?utm_campaign=Facebook+Page&utm_content=Pulse+of+the+Nation+US+Mood+Throughout+the+Day+inferred+from+Twitter&utm_medium=postitat&utm_source=facebook

Mody RN, Willis KS, Kerstein R (2009) WiMo: location-based emotion tagging. In: Proceedings of the 8th international conference on mobile and ubiquitous multimedia, p 14
Němečková M (2014) Komparace vnímání bezpečnosti ve městech Most a Litvínov. Masarykova Univerzita. Retrieved from http://is.muni.cz/th/363798/fss_m/dp_nemeckova.pdf
Nold C (2009) Emotional cartography: technologies of the self. http://WWW.EMOTIONALCARTOGRAPHY.NET. Retrieved from http://emotionalcartography.net/EmotionalCartography.pdf
Oc T, Tiesdell S (1997) Safer city centres: reviving the public realm. Sage, Beverley Hills
Olomouc Region (2015) Prevence kriminality. Retrieved 17 Dec 2015, from https://www.kr-olomoucky.cz/prevence-kriminality-cl-292.html
Pearce MW (2008) Framing the days: place and narrative in cartography. Cartogr Geogr Inf Sci 35 (1):17–32
Pickles J (1995) Ground truth: the social implications of geographic information systems, 1st edn. Guilford Press, New York
Raslan R, Al-hagla K, Bakr A (2014) Integration of emotional behavioural layer "EmoBeL" in city planning. In: REAL CORP 2014, vol 8, pp 309–317
Reeve J (2014) Understanding motivation and emotion. Wiley, New York
Reimann C, Filzmoser P, Garrett RG, Dutter R (2008) Statistical data analysis explained: applied environmental statistics with R. Wiley, New York
Russell JA (1980) A circumplex model of affect. J Pers Soc Psychol 39(6):1161
Santos RB (2012) Crime analysis with crime mapping. Sage, Beverley Hills
Sessar K, Sirotek J (2001) Společnost v období transformace a strach z kriminality. Sociologický Časopis 37(1):7–22
Smith M, Bondi L, Davidson J (2012) Emotional geographies. Ashgate Publishing Ltd., London
Stasíková L (2011) Relevantnost výskumu strachu z kriminality v urbánnej geografii. Geografický Časopis 63(4):325–343
Wilson MW (2011) "Training the eye": formation of the geocoding subject. Soc Cult Geogr 12 (04):357–376

Models for Relocation of Emergency Medical Stations

Ľudmila Jánošíková, Peter Jankovič and Peter Márton

Abstract The paper presents a mathematical programming model for the deployment of stations where the ambulances are kept. The model deals with the reliability of service. Reliability in emergency systems is based on the individual level of service at each demand point. The system operates, if there is an available ambulance that can respond immediately to each emergency call. However, the probability that vehicles are available can hardly be determined before the stations are located. That is why an upper bound on the failure probability is used in presented location model instead of the probability itself. This upper bound is calculated using a Poisson distribution of arriving calls and the maximum time duration of a vehicle service trip. The model maximizes an average reliability that can be achieved on a given territory with a limited number of vehicles. Such a model may be useful when the issue is a reorganization of the current emergency system. A computer simulation model was used to verify vehicle location provided by the mathematical model. Both models were implemented under the conditions of the Slovak Republic. Travel times were calculated using the transportation network data from the OpenStreetMap project. The paper compares the proposed vehicle location with the current one in terms of the objective function and performance characteristics resulting from computer simulation.

Keywords Emergency medical service · Service reliability · Vehicle location · Integer programming · Computer simulation

Ľ. Jánošíková (✉) · P. Jankovič · P. Márton
Department of Mathematical Methods and Operations Research,
Faculty of Management Science and Informatics, University of Žilina,
Univerzitná 1, 010 26 Žilina, Slovakia
e-mail: Ludmila.Janosikova@fri.uniza.sk

P. Jankovič
e-mail: Peter.Jankovic@fri.uniza.sk

P. Márton
e-mail: Peter.Marton@fri.uniza.sk

I. Ivan et al. (eds.), *The Rise of Big Spatial Data*, Lecture Notes
in Geoinformation and Cartography, DOI 10.1007/978-3-319-45123-7_17

1 Introduction

The quality and efficiency of the emergency medical service (EMS) depends mainly on:

- the number of ambulances operating in a given region and
- the deployment of stations where the ambulances are kept.

Specifying the proper number of ambulances is a sensitive issue balancing between two opposing requirements. On one hand, the main task of the EMS—to save lives and reduce human suffering caused by injuries or illnesses—can be fulfilled if the network of emergency stations is sufficiently dense so that each emergency call could receive an immediate response and first aid could be provided in a very short time. On the other hand, there is a justified requirement on the efficiency of public expenditures that reduces the number of ambulances. Reducing the number of ambulances increases their workload and results in a lower availability of urgent health care since the nearest ambulance may be busy at the moment when a new call arrives. The situation when the patient does not receive medical care in a predefined time limit is interpreted as the system failure (Ball and Lin 1993).

This paper concerns the strategic level of the urgent health care management and aims at determining the optimal location and sizing of emergency stations so that the quality of service is as good as possible. Clearly, the main quality output of the EMS production is response time. However, due to the stochastic character of the system, it cannot be modelled directly and various approximation criteria are used instead. We measure the efficiency of the EMS by its reliability, which is defined as the probability that the service will be provided when and where necessary.

The problem of station location and determining the number of vehicles to allocate to each station is denoted as the EMS system design problem (Noyan 2010). The research presented here deals with the existing EMS system and looks for a better station location preserving the current number of vehicles. Therefore the problem to be solved can be expressed by the following question: Given the number of vehicles, where they should be located so that the average reliability of service over all demand nodes could be as high as possible?

We formulate the problem as an optimisation problem and solve it using a mathematical programming approach. The intention is to devise such a model that would be applicable in a wide-spread region containing both urban and rural areas. We have in mind a state territory. The motivation stems from the situation in the Slovak Republic where ambulances are kept in single-purpose stations. Every station houses only one ambulance. The number and locations of stations on the whole state territory are defined by the Regulations of the Ministry of Health of the Slovak Republic No. 10552/2009-OL and 11378/2010-OL. In accordance with the Regulations, 273 stations are currently deployed.

Because stations are supposed to be deployed in a large-scale area, a macroscopic view must be applied. It means that demands are aggregated and customers of the system are not individual patients but whole municipalities. In accordance with Felder and Brinkman (2002), we suppose that the number of calls arising in a municipality is proportional to the number of its inhabitants and so we identify the size of the demand with the municipality population. At the same time, every municipality is regarded as a candidate location for a station. To simplify the model, we allow to place at most one station in a municipality and to allocate multiple vehicles to a station. Such a situation could be resolved in practice so that multiple stations equipped with one ambulance would be dispersed over the municipal area.

2 Literature Review

Ambulance location models are surveyed in Brotcorne et al. (2003). In this review paper, the models are classified into two main categories: deterministic and probabilistic models.

Deterministic models of the EMS systems are based on an implicit assumption that there is always an ambulance available to respond to a call. They can be further divided into covering and allocation models. In covering models, a maximum value is preset for either distance or travel time. If a service is provided by a facility located within this limit, then the service is considered acceptable, and a customer is considered covered by the service, if he or she has a facility sited within the preset distance or time. In the allocation models, demand zones are assigned to ambulance locations. One of the possible models of this type is the p-median model that minimizes the average travel time from the ambulance locations to potential patients. The objective function is a surrogate of the efficiency criterion which aims at providing the best possible level of service to as many people as possible with a limited number of resources. Besides efficiency, equity (or fairness) is a core performance dimension in a health care system (see e.g. Kvet and Janáček 2014; Janáček and Kvet 2015). Fairness is achieved when each customer receives the service of required and/or acceptable quality. Although an equity principle is usually favoured in health care provision, study by Felder and Brinkmann (2002) proves that efficiency maximizes the number of lives saved and is often applied by policy makers.

Probabilistic models take into account the reliability of service provision that may be affected by uncertainty present in a real system. Arriving calls are stochastic events, and treating a patient is also a random variable. Moreover, the travel time of an ambulance may be affected by the traffic and weather conditions. Therefore, the nearest ambulance may happen to be busy when an accident occurs. Then another ambulance must be dispatched to serve the call, or the service must be postponed. Probabilistic models incorporate so called busy fraction of ambulances, which is the probability that the ambulance will be busy when a new call arrives and it will not be able to respond to the call. For a thorough discussion of the probabilistic models see, for example, a chapter by ReVelle and Williams (2004). Sorenses and Church (2010) distinguish three generations of probabilistic models. First-generation probabilistic models suppose that all ambulances are equally busy. Models are variants of the famous Maximum Expected Coverage Problem (MEXCLP) formulation proposed by Daskin (1983). The Maximum Availability Location Problem (MALP), introduced by ReVelle and Hogan (1989) represents the second generation of probabilistic models. These models abandon the assumption of the uniform distribution of workload among ambulances and take into account the workload of individual ambulances. The probability of an ambulance being busy is calculated as the ratio of the total demand (in hours) within the local service area surrounding the server and the available service time of all servers located within the same area. The number of calls that can be responded to with a predefined level of reliability is maximized. Models of the third generation [represented by the Queueing-based Maximum Availability Location Problem (Marianov and ReVelle 1996)] maintain the concept of individual busy-fractions but estimate them using the queueing theory.

Common drawbacks of these models are simplifying assumptions that are necessary to approximate probabilistic behaviour within linear equations. Some of the more significant assumptions include (Sorensen and Church 2010):

1. Server independence meaning that the probability of one ambulance being available to handle a call is independent on the status of other ambulances in the region.
2. Locally-constrained service areas meaning that ambulances within a local area only serve calls within the same area and vice versa demand nodes located within the local area are not served by ambulances from outside the area.
3. Location-independent service times meaning that service times do not depend on the distances between stations and call locations. This assumption prevents the application of the mentioned models in a large-scale network, where distances between stations and incident sites differ significantly.
4. Uniform service times that do not distinguish categories of patients, severity of their diagnosis nor a possible need of their transport to a hospital. Obviously individual demands cannot be distinguished in the system design, however if the models are based on an average service time, they need not give realistic outputs.

Exact service times are not included in the reliability model introduced by Ball and Lin (1993), just the upper limit of the service time inputs the model and is used to calculation of the upper bound on the failure probability. This upper bound is based on the assumption that nobody outside the service area can help to serve demands within the area. So the upper bound remains in force also in the case of overlapping areas where the servers can help each other. If locally-constrained service areas do not represent an obstacle to the applicability of the reliability upper bound, neither does the server independence assumption. The consequence of the first and second simplifications is that the model takes into account a more pessimistic failure probability. However, the model does not need other parameters estimation of which would be difficult, and that is why we took it as a basis for our vehicle relocation model. Although Ball and Lin's model is often classified only as a variant of the MALP, we think that they propose a principally new approach. They do not calculate the probability of an ambulance's unavailability as a ratio of demanded and offered service times but as the probability that the number of calls arising in the service areas of the servers located in the coverage area of the given demand point exceeds the number of available ambulances. Ball and Lin solve the Reliability Location Set Covering Problem that minimizes the number of vehicles needed to ensure a desired level of reliability for each demand node. The model was further elaborated by Borrás and Pastor (2002) who regarded the EMS system as a system with losses of calls in case of congestion. The reliability constraints proposed by Ball and Lin were also used by Beraldi and Bruni (2009) to improve the two-stage EMS model that incorporates first-stage strategic decisions about location and capacities of the EMS stations and second-stage tactical decisions concerning assignment of demand nodes to stations.

3 Mathematical Programming Model

In this section we describe a mathematical programming model that aims at finding a new location of the current number of ambulances. The model is based on the definition of the failure probability introduced by Ball and Lin (1993).

It is supposed that potential station locations have been predetermined due to some previous studies. Let us denote the set of location candidates by symbol I. Its elements are indexed by symbol i. Potential patients live in municipalities spatially spread in the given territory. We denote the set of municipalities (demand points) by J and the population of municipality j by b_j. Members of both sets I and J correspond to nodes of a road network. The network segments are associated with the travel times that depend on the length of the segment, category of the road and average speed of an ambulance. Based on the segment travel times, the shortest travel time t_{ij} from node $i \in I$ to node $j \in J$ can be calculated using, for example, the Dijkstra's algorithm. The travel times are needed to develop a set of candidate locations that cover each municipality j, it means, that are within the specified response time T^{max} of municipality j. We denote this set by N_j and define $N_j = \{i \in I: t_{ij} \leq T^{max}\}$. This specification of the problem shows that we face a network location problem (Daskin 2013).

The acceptable level of service is ensured by imposing an upper limit on the system failure for each demand point. The system fails if there are no feasible vehicles available for the service when a demand call occurs. The failure probability should be less that $1 - \alpha$, where α is a minimum level of reliability. Due to the difficulty of computing the exact failure probability, Ball and Lin developed an upper bound on it. A patient cannot be served immediately if there is no available vehicle in covering stations at time t when the emergency call arrives. This situation occurs if all vehicles have been servicing other demand calls. To calculate the probability that all feasible vehicles are busy at time t we need to know the duration of service trips. Since an upper bound on the failure probability is the issue, a time amount T equal to an upper bound on service time may be used. The probability that the number of calls that have occurred in the previous interval of size T in the service area of station i exceeds the number of available vehicles can be denoted as $\Pr(D_i \geq K_i)$, where D_i is the number of calls that have occurred in the service area of station i during time interval $\langle t - T, t \rangle$ and K_i is the number of vehicles housed at station i. If calls arise from each demand point according to the Poisson distribution, then the upper limit of the system failure for demand point j is defined by the term on the left hand side of the following relation:

$$\prod_{i \in N_j} \Pr(D_i \geq K_i) \leq 1 - \alpha \tag{1}$$

Although the product form suggests that service areas of the stations do not overlap, it still remains an upper bound in a more realistic case of overlapping areas.

In the location model, the number of vehicles housed at each station is a decision variable that can be defined as follows.

$$x_{ik} = \begin{cases} 1 & \text{if } k \text{ vehicles are stationed at location } i, \\ 0 & \text{otherwise,} \end{cases}$$

for $i \in I$ and $k = 1, 2, \ldots, M_i$, where M_i is the maximum number of vehicles that can be located at i. Using these variables the restriction (1) can be written as

$$\prod_{i \in N_j} \prod_{k=1}^{M_i} [\Pr(D_i \geq k)]^{x_{ik}} \leq 1 - \alpha \tag{2}$$

By taking the logarithm function on both sides and changing the signs, (2) can be transformed into a linear constraint:

$$\sum_{i \in N_j} \sum_{k=1}^{M_i} a_{ik} x_{ik} \geq -\log(1 - \alpha) \quad \text{for } j \in J \tag{3}$$

where $a_{ik} = -\log(\Pr(D_i \geq k))$.

The optimization criterion will be the weighted average of failure probabilities over all municipalities where the weight is the population of municipality j. Let α_j be the unknown reliability in municipality j. To simplify the notation we introduce a new variable y_j that substitutes for $-\log(1 - \alpha_j)$. It represents an upper bound on the failure probability in municipality j on the logarithmic scale.

The complete integer programming formulation of the model is as follows:

$$\text{maximize} \quad \frac{1}{\sum_{j \in J} b_j} \sum_{j \in J} b_j y_j \tag{4}$$

$$\text{subject to} \quad \sum_{k=1}^{M_i} x_{ik} \leq 1 \quad \text{for } i \in I \tag{5}$$

$$\sum_{i \in N_j} \sum_{k=1}^{M_i} a_{ik} x_{ik} \geq y_j \quad \text{for } j \in J \tag{6}$$

$$\sum_{i \in I} \sum_{k=1}^{M_i} k x_{ik} \leq p \tag{7}$$

$$y_j \geq -\log(1-\alpha) \quad \text{for } j \in J \tag{8}$$

$$x_{ik} \in \{0,1\} \quad \text{for } i \in I, k = 1, 2, \ldots, M_i \tag{9}$$

Constraints (5) state that each location i either is not chosen or houses exactly one number of k vehicles. Constraints (6) bound new variables $\boldsymbol{y}$ with location variables $\boldsymbol{x}$. Constraint (7) ensures that the number of allocated vehicles does not exceed a predefined number p. Constraints (8) are lower bounds on variables $\boldsymbol{y}$ ensuring a minimum level of reliability α at each demand node j. Constraints (9) are obligatory constraints on bivalent variables.

To calculate the average failure probability we must use the exponentiation to return the computation to the original scale. Finally instead of failure probability we can express the average lower bound on reliability $\bar{\alpha}$ per inhabitant using the following formula:

$$\bar{\alpha} = 1 - \exp\left(-\frac{1}{\sum_{j \in J} b_j} \sum_{j \in J} b_j y_j^*\right) \tag{10}$$

where y_j^* are optimal values of variables $\boldsymbol{y}$ in the solution of the model (4)–(9).

The average lower bound on reliability per municipality is:

$$\bar{\alpha} = 1 - \exp\left(-\frac{1}{n} \sum_{j \in J} y_j^*\right) \tag{11}$$

where n is the number of municipalities ($n = |J|$).

The fact that a station houses multiple vehicles can be interpreted in such a way that multiple stations equipped with one ambulance are dispersed over the town area. We will use this interpretation in the following sections describing a case study of the Slovak Republic. Moreover we will use a simplified term reliability instead of a lower bound on reliability.

4 Computer Simulation Model

For the verification of the mathematical model we created a microscopic simulation model. The simulation allows testing many variants of the location of stations. The agent based simulation model simulates all processes of emergency medical service. It was built in AnyLogic simulation software (Borschev 2013; Grigoryev 2014). AnyLogic is developed on Java simulation core. We implemented a library of classes and functions in Java for the simulation support. The main advantage of our simulation model is a very precise simulation of moving the ambulance. The OpenStreetMap (OSM) data (https://www.openstreetmap.org) are used for creating the transportation network (explained in Sect. 5).

During the simulation run, the shortest path in terms of travel time is calculated for every ambulance trip. The travel time through a road segment depends on the length of the segment and the speed of the ambulance. The OSM data distinguish six road categories and for every category we define a speed limit. The ambulance can move in two modes: when it travels to a patient or transports the patient to a hospital, then it drives at a higher speed as usually. We also take into account maximal speed limits from the OSM data file. The result of this solution is a very realistic model of the ambulance movement.

For modeling a patient position we apply a sophisticated strategy using data from the LandScan database (http://web.ornl.gov/sci/landscan). LandScan data represent an ambient population (average over 24 h) over the whole state territory. The territory of the Slovak Republic is covered by 70,324 grid elements.

A special population tree is built from the population grid. It helps to select one node of the transportation network as a source of the emergency call. The proposed population tree has six levels. From leaves (bottom) to the root (top) there are levels with OSM nodes, population grids, municipalities, districts, regions and the state (the root of the tree). A weight is associated with every node of the tree. The OSM nodes at the bottom level have zero weight. They are assigned to population grids according to their geographic position. The weight of a population grid is set to the number of people living in the grid (the value of the corresponding element in the LandScan database). On the next level we assign population grids to municipalities also according to a geographic position. The weight of a tree node representing a municipality is a sum of the weights of the child nodes. Using the same algorithm we build levels with districts and regions. When we generate a patient position, we traverse from the root of the tree to the leaves. On every level we need to decide which child of the current node should be used. We generate a random number from the discrete uniform probability distribution (from zero to the weight of the current node) which is interpreted as a cumulative weight of some of its children. This way a child is chosen randomly but with regard to the population density. On the bottom level we take a random leaf. This strategy for location of patients in the space is very realistic.

For every simulation model the maximal simulation speed is very important. When the simulation run is slow, it is difficult to take enough data from it. Our model works with big data files containing population distribution, transportation network and borders (municipalities, regions). We maximize the speed of simulation by using our special algorithms and data collectors.

For example for searching the shortest path on the network we performed many experiments with variations of the Dijkstra's algorithm. We tested several data structures for storing the nodes and we measured their performance during the simulation run. The outcomes from the profiler for 50,000 calculations of the shortest path are presented in Table 1. In the first three rows there are computation times for the implementations where references to nodes are stored in HashMap, ArrayList, and LinkedList data structures, respectively. Here all nodes were processed during the initialization phase of the algorithm. However, the initialization of all nodes is time consuming. Therefore we proposed an efficient implementation, where BitSet is used for storing information about the node initialization. Every bit in BitSet represents a boolean value indicating, whether the corresponding node has already been initialized. The BitSet is cleared at the beginning. It is extremely fast operation. When the Dijkstra's algorithm encounters a node, it checks in the BitSet, whether the node is initialized. If not, the algorithm initializes it and sets the corresponding bit in the BitSet. This solution allows to initialize only visited nodes. In the last row of the table the computation time for the combination of the BitSet and HashMap is reported.

From the simulation we take a lot of statistical information:

- reliability for cities, regions, whole state,
- statistics about response time, ambulance workload, duration of the ambulance move (all for cities, regions, whole state),
- statistics about utilization of rescue stations,
- many statistics for hospitals and rescue stations.

The simulation model was validated using actual positions of rescue stations. On the basis of published data about emergency medical service in the Slovak Republic

Table 1 Total time spent for searching the shortest path

Data structure	Total time (ms)
HashMap	382,626
ArrayList	325,839
LinkedList	247,268
BitSet + HashMap	230,304

we can compare results from the model with the real data. Animation during the simulation allows the user to verify behaviour of the ambulance. The user can see also many statistics, states and positions of ambulances, patients etc. Our model is fully valid and it produces relevant data. The results from the simulation model are more relevant than from the optimization model, thanks modelling occupation of the ambulance.

5 OpenStreetMap

The OpenStreetMap is a service, which allows users all over the world to create their own maps. All maps are editable and users can use these geographical data for their own purposes (data are under Open Database License). The OpenStreetMap data are exported in xml (extensible markup language) format. The basic components of the OSM are nodes (defining points in space), ways (defining linear features and area boundaries), and relations (which are sometimes used to explain how other elements work together). We use nodes and ways for the simulation model. A node represents a specific point on the earth surface defined by its latitude and longitude. Each node comprises at least an id number and a pair of coordinates. A way is an ordered list of nodes that define a polyline. Ways are used to represent linear features such as roads.

The main reasons for the use of the OSM data are:

- open and free data with high precision,
- the AnyLogic software supports the OSM infrastructure,
- the database contains a lot of supplementary information (type of the road, maximal speed, one-way direction etc.).

From the OSM we can take also information about borders of municipalities, regions etc. Before the start of the simulation run, the program loads all data into internal data structures. The whole transportation network saved in the OSM data file includes 38,082 ways. To build the model of the transportation network, we transform the ways to the edges. The model is then a graph with 359,947 nodes and 364,862 edges.

Map data from OSM for the Slovak Republic contain some errors and we had to correct them. The main correction consisted in creating one continuous graph representing the transportation network. Some roads are not connected properly. If such a situation occurred, the simulation model could not find the route between two nodes. So we checked the data and created a continuous graph. The second problem is that some roads have a wrong one-way direction setting. These data cause an error, that the algorithm can't find the route to the village. If the shortest path is not found, the algorithm ignores one-way direction on this critical section.

6 Case Study

The mathematical programming model (4)–(9) was implemented by using the real data of the Slovak transportation network. The *Xpress* (2011) optimisation tool was used for optimisation.

Currently 273 stations are deployed in the area of the Slovak Republic. They are located in 211 towns and villages including urban districts of the capital Bratislava and the second largest Slovak town Košice. They altogether serve 2,926 municipalities populated with 5,410,827 inhabitants. Large towns have multiple stations. The most stations (5) are in Bratislava—Petržalka, an urban district of Bratislava, and in Prešov, a highly populated town in the eastern Slovakia.

The mathematical model preserves the total number of stations and looks for a better location of them in terms of reliability. The coefficients in the model were set in the following way:

Potential locations (candidates) for stations consist of all municipalities and all the nodes with existing EMS stations defined by the official regulations.

To create the set N_j of candidate locations that are in the neighbourhood of municipality j, a requested response time T^{max} must be defined. The Slovak legislation does not specify performance standards, there is only a recommendation to deliver medical care in 15 min from when a call for rescue arrives. If we neglect the time taken by a dispatcher to handle a call, T^{max} may be set to 15 min.

To calculate coefficients a_{ik} we need to know the maximum service time T and the arrival rate of calls in municipalities located in the service area of a candidate i. On the basis of the news and published analysis on the EMS performance in Slovakia, we consider $T = 100$ min as a reasonable estimation of the maximum time when an ambulance is busy. The number of calls that have occurred in the previous interval of size T in a municipality j is calculated using the Poisson distribution with arrival rate λ_j (calls per T minutes). The statistics describing the number of calls with particular municipalities were not available to us. Therefore, we suppose that the number of calls in a municipality is proportional to the number of its inhabitants. We stem from the aggregated statistics for the Slovak Republic mapping the year 2013, reporting that the overall number of patients was 465,076. We calculate the rate λ per one inhabitant, and use it to estimate $\lambda_j = \lambda b_j$, where b_j is the number of inhabitants registered in the municipality j.

The maximum number of vehicles that can be located at candidate i takes into account the population of the candidate and the size of the service area of one ambulance. According to the analysis of the EMS system (Bahelka 2008) the capacity limit of one ambulance is 25,000 inhabitants. It follows that $M_i = \lceil b_i/25000 \rceil$.

The remaining coefficients are $p = 273$ and the minimum reliability $\alpha = 0.3$.

A new location of 273 stations was calculated using the model (4)–(9). Currently there are 360 villages whose reliability is 0 because there is no station in their

15 min vicinity. In total 472 municipalities have reliability under the threshold value 0.3. The average reliability per municipality is 0.776, average reliability per inhabitant is 0.772. The model calculates the average reliability per municipality of 0.779 and the average reliability per inhabitant of 0.754. The model ensures that a minimum reliability of 0.3 will be maintained in every municipality. There are only 32 municipalities with this minimum reliability. The expected reliability of the particular municipalities is calculated by using optimal values of variables $\boldsymbol{x}$ in the left hand side of the inequality (3). Expected reliabilities are visualized on the map of the Slovak Republic in Fig. 1 for the current and proposed location of the EMS stations, respectively.

We used computer simulation to estimate performance characteristics of the proposed system more realistically. Out of the performance measures in the emergency systems we are most interested in those reported in Table 2. Reliability is calculated as the percentage of calls responded to within 15 min. Simulated reliability for the current and proposed location of the EMS stations is in Fig. 2.

The simulation reveals that the proposed mathematical programming model does not outperform the current design of the EMS system in the average response time, the number of municipalities with the average response time greater than 15 min,

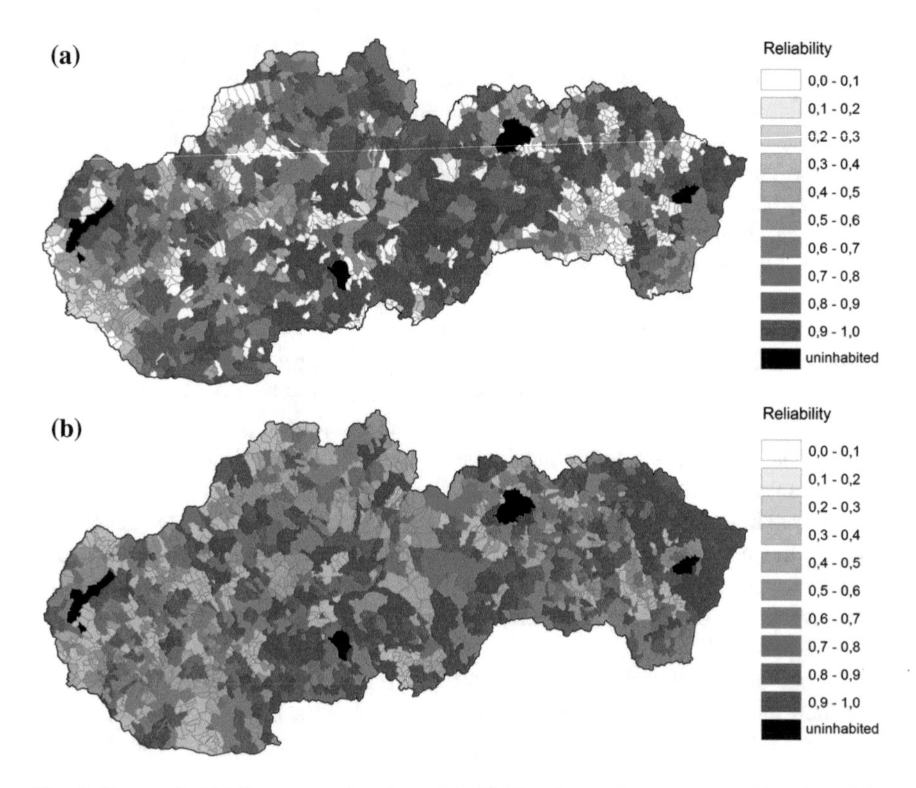

Fig. 1 Expected reliability: current location of the EMS stations (*above*), proposed location of the EMS stations (*below*)

Table 2 Performance characteristics from simulation

Output	Current location	Proposed location
Average response time	10.50	12.58
Number of municipalities with average response time >15 min	422	473
Average reliability	0.785	0.695
Number of municipalities with reliability <0.3	769	696

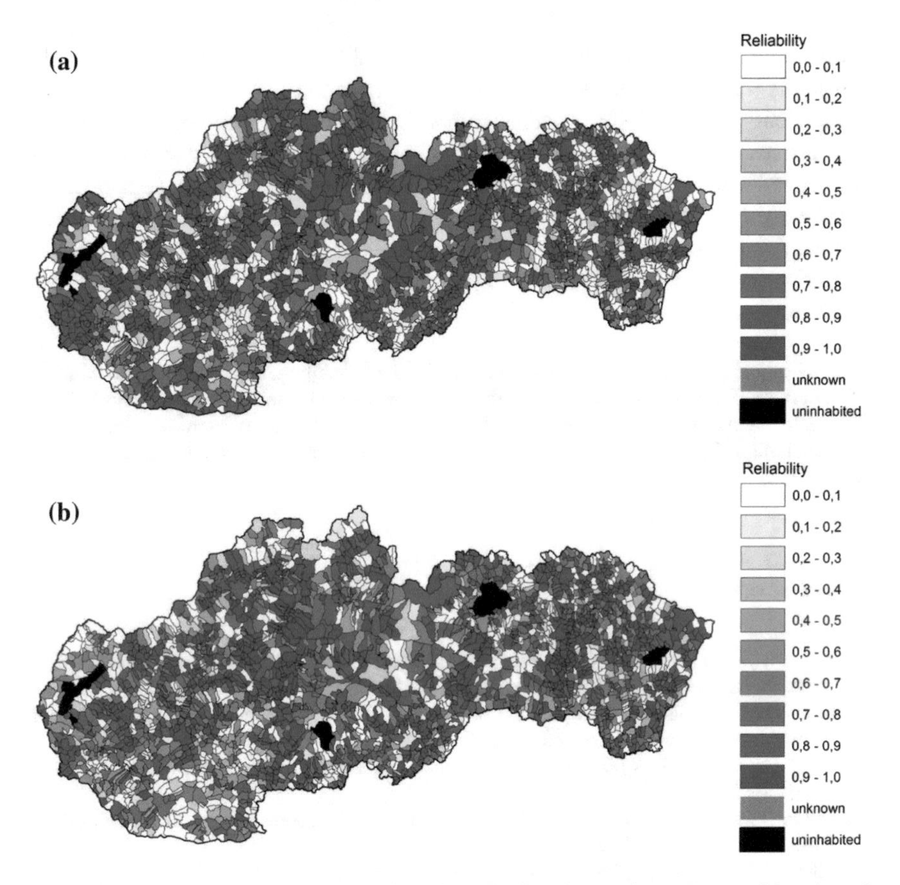

Fig. 2 Simulated reliability: current location of the EMS stations (*above*), Proposed location of the EMS stations (*below*)

neither in average reliability. However the model accentuates fairness as can be seen from the figures. It ensures that the reliability in more villages will be above the minimum level of 0.3.

7 Conclusions

In the paper, the problem of EMS ambulance location is formulated as a probabilistic problem that maximizes average reliability of the service over the state territory. We suggest that probabilistic models are more suitable than deterministic models in public service systems because they enable to model to some extent the uncertainty in the service provision. However in a real system there still exist many sources of the uncertainty that cannot be captured in a mathematical model. In such a situation a computer simulation model plays an undoubtedly significant role in the estimation of the performance characteristics of the system. The simulation model itself is not able to propose the best station location, however it is useful in the sensitivity analysis to answer the "what if" questions. It means it enables to experiment with changing system parameters such as the number of available ambulances in the proposed stations, demographic characteristics that influence the rate of calls in some regions and so on. Both models can serve as a decision supporting tool for an authority responsible for the emergency medical system design. A crucial issue that limits their usage is the input data relevant to the system operation. Reliable input data may improve the calibration of the models and their validation.

Acknowledgments This research was supported by the Scientific Grant Agency of the Ministry of Education of the Slovak Republic and the Slovak Academy of Sciences under Project VEGA 1/0518/15 "Resilient rescue systems with uncertain accessibility of service" and by the Project ITMS 26220120050 "Centre of excellence for systems and services of intelligent transport II".

References

Bahelka M (2008) Analysis of the emergency medical system after reform (Analýza systému záchrannej zdravotnej služby po reforme) [in Slovak] [online]. Health Policy Institute, Bratislava. http://www.hpi.sk/hpi/sk/view/3795/analyza-systemu-zachrannej-zdravotnej-sluzby-po-reforme.html. Accessed 3 Mar 2014

Ball MO, Lin FL (1993) A reliability model applied to emergency service vehicle location. Oper Res 41:18–36

Beraldi P, Bruni ME (2009) A probabilistic model applied to emergency service vehicle location. Eur J Oper Res 196:323–331

Borschev A (2013) The big book of simulation, USA. 978-0-9895731-7-7

Brotcorne L, Laporte G, Semet F (2003) Ambulance location and relocation models. Eur J Oper Res 147:451–463

Borrás F, Pastor JT (2002) The ex-post evaluation of the minimum local reliability level: an enhanced probabilistic location set covering model. Ann Oper Res 111:51–74

Daskin M (1983) A maximum expected covering location model: formulation, properties, and heuristic solution. Transp Sci 17:48–70

Daskin M (2013) Network and discrete location: models, algorithms, and applications. Wileys, Hoboken

Felder S, Brinkmann H (2002) Spatial allocation of emergency medical services: minimising the death rate or providing equal access? Reg Sci Urban Econ 32:27–45

FICOTM Xpress Optimization Suite [online]. http://www.fico.com/en/products/fico-xpress-optimization-suite. Accessed 10 Oct 2011

Grigoryev I (2014) AnyLogic 7 in three days: a quick course in simulation modeling [online]. http://www.anylogic.com/free-simulation-book-and-modeling-tutorials. Accessed 2 Nov 2015

Janáček J, Kvet M (2015) Min-max optimization of emergency service system by exposing constraints. Komunikacie 17:15–22

Kvet M, Janáček J (2014) Price of fairness in public service system design. In: Proceedings of the 32nd international conference on mathematical methods in economics, Olomouc, Czech Republic, Sept 10–12, pp 554–559

Marianov V, ReVelle C (1996) The queueing maximal availability location problem: a model for the siting of emergency vehicles. Eur J Oper Res 93:110–120

Noyan N (2010) Alternate risk measures for emergency medical service system design. Ann Oper Res 181:559–589

ReVelle C, Hogan K (1989) The maximum availability location problem. Transp Sci 23:195–200

ReVelle C, Williams JC (2004) Reserve design and facility sitting. In: Drezner Z, Hamacher HW (eds) Facility location: applications and theory. Springer, Berlin

Sorensen P, Church R (2010) Integrating expected coverage and local reliability for emergency medical service location problems. Soc Econ Plan Sci 44:8–18

Spatio-Temporal Variation of Accessibility by Public Transport—The Equity Perspective

Marcin Stępniak and Sławomir Goliszek

Abstract The growth of large, open datasets coupled with an acceleration of technical developments, including GIS solutions, opens the door to new challenges in transport research. One of the emerging fields of research is the temporal dynamics of accessibility. The increase in availability of General Transit Feed Specification (GTFS) data permits the inclusion of very detailed, schedule-based travel time information. In the study presented we focus on the spatial and temporal variation in accessibility by public transport in the city of Szczecin (Poland). This paper advocates the necessity of incorporating a temporal component in accessibility analysis. We conducted a full day analysis for 1 day using averaged 15-min-long time periods at a very detailed spatial scale (enumeration districts). Based on the calculated origin-destination matrix in 96 time-profiles we calculated the potential accessibility indicator. Then we investigated spatial disparities and their variability during the day-long observation. Apart from the well-known spatial disparities in accessibility level, our findings underline the uncertainty of the accessibility pattern. Moreover, the results show that less accessible areas are also more affected by the daily variation in accessibility level. The findings provide a more realistic insight into accessibility patterns which will be useful for transport planners and policy makers.

Keywords Accessibility · Public transport · Open data · GTFS · Spatial and temporal analysis · Szczecin

M. Stępniak (✉) · S. Goliszek
Institute of Geography and Spatial Organization, Polish Academy of Sciences,
ul. Twarda 51/55, 00-818 Warsaw, Poland
e-mail: stepniak@twarda.pan.pl

S. Goliszek
e-mail: sgoliszek@twarda.pan.pl

I. Ivan et al. (eds.), *The Rise of Big Spatial Data*, Lecture Notes
in Geoinformation and Cartography, DOI 10.1007/978-3-319-45123-7_18

1 Introduction

Accessibility plays a decisive role in shaping our contemporary cities and regions. Its evaluation has become a key issue in spatial planning and modelling. For several decades urban planners, transportation experts and researchers use accessibility in order to examine transport efficiency (Benenson et al. 2011; O'Sullivan et al. 2000), risk of social exclusion (Church et al. 2000; Lucas 2011) and the potential for equitable transport systems (Delmelle and Casas 2012; Foth et al. 2013) or the assessment of new transport investments (Golub and Martens 2014; Manaugh and El-Geneidy 2012; Martens and Hurvitz 2011).

The emergence of new, large datasets and progress in computational capacities offers new opportunities, and enables researchers to increase the temporal and spatial resolution of accessibility analyses. The rapid development of open databases has provided valuable open data sources which are critically useful, especially for transport analyses. For example the OpenStreetMap database collects a huge number of gigabytes of data on transport networks, while GTFS datasets contain all the information required for analyses that investigate public transport. Taking all these new capabilities into account, we can face new research challenges, or investigate established ones from a new perspective.

One of the new fields of research is related to the temporal dynamics of accessibility (Geurs et al. 2015; van Wee 2016), which may provide a valuable, previously hidden input for policy makers and practitioners. Given that the level of accessibility (Farber et al. 2014) and its equity vary through the day (El-Geneidy et al. 2015), it is inequitable to include a temporal dimension into accessibility analysis (Baradaran and Ramjerdi 2001). The present study is focused on the temporal variation of the spatial pattern of accessibility by public transport in the city of Szczecin, Poland. We investigate this variability from the equity perspective focusing on its spatial and temporal dimensions. Thus, in the next section we describe two core issues that are used as a point of departure for the research conducted: accessibility and equity. Next, we present a selected case study, data sources and the methods applied. The results of our analyses, illustrated by maps, graphs and tables, are then presented. The final section presents our conclusions.

2 Accessibility and Equity—A Multidimensional Perspective

Accessibility, following the classical definition by Hansen (1959), is commonly understood as the potential of opportunities for interaction. Limited accessibility may therefore be considered as one of the main limitations of full participation in the social and economic activity of a given society, and in consequence—as one of the main factors responsible for social exclusion (Church et al. 2000; Lucas 2011; Preston and Rajé 2007). Given that the link between accessibility and exclusion is receiving growing attention, the focus of transport studies is evolving towards equity approach (Manaugh et al. 2015; Martens 2012; van Wee and Geurs 2011) and this is becoming more frequently included in accessibility instruments for planning practice (Papa et al. 2016).

Accessibility is a multidimensional phenomenon which covers a wide range of issues. Geurs and van Wee (2004) established fourfold typology of accessibility components, which includes: land use, transport, temporal and individual components. In a given study, we focus on the first three components, leaving an individual one behind the scope of this study. Thus, we assume, that accessibility level is a by-product of spatial distribution of potential origins and destinations (i.e. land use component, hereby population distribution), reflected by an existing transport system (i.e. transport component, hereby public transport service), which varies in time (i.e. temporal component, hereby daily variability). As accessibility determines human opportunities—for jobs, for economic or social interaction, for access to goods, or for the use of social or economic resources—the differentiation in the level of accessibility creates variations in opportunities and when differentiation comes onto the stage, equity should also be considered. The discussion about accessibility should therefore include an equity perspective, with equity understood as equality in accessibility level.

Equity in transport appears to be a very complex phenomenon, which results from the fact that there is no standard definition for it (Martens et al. 2012). Nevertheless, a common framework established to analyse equity from the accessibility perspective does exist, and this is based on two main dimensions: the vertical and the horizontal (Litman 2002). The vertical focuses on the differences between individuals, including their social and economic status as well as their mobility, with regard to their needs and abilities. The horizontal dimension has its roots in egalitarian theories of social justice (El-Geneidy et al. 2015; van Wee and Geurs 2011), i.e. it assumes that no one should be disfavoured, no matter who they are and where they live. Common practice usually identifies the horizontal dimension as spatial disparities (El-Geneidy et al. 2016) underlying the importance of a uniform distribution of accessibility levels across space (Chang and Liao 2011; Thomopoulos et al. 2009). The equal level of access is obviously an utopia, as a spatial pattern of any city or region creates conditions for unequal distribution of potential destinations, e.g. due to their centre-peripheral division, as stated by Martens et al. (2012). Nevertheless, we assume that one of the main aim of transport system is to diminish existing differences, by providing a proper level of public transport service.

Another equity dimension, which is investigated with regard to accessibility is modal equity, i.e. the comparison between accessibility by private car and public transport modes (Benenson et al. 2011; Golub and Martens 2014; Martens et al. 2012). The scale of disparities of accessibility level between both transport modes, called public transport gaps (Fransen et al. 2015), can be used as a proxy of car dependency and results in higher inequality levels (Martens 2012). Furthermore, in recent studies a next, intergenerational dimension of equity is identified (El-Geneidy et al. 2015; Foth et al. 2013; Kaplan et al. 2014), but this seems to be a special type of Litman's vertical dimension, where demographic groups take the place of social classes, rather than a separate equity dimension. Finally, apart from the spatial (identified with the horizontal) dimension and the socio-demographic (related to the vertical) dimension, the third, temporal dimension is beginning to attract more attention (Jones and Lucas 2012; Kawabata 2009), and this can be linked to the increasing availability of real-time data on traffic conditions and General Transit Feed Specification (GTFS) data (Geurs et al. 2015).

Given the complexity of the equity issues, we are limiting our interest to two of its dimensions: the horizontal and temporal ones. We consider spatial disparities to be applying the horizontal dimension of equity. We have investigated these disparities in accessibility level by public transport in the city area. Additionally, we have included temporal equity in our analysis. With regard to the latter, our analysis is twofold: First, we investigate the extent to which accessibility and equity levels vary during the day. Second, we estimate temporal disparities in accessibility level in given spatial units (enumeration districts). Taking the joint perspective of concepts of accessibility and resilience (Östh et al. 2015) as a point of departure, we argue that, for a given areal unit, it is not only low accessibility that may be considered 'problematic'. The constancy of travel time to a given destination, independent of the time period, also determines quality of life and travel behaviour.

3 Data and Methods

3.1 Case Study

The study covers the area of the city of Szczecin, north-west Poland. It is a subregional centre, one of the capitals of the Polish regions (voivodship, NUTS-2 region). The city area of 300 km^2 has a population of about 410,000. The city is divided by the River Oder and the two river banks are connected by only two main bridges. The main settlements of the city are located on the left-bank of the river, including the city and regional administration, main railway station as well as most of the population (Fig. 1) and employment. Moreover, almost a quarter of the city area consists of uninhabited natural zones, mainly wetlands. The public transport network consists of a radial tramway network (12 routes, mainly in the city centre) supplemented by a bus network, which connects most of the residential areas with the city centre (Fig. 2). The night-time public transport is operated by 16 dedicated night-bus routes. Only a few routes connect both river banks (including 3 night-time ones).

The study covers a complete day long accessibility analysis, applied for a typical weekday. We have used a complete day measurement (timespan 00:00–23:59) using 15-min intervals, i.e. we obtained 96 origin-destination travel time matrixes. Given that the total number of origin-destination nodes in the accessibility model is equal to 1745, each of the 96 averaged travel time origin-destination matrices

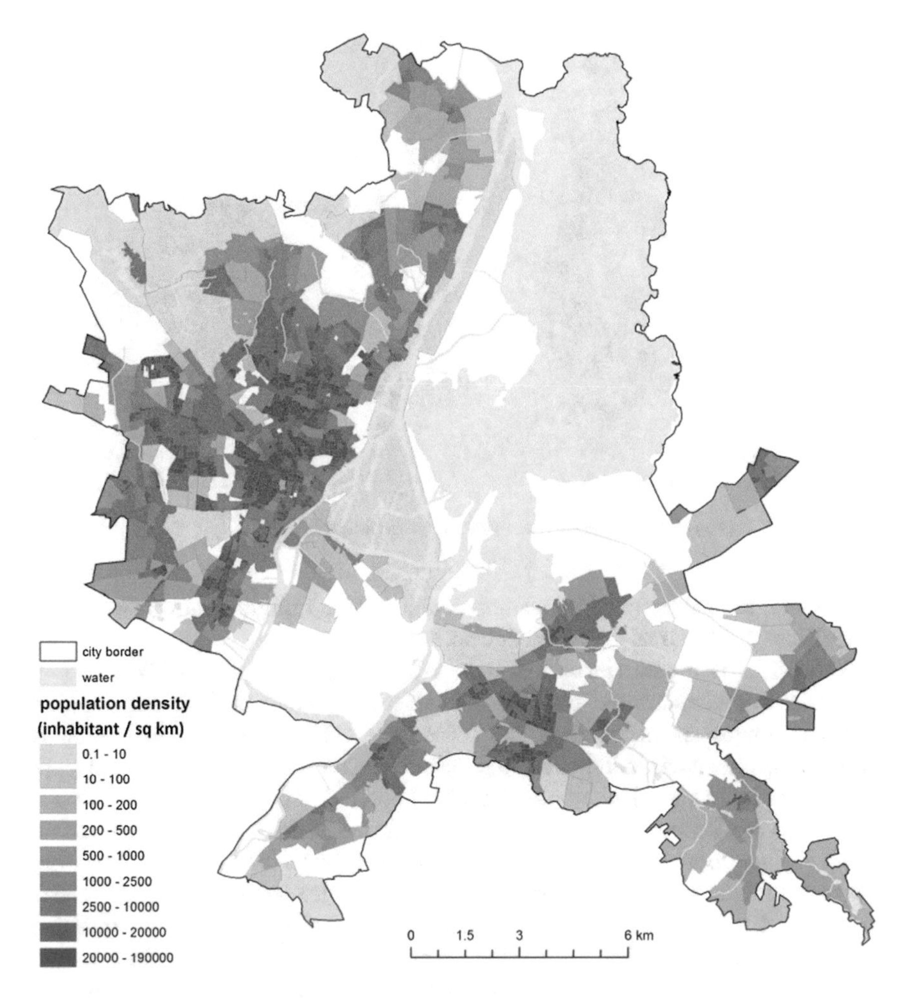

Fig. 1 Population density in Szczecin (according to National Census 2011)

contains about 3 million records, and the whole dataset contains almost 300 million individual travel time measurements. All of these are used to estimate the accessibility level and its daily variability for each individual spatial unit.

3.2 Open Data

The assumption that lies behind this study is that it should be fully replicable, requiring only data (including any spatial data) that are available in open access. We take advantage of the growing potential of the GTFS data format, and the freely available datasets containing the pedestrian network (derived from the

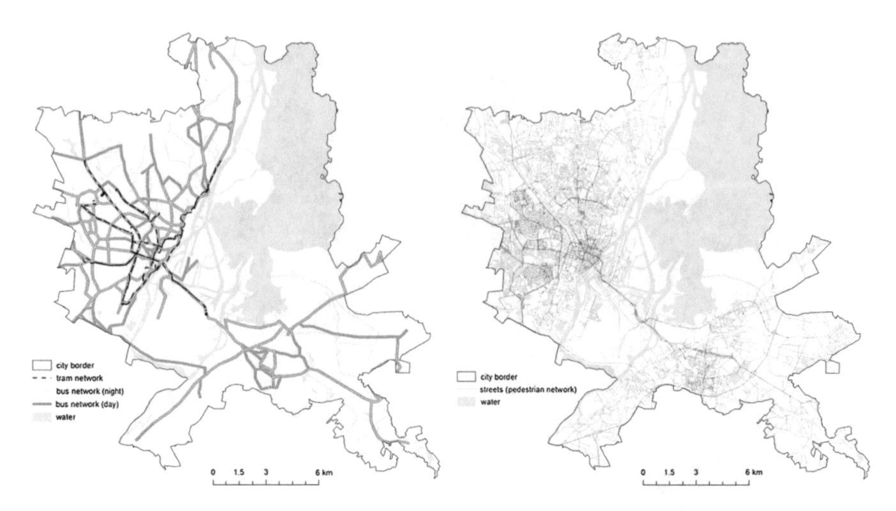

Fig. 2 Public transport and pedestrian network in Szczecin (July 2015)

OpenStreetMap dataset) and population data available from the Information Portal of the Central Statistical Office of Poland.

The GTFS is a data standard which is used to describe the public transport system and it contains the public transport stops (including their geographical location), routes and schedules. The subsequent GTFS files used in this study include stops (i.e. their individual geographical location), routes (i.e. the set of stops which constitutes a particular public transport line), trips (i.e. a sequence of two or more stops that occurs at specific time), stop-times (i.e. times that a vehicle arrives at and departs from individual stops for each trip) and calendar (i.e. dates for service IDs using a weekly schedule) (https://developers.google.com/transit/gtfs/reference). In result, it enables to calculate a precise, real travel time by public transport between any pair of origin-destination points. Due to the fact it enables one to investigate real-time accessibility by public transport, it is attracting growing attention from practitioners and the scientific community. It is used to evaluate the efficiency of the public transport network in general (Hadas 2013) and in comparison with travel times by private car (Benenson et al. 2011; Salonen and Toivonen 2013), providing practical conclusions that enhance the effectiveness of public transport systems (Tao et al. 2014). In addition, GTFS data enables a more precise assessment to be made of the level of socio-spatial disparities in the metropolitan area (El-Geneidy et al. 2016) with special attention being paid to socially disadvantaged groups (El-Geneidy et al. 2015). The precise information on the public transport schedules permits a better understanding of temporal changes of accessibility patterns, e.g. in the case of accessibility to health-care services (Fransen et al. 2015) or supermarkets (Farber et al. 2014).

The original GTFS dataset, which is applied in the present study, is published by the Szczecin public transport authority (http://zditm.szczecin.pl/rozklady/GTFS/latest/) and it covers the whole city area. We have used the most current dataset available at the time of the study (i.e. July, 2015). The dataset contains detailed information on 88 transit lines (60 daytime and 16 night buses, supplemented by 12 tram lines), 1888 edges in the geodatabase and 1394 points that represent particular public transport stops (more about GTFS standard may be found here: https://developers.google.com/transit/gtfs/reference), among them 1112 located within the city borders and the rest in adjoining areas. The Add GTFS datasets to a Network tool for ArcGIS (available at: http://www.transit.melindamorang.com/overview_AddGTFStoND.html) is then applied to create a routable network and the ArcGIS Network Analyst extension (10.2) is used for travel time calculations [Similar calculations may be also executed using open software instead of commercial one provided by ESRI, e.g. Open-TripPlanner (http://www.opentripplanner.org/)]. The travel time from origin i to destination j is calculated for a specific departure time and includes: walking to and from the public transport stop, waiting, boarding (0.25 min) and duration of the ride. The last of these includes intermediate boarding and waiting times, as well as walking between particular stops (where this applies).

The OpenStreetMap data are used to characterise the pedestrian network, required to connect enumeration district centroids to the nearest public transport stop, as well as to connect different public transport stops in the case of transfer. The full dataset consists of almost 139,000 edges. The distance travelled is derived from GIS data using the shortest path algorithm along the pedestrian network. Previous studies apply a wide range of walking speeds, starting from 3.2 up to 5.4 km/h (Table 1). Based on these examples we have applied an average walking

Table 1 Examples of walking speeds applied

Study	Walking speed (km/h)	Comment
Reyes et al. (2014)	3.2	Minimum typical speed for children aged 5–11
Fransen et al. (2015)	4.0	Adult's average
Ritsema van Eck et al. (2005)	4.0	Distance as the crow flies
Hadas (2013)	4.0	–
Nettleton et al. (2007)	4.8	–
Farber et al. (2014)	4.8	–
Willis et al. (2004)	5.3	Mean walking speed of individuals
Reyes et al. (2014)	5.4	Maximum typical speed for children aged 5–11
Krizek et al. (2012)	5.4	Average walking speed for 14–64 year old

travel speed of 4.5 km/h. Both of the datasets, i.e. GTFS data on the public transport network and OpenStreetMap create a very complex, multimodal transit network, which enables connections to be made between any pair of origin-destination nodes.

The nodes are located in the centroids of all the inhabited enumeration districts (census tracks) in the case study area (1745 units connected to the multimodal transport network). The average area of these units is equal to 98.8 ha and they are inhabited by 235 persons on average (in a range 3–899 persons). Such detailed, spatially disaggregated data are available from the geoportal of the Central Statistical Office of Poland (see http://geo.stat.gov.pl/inspire), established in the framework of the European INSPIRE directive (2007).

3.3 Accessibility calculations

As accessibility is a multidimensional phenomenon, there exist many measures of accessibility. The one applied in the present study is potential accessibility. Geurs and van Wee (2004) note that the gravity-based measure of potential accessibility provides a suitable framework as it combines two essential elements: land use and transportation. The measure is based on travel time calculations between all pairs of origin-destination nodes within the case study area. It is based on the framework established by the 'first law of geography' (Tobler 1970) and the Huff model (Griffith 1982; Huff 1963). According to the former, everything is related to everything else, but near things have a stronger relationship than distant ones. This means that closer destinations have more influence than distantly located ones, hence they appear as more 'attractive' destinations. According to the latter, the more important (e.g. larger) the destination, the farther its influence extends. Thus, accessibility level is determined by travel time to a diverse range of destinations and it assigns more importance to larger centres than to smaller ones with diminishing attractiveness with increasing travel time. Let M_j be the attractiveness of destination j and $f(t_{ij})$ be the impedance component. From this the accessibility A_i of the spatial unit i is then calculated using the following formula:

$$A_i = M_i f(t_{ii}) + \sum_j M_j f(t_{ij})$$

The first component of the formula is the self-potential of a given enumeration district, i.e. the potential produced by the area itself. Its value is estimated based on the area of a given spatial unit, following the method proposed by Rich (1978) (cf. Bröcker 1989; Frost and Spence 1995), i.e. internal travel distance is equal to half the radius of the area and the average internal travel speed is the typical walking speed applied in the study (4.5 km/h). The second component is then the 'external' potential, i.e. the potential produced by all other destinations included in the study.

We use population size, which is assumed to be a proxy of destination attractiveness. The impedance component is determined by time, with the negative exponential function as an impedance form (for a review on the selection of the proper distance decay function consult: De Vries et al. 2009; Kwan 1998; Martínez and Viegas 2013; Reggiani et al. 2011; Rosik et al. 2015). In order to estimate the β parameter which influences the slope of the distance decay function, we apply the 'half-life' approach (Östh et al. 2014), i.e. we assume that the destination loses half its attractiveness at the observed median travel time for a given trip purpose. According to the data derived from the Comprehensive Traffic Study conducted in Szczecin (2010) the average travel time is 27 min which gives a β equal to 0.02567.

Daily accessibility values are calculated using the weighted average, where the share in daily flows by public transport is used as a weighting factor. As a result, the accessibility level during the peak-hours has more influence on the aggregated daily value than night-time accessibility. The same weighting procedure is applied to five time-periods during the day. The share of flows in a given time period is derived from the Comprehensive Traffic Study of Szczecin (2010). The extract from the Study is based on individual travel patterns during a weekday. It does not differentiate trip motivations (i.e. it includes work- as well as non-work-related trips) and it is limited only to the walking and public transport modes (buses and tramways), while private cars, taxis and regional trains are excluded from the weighting procedure. Nevertheless, the variability of the share of the departure times of journeys during the subsequent periods of the day is quite similar for different transport modes (Table 2).

3.4 *Equity and Spatio-Temporal Variation*

Similar to the broad range of accessibility measures, there also exist several equity measures. Due to the fact that each of these reflects a different perception of (in) equality, it may be difficult to assess the degree of equality on the basis of a single measure (Ramjerdi 2006). Nevertheless, all the equality indicators already devised and tested (i.e. Gini coefficient, coefficient of variation, standard deviation and Theil's entropy index) are strongly correlated with each other, as underlined in the cited paper (Ramjerdi 2006), as well as in our case study area (Table 3). thus we selected the Gini coefficient for further analysis (cf. Kaplan et al. 2014; Neutens et al. 2010; van Wee and Geurs 2011). The Gini coefficient is based on the Lorenz curve, and it is calculated as a ratio between the line of the perfect equality and an observed Lorenz curve. It can assume values between 0 (a perfectly even distribution) and 1 (a maximal concentration). We use 'ineq' package for RStudio environment (ineq package: Measuring Inequality, Concentration, and Poverty https://cran.r-project.org/web/packages/ineq/index.html) to calculate Gini coefficient.

Taking advantage of the completeness of the GTFS data, we investigated the daily variability of both accessibility levels and equity measures. In the case of the latter we can investigate both of its dimensions, i.e. the horizontal dimension, using

Table 2 Share of trips by time-of-day (journeys started, all trip motivations)

%	0:01–6:00	6:01–7:00	7:01–8:00	8:01–9:00	9:01–11:00	11:01–12:00	12:01–13:00	13:01–14:00	14:01–15:00	15:01–16:00	16:01–17:00	17:01–18:00	18:01–20:00	20:01–24:00
Total (n = 8039)	3.2	7.5	13.5	6.8	12.9	5.1	4.7	5.9	8.0	10.7	7.7	5.1	6.2	2.7
Walking (n = 1511)	1.8	3.6	11.6	8.8	17.8	7.4	6.0	7.5	8.3	7.6	5.8	4.0	6.6	3.2
Public transport (n = 2696)	3.5	8.1	12.0	6.8	14.6	5.5	5.7	6.5	8.4	10.2	7.4	3.9	5.2	2.2
Walking and public transport	**2.9**	**6.5**	**11.8**	**7.5**	**15.7**	**6.2**	**5.8**	**6.9**	**8.4**	**9.3**	**6.8**	**4.0**	**5.7**	**2.5**

Source Comprehensive Traffic Study of Szczecin (2010)

Table 3 Correlation matrix for the calculated equity measures

	SD	Coefficient of variation	Gini	Theil
SD	1.00	0.91	0.91	0.85
Coefficient of variation	0.91	1.00	1.00	0.98
Gini	0.91	1.00	1.00	0.98
Theil	0.85	0.98	0.98	1.00

the Gini coefficient calculated for all spatial units (enumeration districts), and the temporal dimension, focusing on the daily variability of the accessibility level noted in particular spatial units.

4 Results

4.1 Spatial Disparities

The spatial pattern of the daily weighted average of potential accessibility level replicates the distribution of population including the modification resulting from the central-peripheral division of the city (Fig. 3). Further located areas are obviously less accessible (Martens 2012), but those located on the right bank are even less accessible than left-bank peripheries. The separation of the city produced by the river crossing from south to north and limited connectivity of both parts of the city is then clearly visible. Moreover, the fact that almost 80 % of city's population inhabits residential areas located in the western part of the city is also reflected by the spatial pattern of accessibility values, which are significantly higher on the left bank of the River Oder.

Nevertheless, we argue that public transport, to some extent, fulfils its role in diminishing disparities in accessibility level. The spatial disparities in walking accessibility (Fig. 4a) are far more intense than those resulting from public transport accessibility. Ai values in the western part of the city are significantly higher than those in the other part and these differences are more extreme than in the case of accessibility by public transport. The evenness of the spatial distribution of potential accessibility values is also significantly higher for accessibility by public transport—in the case of the latter the Gini coefficient is equal to 0.13, while the Gini coefficient of potential accessibility through the pedestrian network increases up to 0.28. In case of walking accessibility, the most accessible areas have Ai values almost 45 times higher than the lowest ones, comparing to 5.5 in the case of public transport accessibility. The public transport system compensates, to some extent at least, the accessibility imbalance resulted from the spatial distribution of population and terrain constraints. The areas that gain the most from public transport are located on the east-bank as well as in the city's peripheries in general (Fig. 4b).

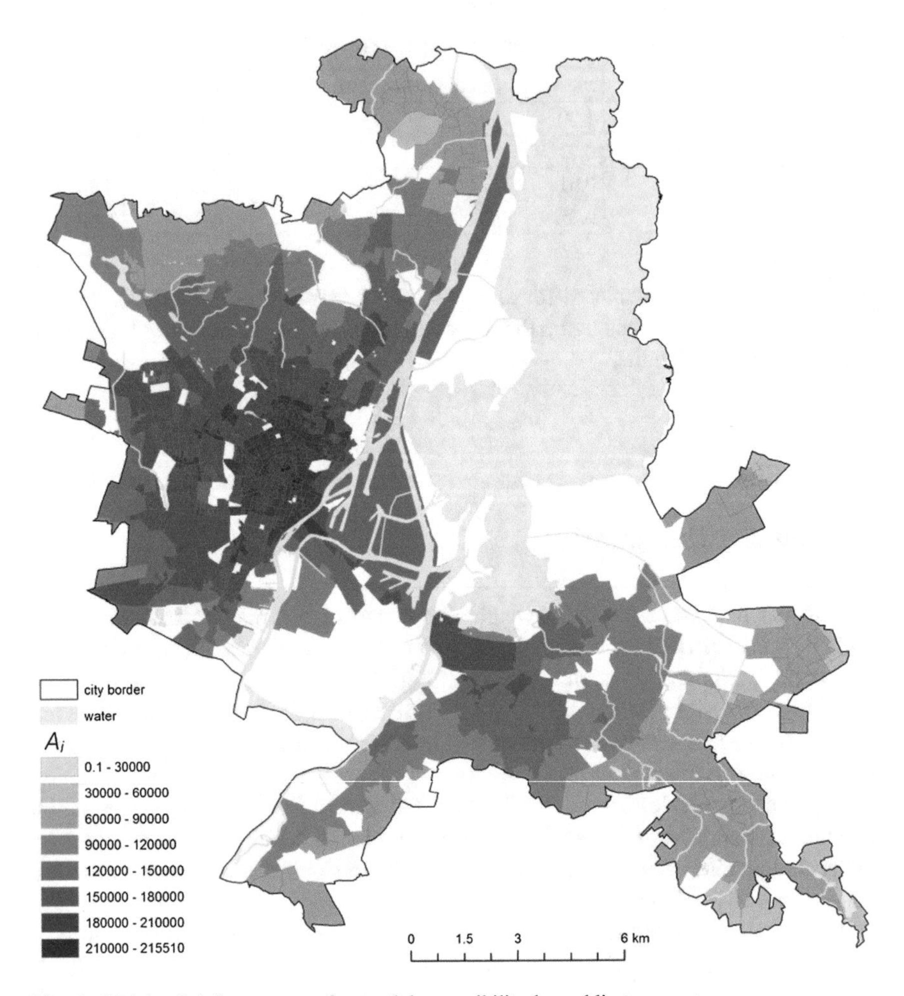

Fig. 3 Weighted daily average of potential accessibility by public transport

4.2 Temporal Variability

One would expect potential accessibility to vary during the day, and that the lowest values would be noted during the night (Fig. 5). The night-time service is less frequent and does not cover the whole city area, thus some neighbourhoods are reachable only by walking. However, it is surprising that the peak hours (especially the afternoon ones), when the public transport frequency is the highest, are almost invisible. The difference in the overall potential accessibility values between morning peak-hours and the daily off-peak does not exceed 4 %, and for the less-concentrated evening peak the difference is even smaller. If we look at the averaged potential accessibility values during successive periods during the day, we find hardly any differences (Fig. 6). The only exception is night-time accessibility.

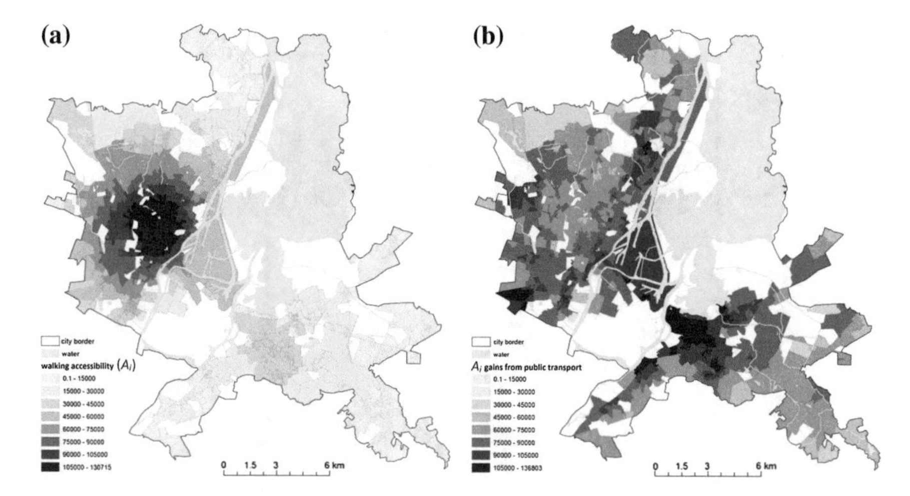

Fig. 4 Weighted daily average of potential accessibility by walking (**a**) and gains from public transport (**b**)

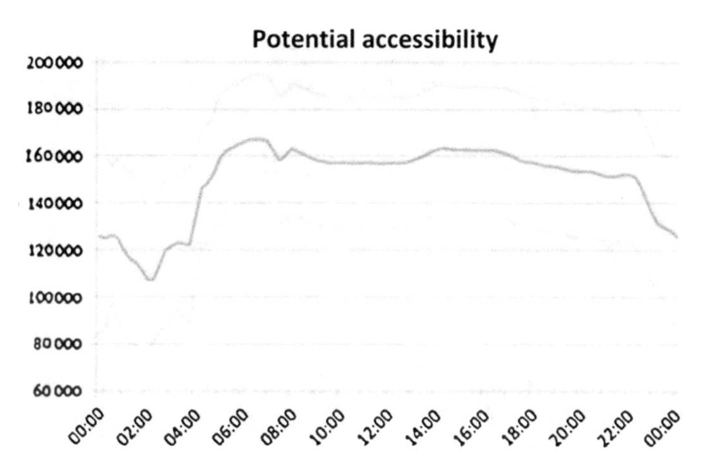

Fig. 5 Daily temporal fluctuations (30-min averages) of population weighted accessibility level at the city scale

The growth of spatial disparities during the night-time is reflected by an increase in standard deviation values (Fig. 5). Night-time inequality is also illustrated by an almost doubled Gini coefficient (Fig. 7). The higher the accessibility is by public transport, the lower the level of inequalities that is noted. This supports the statement that efficient public transport plays a decisive role in reducing spatial disparities in accessibility level.

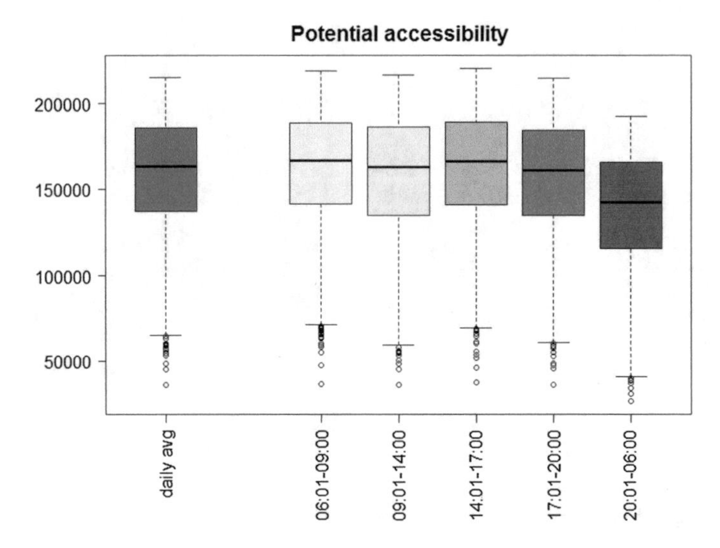

Fig. 6 Boxplots of potential accessibility values

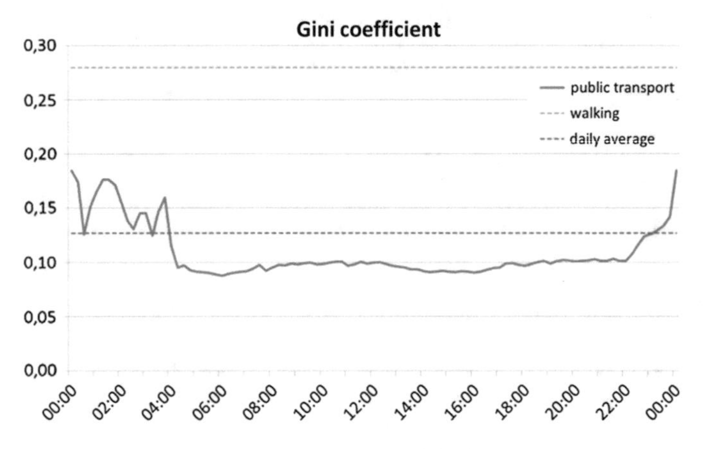

Fig. 7 Daily temporal fluctuations (30-min averages) of Gini coefficient at the city scale

4.3 Spatio-Temporal Arrangements

Potential accessibility values calculated for particular enumeration districts during different periods of a the day are still highly correlated to each other, as well as to the daily average (Table 4). The comap shows the changes of absolute values and their spatial pattern between different periods of the day (Fig. 8). Although the picture underlines slightly higher accessibility values during the morning peak-hours, the spatial patterns remain generally the same throughout the day. Again, the only exception is found in the night-time period and the most affected

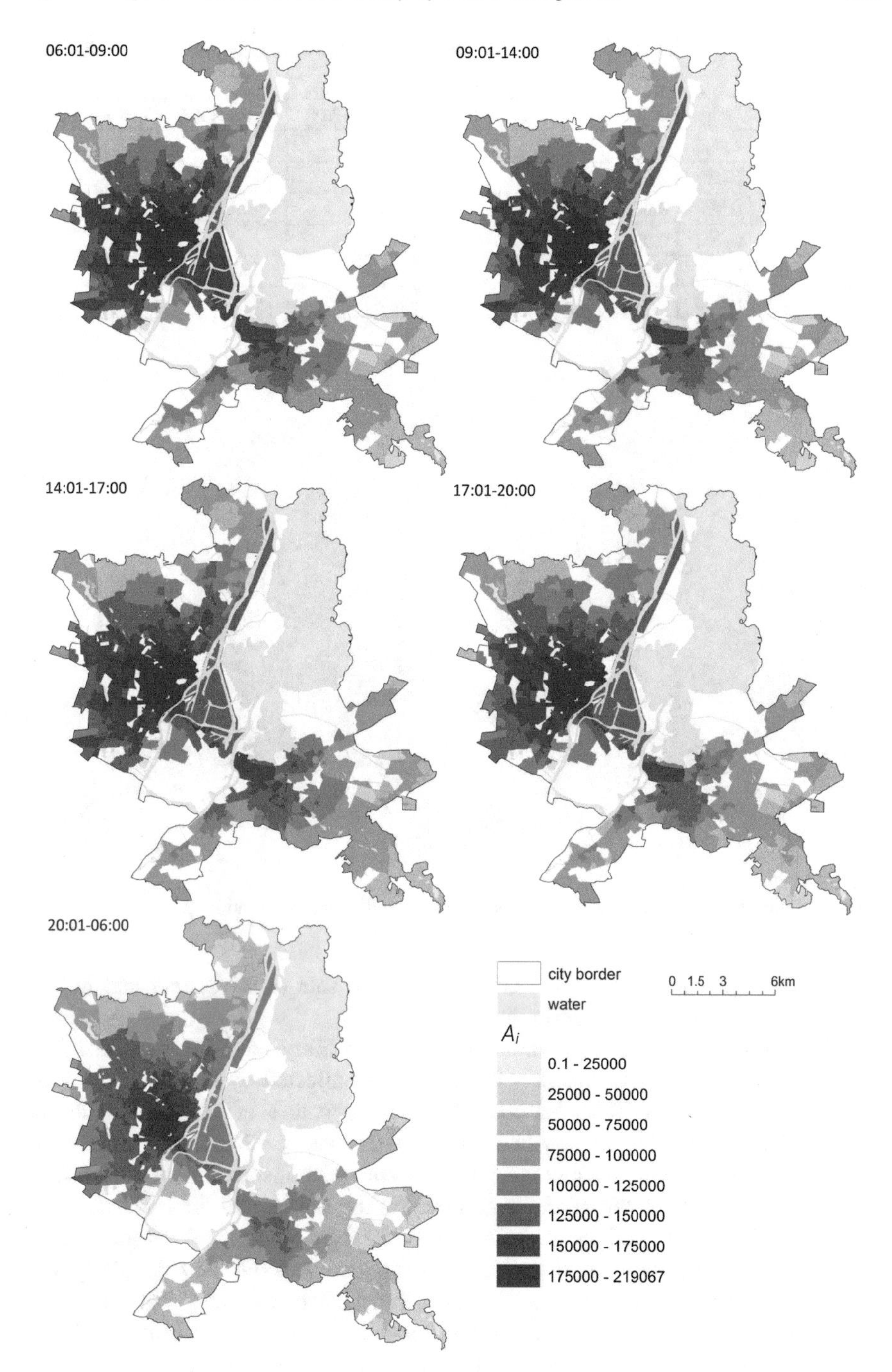

Fig. 8 Comap for potential accessibility values (by times of day)

Table 4 Correlation matrix of weighted averages of potential accessibility for different times of a day

	Daily	06:01–09:00	09:01–14:00	14:01–17:00	17:01–20:00	20:01–06:00
Daily	1.000					
06:01–09:00	0.999	1.000				
09:01–14:00	0.998	0.996	1.000			
14:01–17:00	0.998	0.996	0.994	1.000		
17:01–20:00	0.998	0.995	0.996	0.994	1.000	
20:01–06:00	0.990	0.989	0.986	0.987	0.989	1.000

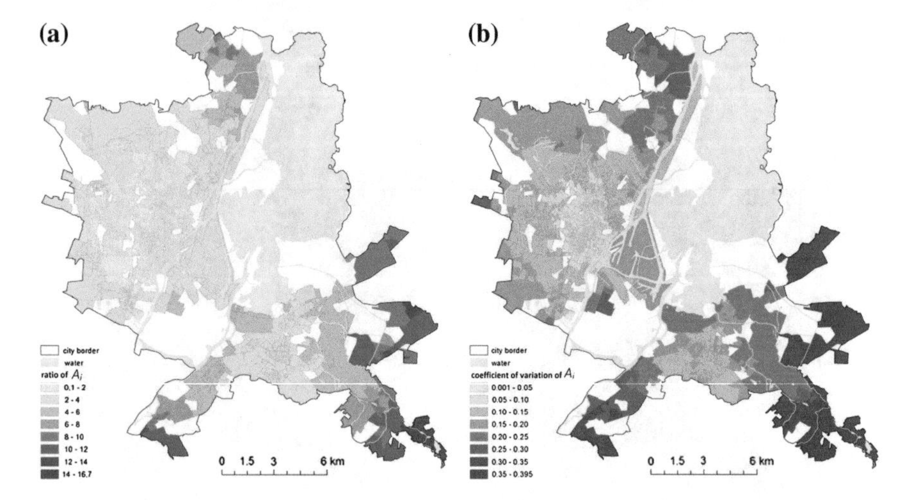

Fig. 9 Ratio of potential accessibility values (**a**) and coefficient of variation (**b**)

area is the right-bank part of the city. The crucial role of the terrain and infrastructure constraints is again underlined.

The spatial pattern of daily variability of accessibility level indicates that less accessible areas are simultaneously the areas most affected by the inconstancy of accessibility level. The ratio between the extreme values of accessibility levels reflects this inconstancy in absolute terms, while the coefficient of variation of accessibility level during the day indicates the same relationship in relative terms. In the case of the former, the relationship between the lowest and the highest accessibility level in some, peripheral areas exceeds a factor of ten, while in the case of the highly accessible city centre the ratio is lower than two (Fig. 9a). Similarly, the latter underlines the same areas as the most imbalanced ones (Fig. 9b). The scatterplots depict a strong relationship between a daily average accessibility level and its daily variation in both absolute and relative terms (Fig. 10).

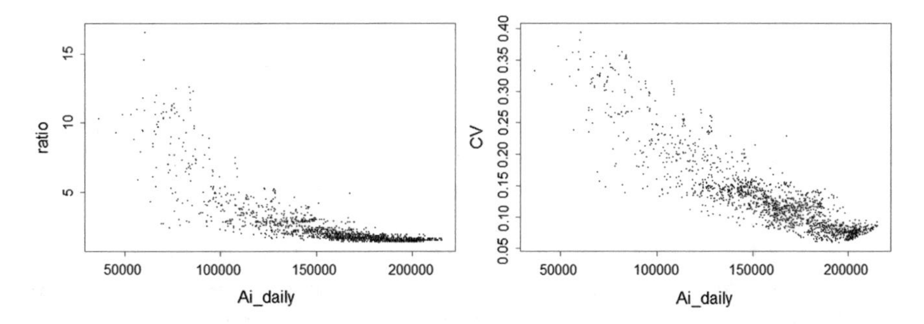

Fig. 10 Scatterplot of weighted average of potential accessibility versus ratio and coefficient of variation

5 Conclusions

The present study tends to shed more light on the less investigated, temporal dimension of equity in the distribution of accessibility level across the city. Increasing computational efficiency, and the rise of large-scale open data, including GTFS standard, enables an unprecedentedly detailed analysis of accessibility disparities. Using real travel times by public transport calculated on the basis of up-to-date schedules, we built up an original, time-varying origin-destination matrix which is then converted into potential accessibility values. Highly disaggregated spatial units (1745 enumeration districts) provide very detailed information about spatial disparities in Szczecin. The effectiveness of combination of GTFS large datasets and GIS solutions, provides tools for highly detail accessibility analysis for a large case study areas, which have been unavailable before.

Our results show, that the public transport system in Szczecin at least partly compensates the accessibility imbalance resulted from the unusual spatial pattern of land use. The spatial disparities in accessibility levels by public transport are not as heavily affected by the city's division by Oder river, as it might be expected. The accessibility by walking in the eastern part of the city is significantly lower than in case of daily average of accessibility by public transport. The central-peripheral division of accessibility disparities is still visible, but it is quite typical for any urban area (cf. Martens et al. 2012).

We acknowledge the previous findings about the existence of great temporal variability in accessibility level by public transport (Farber et al. 2014), even though the dissimilarities decrease when using temporally aggregated values (Fransen et al. 2015). Investigating equity indicators during the course of the day, we add a better understanding to previous studies regarding the temporal component of spatial disparities in terms of accessibility. The average accessibility level does not vary significantly between peak and off-peak hours, nor does the equity indicator (Gini coefficient). Nevertheless, the limited service during the night-time period strongly affects both the overall accessibility level and its degree of equity. It should be clearly pointed out that public transport frequency and coverage not only has an influence on the population accessibility level, but also has a significant impact on spatial inequalities. Accessibility inequality is strongly inversely related to accessibility level.

A major finding of this study is that a low accessibility level is associated with a higher inconstancy of accessibility level. In the results some areas, mainly the peripheral ones, are doubly affected by the limited public transport service. Not only do their inhabitants need more time to reach their desired destination, but also their expected travel time is strongly influenced by uncertainty and greatly depends on a particular time of a departure. Apart from public transit gaps, this uncertainty may be the most important factor in car dependency and transport inequalities.

In this study we focus on the spatial and temporal dimension of equity in accessibility level. Neither the distribution of accessibility between the different population groups (vertical equity), nor modal equity remain unaffected and they should be included in further investigations. We identify a compelling need for a complex, multidimensional analysis of equity, which includes all of its identified dimensions (i.e. horizontal, vertical, temporal and modal). Such an approach requires comprehensive data which describe several fields of population characteristics and travel behaviour including their very detailed spatial representation. Further, in the presented study we apply an average walking speed, equal for the whole population. The combination of vertical and horizontal dimension of equity and availability of supporting data provides an opportunity to include, among others, variability of walking speed depending of the characteristic of individual (e.g. age, potential disabilities etc.). We believe that our findings on the temporal and equity dimensions supplement the existing body of knowledge, and the rise of 'big data', supported by increasing technical capacities, provides a stable ground reference for future investigations. The proposed methodology enables to combine various equity dimensions into one analysis, which can be easily adapted for any other case study area. Moreover, similar approach may be implemented when comparing any other equity dimensions (e.g. vertical and/ or modal equity dimensions). Thus, the described research procedure may be easily implemented e.g. for an ex-ante evaluation of changes in organization of public transport services.

Finally, our analysis suggests, that some of neighbourhoods (and their inhabitants) are "double-affected", i.e. not only by low average level of accessibility by public transport, but also—by the high daily variation in accessibility level. It should be further evaluated, whether this multiplication of a negative impact of temporal and horizontal inequalities is something typical or it is a by-product of the unusual spatial pattern of the land use (city divided into two parts by natural constraints) in the particular case study.

Acknowledgments This research was funded by the National Science Centre allocated on the basis of the Decision No. DEC-2013/09/D/HS4/02679.

References

Baradaran S, Ramjerdi F (2001) Performance of accessibility measures in Europe. J Transp Stat 4:31–48

Benenson I, Martens K, Rofé Y et al (2011) Public transport versus private car GIS-based estimation of accessibility applied to the Tel Aviv metropolitan area. Ann Reg Sci 47:499–515

Bröcker, J. (1989). How to eliminate certain defects of the potential formula. Environ Plan A 21 (6):817–830

Chang H-S, Liao C-H (2011) Exploring an integrated method for measuring the relative spatial equity in public facilities in the context of urban parks. Cities 28:361–371

Church A, Frost M, Sullivan K (2000) Transport and social exclusion in London. Transp Policy 7:195–205

de Vries JJ, Nijkamp P, Rietveld P (2009) Exponential or power distance-decay for commuting? An alternative specification. Environ Plan A 41:461–480

Delmelle EC, Casas I (2012) Evaluating the spatial equity of bus rapid transit-based accessibility patterns in a developing country: the case of Cali, Colombia. Transp Policy 20:36–46

El-Geneidy A, Buliung R, Diab E et al (2015) Non-stop equity: assessing daily intersections between transit accessibility and social disparity across the Greater Toronto and Hamilton Area (GTHA). Environ Plan B Plan Des 43:540–560

El-Geneidy A, Levinson D, Diab E, et al (2016) The cost of equity: assessing transit accessibility and social disparity using total travel cost. In: 95th annual meeting of the transportation research board, Washington DC, USA, pp 1–34

Farber S, Morang MZ, Widener MJ (2014) Temporal variability in transit-based accessibility to supermarkets. Appl Geogr 53:149–159

Foth N, Manaugh K, El-Geneidy AM (2013) Towards equitable transit: examining transit accessibility and social need in Toronto, Canada, 1996–2006. J Transp Geogr 29:1–10

Fransen K, Neutens T, Farber S et al (2015) Identifying public transport gaps using time-dependent accessibility levels. J Transp Geogr 48:176–187

Frost ME, Spence NA (1995) The rediscovery of accessibility and economic potential: the critical issue of self-potential. Environ Plan A 27:1833–1848

Geurs KT, van Wee B (2004) Accessibility evaluation of land-use and transport strategies: review and research directions. J Transp Geogr 12:127–140

Geurs KT, De Montis A, Reggiani A (2015) Recent advances and applications in accessibility modelling. Comput Environ Urban Syst 49:82–85

Golub A, Martens K (2014) Using principles of justice to assess the modal equity of regional transportation plans. J Transp Geogr 41:10–20

Griffith D (1982) A generalized Huff model. Geogr Anal 14:135–144

Hadas Y (2013) Assessing public transport systems connectivity based on Google Transit data. J Transp Geogr 33:105–116

Hansen WG (1959) How accessibility shapes land-use. J Am Inst Plan 25:73–76

Huff DL (1963) A probabilistic analysis of shopping center trade areas. Land Econ 39:81–90

INSPIRE (2007) Directive 2007/2/ EC OF the European Parliament and of the Council of 14 March 2007. Establishing an Infrastructure for Spatial Information in the European Community (INSPIRE)

Jones P, Lucas K (2012) The social consequences of transport decision-making: clarifying concepts, synthesising knowledge and assessing implications. J Transp Geogr 21:4–16

Kaplan S, Popoks D, Prato CG et al (2014) Using connectivity for measuring equity in transit provision. J Transp Geogr 37:82–92

Kawabata M (2009) Spatiotemporal dimensions of modal accessibility disparity in Boston and San Francisko. Environ Plan A 41:183–198

Krizek KJ, Horning J, El-Geneidy A (2012) Perceptions of accessibility to neighbourhood retail and other public services. In: Geurs KT, Krizek KJ, Reggiani A (eds) Accessibility analysis

and transport planning challenges for Europe and North America. Edward Elgar, Cheltenham, pp 96–117
Kwan M-P (1998) Space-time and integral measures of individual accessibility: a comparative analysis using a point-based framework. Geogr Anal 30:191–216
Litman T (2002) Evaluating transportation equity. World Transp Policy Pract 8:50–65
Lucas K (2011) Making the connections between transport disadvantage and the social exclusion of low income populations in the Tshwane Region of South Africa. J Transp Geogr 19:1320–1334
Manaugh K, El-Geneidy A (2012) Who benefits from new transportation infrastructure? Using accessibility measures to evaluate social equity in public transport provision. In: Geurs KT, Krizek KJ, Reggiani A (eds) Accessibility analysis and transport planning challenges for Europe and North America. Edward Elgar, Cheltenham, pp 211–227
Manaugh K, Badami MG, El-Geneidy AM (2015) Integrating social equity into urban transportation planning: a critical evaluation of equity objectives and measures in transportation plans in North America. Transp Policy 37:167–176
Martens K (2012) Justice in transport as justice in accessibility: applying Walzer's "spheres of justice" to the transport sector. Transp Amst 39:1035–1053
Martens K, Hurvitz E (2011) Distributive impacts of demand-based modelling. Transportmetrica 7:181–200
Martens K, Golub A, Robinson G (2012) A justice-theoretic approach to the distribution of transportation benefits: implications for transportation planning practice in the United States. Transp Res A Policy Pract 46:684–695
Martínez LM, Viegas JM (2013) A new approach to modelling distance-decay functions for accessibility assessment in transport studies. J Transp Geogr 26:87–96
Nettleton M, Pass DJ, Walters GW et al (2007) Public transport accessibility map of access to general practitioners surgeries in longbridge, Birmingham, UK. J Maps 3:64–75
Neutens T, Schwanen T, Witlox F et al (2010) Equity of urban service delivery: a comparison of different accessibility measures. Environ Plan A 42:1613–1635
O'Sullivan D, Morrison A, Shearer J (2000) Using desktop GIS for the investigation of accessibility by public transport: an isochrone approach. Int J Geogr Inf Sci 14:85–104
Östh J, Reggiani A, Galiazzo G (2014) Novel methods for the estimation of cost–distance decay in potential accessibility models. In: Condeço-Melhorado A, Reggiani A, Gutiérrez J (eds) Accessibility and spatial interaction. Edward Elgar, Cheltenham, pp 15–37
Östh J, Reggiani A, Galiazzo G (2015) Spatial economic resilience and accessibility: a joint perspective. Comput Environ Urban Syst 49:148–159
Papa E, Silva C, Brömmelstroet M et al (2016) Accessibility instruments for planning practice: a review of European experiences. J Transp Land Use 9:1–20
Preston J, Rajé F (2007) Accessibility, mobility and transport-related social exclusion. J Transp Geogr 15:151–160
Ramjerdi F (2006) Equity measures and their performance in transportation. Transp Res Rec J Transp Res Board 1983:67–74
Reggiani A, Bucci P, Russo G (2011) Accessibility and impedance forms: empirical applications to the german commuting network. Int Reg Sci Rev 34:230–252
Reyes M, Páez A, Morency C (2014) Walking accessibility to urban parks by children: a case study of Montreal. Landsc Urban Plan 125:38–47
Rich DC (1978) Population potential, potential transportation cost and industrial location. Area 10:222–226
Ritsema van Eck J, Burghouwt G, Dijst M (2005) Lifestyles, spatial configurations and quality of life in daily travel: an explorative simulation study. J Transp Geogr 13:123–134
Rosik P, Stępniak M, Komornicki T (2015) The decade of the big push to roads in Poland: Impact on improvement in accessibility and territorial cohesion from a policy perspective. Transp Policy 37:134–146. http://doi.org/10.1016/j.tranpol.2014.10.007
Salonen M, Toivonen T (2013) Modelling travel time in urban networks: comparable measures for private car and public transport. J Transp Geogr 31:143–153

Tao S, Rohde D, Corcoran J (2014) Examining the spatial–temporal dynamics of bus passenger travel behaviour using smart card data and the flow-comap. J Transp Geogr 41:21–36
Thomopoulos N, Grant-Muller S, Tight MR (2009) Incorporating equity considerations in transport infrastructure evaluation: current practice and a proposed methodology. Eval Program Plan 32:351–359
Tobler WR (1970) Computer movie simulating urban growth in Detroit region. Econ Geogr 46:234–240
Urząd Miasta Szczecin (2010) Comprehensive traffic study—Szczecin 2010
Van Wee B (2016) Accessible accessibility research challenges. J Transp Geogr 51:9–16
Van Wee B, Geurs KT (2011) Discussing equity and social exclusion in accessibility evaluations. Eur J Transp Infrastruct Res 11:350–367
Willis A, Gjersoe N, Havard C et al (2004) Human movement behaviour in urban spaces: implications for the design and modelling of effective pedestrian environments. Environ Plan B Plan Des 31:805–828

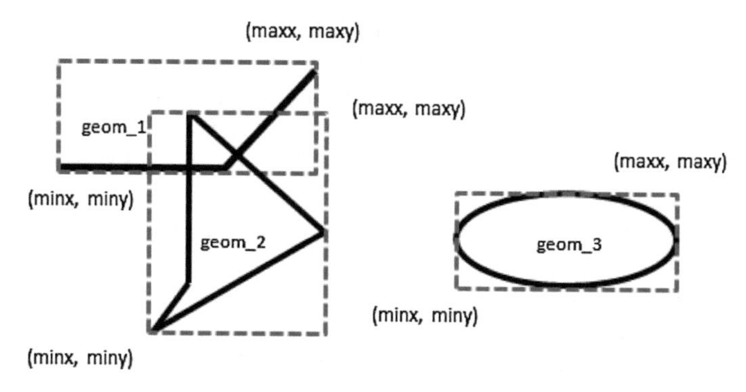

Fig. 1 MBRs of different types of spatial objects

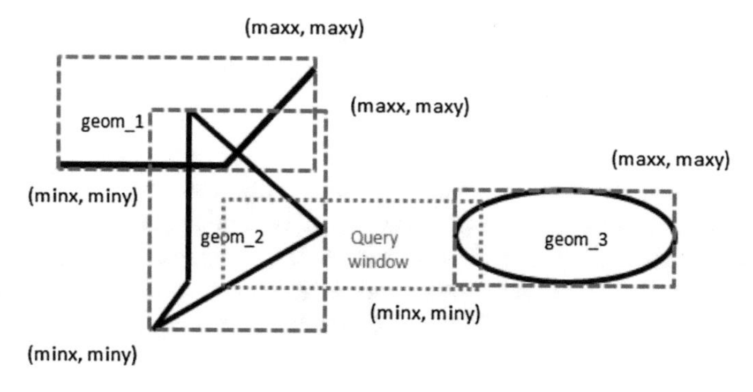

Fig. 2 An example of range query

Figure 2 illustrates the problem regarding range query for different spatial data. In Fig. 2, dotted red rectangle represents query window (range query) and others represent different spatial objects. Our goal is to find rectangles from millions rectangles coincident with the query rectangle by means of distributed programming framework. According to Fig. 2, two MBRs (geom_2 and geom_3) are intersected with range query.

5 Scalable Range Query Architecture

Details of the proposed range query architecture are as following: All rectangles and their bottom left and upper right coordinates are stored in HDFS. Region (query region) is specified by users. Spatial region query could be resolved with one

MapReduce job. This job includes Map and Reduce functions. In the Map function, the filtering strategy can be used to find the rectangles intersected with of region query. The results of the Map stage are stored in the distributed file system directly. In filtering phase, bottom left and upper right coordinates of every mosaic are examined whether intersects with the query region or not. Architecture for range query is illustrated in Fig. 3.

Pseudo-code for the algorithm to filter step is as following:

Algorithm 1 Map

Input: {**Init:** *q*: query; *r*: rectangle; **Key:** line number of the file; **Value:** line content of the file}

Output: {**Key:** Coordinates of intersected rectangle; **Value:** 1}

1. **begin**
2. splits a line and extracts coordinates of a rectangle as r.minx, r.miny, r.maxx, and r.maxy
3. **if**(!(q.mix>r.maxx)) && !(r.minx>q.maxx) && !(q.miny>r.maxy && !(r.miny>q.maxy)) **then output** (r, 1)
4. **end**

where (q.minx, q.miny) and (q.maxx, q.maxy) show bottom left and upper right coordinates of query region, respectively. (r.minx, r.miny) and (r.maxx, r.maxy) show bottom left and upper right coordinates of a rectangle in rectangle dataset, respectively. Each Mapper processes a file, extracts rectangles (r) intersected with range query (q) and emits the following key/value pair: <r,1>.

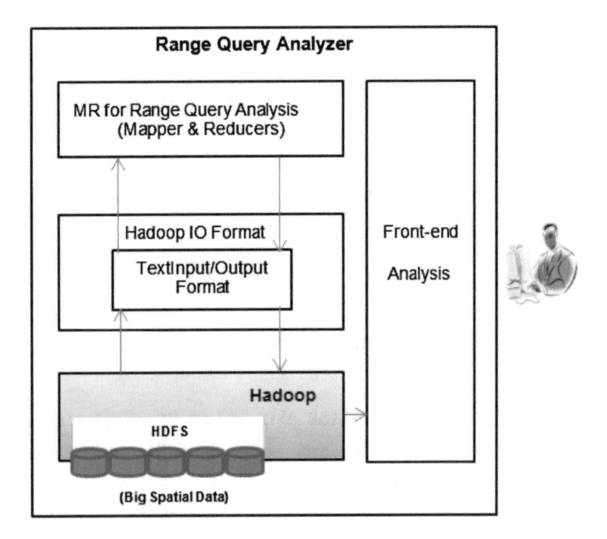

Fig. 3 Range query architecture

Pseudo-code for Reduce function can be defined as follows:

Algorithm 2 Reduce

Input: {**Init:** *sum*: total number of rectangles intersected with range query; **Key:** Coordinates of intersected rectangle; **Value:** 1}

Output: {**Key:** unused, **Value:** unused}

1. **begin**
2. **for each** (rectangle r **in** intersected rectangle list) **do**
3. sum+=1
4. **end for each**
5. **output** (sum)
6. **end**

The Reducer receives key/value pairs that have the following format: <r, 1>. The Reducer simply add up the 1 s to provide a final count of the rectangles and send the result to the output as the following value <count of rectangles intersected with range query>. After defining map and reduce functions, jobs are executed on worker nodes as MapTask/ReduceTask. JobTracker is the main process of Hadoop for controlling and scheduling tasks. JobTracker gives roles to the worker nodes as Mapper or Reducer task by initializing TaskTrackers in worker nodes. TaskTracker runs the Mapper or Reducer task and reports the progress to JobTracker.

To test the system and evaluate the results, we have set up an HDFS cluster with two nodes of Hewlett–Packard. Each node has an Intel(R) Core(TM) i7-3610QM CPU @ 2.30 GHz, 8 GB memory, 160 GB SATA disk. The operating system is Ubuntu with kernel 3.13.0-37-generic. Hadoop version is 2.6.0 and java version is 1.7.0. We make four different test platforms using these two nodes: (i) one NameNode and one DataNode worked in same node, (ii) one NameNode and two DataNodes worked in same node, (iii) one NameNode and two DataNodes worked in one node and one DataNode worked in another node (totally 3 DataNodes), (iv) traditional java implementation instead of distributed framework. Each node has a Hadoop framework installed on a virtual machine. Although virtualization causes some performance loss in total execution efficiency, installation and management of Hadoop become easier by cloning virtual machines. In order to verify the efficiency of the proposed approach, three datasets are created with different size (1, 3, and 5 GB). Each dataset is composed of millions of rectangle names and their bottom left and upper right coordinates. According to experimental results, when NameNode and DataNode are in same computer, they spend more time than traditional java implementation. Because, coordination and data flow between NameNode and DataNode require more time. For example, average process time of 300 million of MRBs with traditional java implementation is 9.57 min and average process time of 300 million of MRBs with NameNode and DataNode in same computer is 11.01 min. Also, experimental results show that average processing times reduce

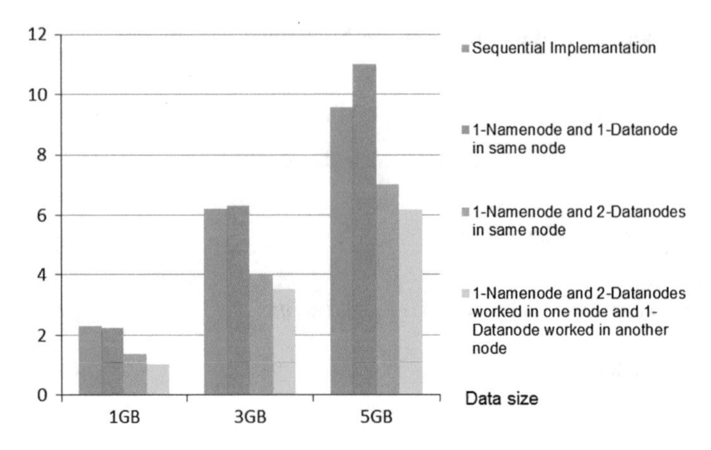

Fig. 4 Comparison of process times of different test platforms

with increase of conventional machines as seen Fig. 4. For example, average process time of 300 million of MRBs with the third test platform is 7.01 min and average process time of 300 million of MRBs with the fourth test platform is 6.18 min.

6 Results and Future Works

Range query, both on point and general geographic object datasets, has received important attention in the literature. In this paper, we have shown that the MapReduce parallel programming paradigm can be used to process range queries for big spatial data. The performance evaluation demonstrates the feasibility of processing range queries with MapReduce. The proposed approaches can be used in object extraction, object recognition, and image stitching as a preprocessing step. In the near future, we plan to extend the proposed system with other operations for example union, difference, etc. for implementing polygon coverage problem on big spatial datasets.

Acknowledgments This work has been supported by the TUBITAK under Grant 215E189.

References

Akdogan A, Demiryurek U, Banaei-Kashani F, Shahabi C (2010) Voronoi-based geospatial query processing with MapReduce. In: IEEE second international conference on cloud computing technology and science, pp. 9–16

Andreica MI, Tapus N (2010) Sequential and MapReduce-based algorithms for constructing an in-place multidimensional quad-tree index for answering fixed-radius nearest neighbor queries. Acta Univ Apulensis-Mathematics-Informatics (ISSN: 1582–5329), 131–151

Dean J, Ghemawat S (2008) MapReduce: simplified data processing on large clusters. Commun ACM 51(1):107–113
Demir İ, Sayar A (2012) Hadoop plugin for distributed and parallel image processing. In: 20th signal processing and communications applications conference, Mugla, Turkey, pp. 1–4
Demir İ, Sayar A (2014) Hadoop optimization for massive image processing: case study face detection. Int J Comput Commun Control 9(6):664–671
Eken S, Sayar A (2015a) An automated technique to determine spatio-temporal changes in satellite island images with vectorization and spatial queries. Sadhana 40(1):121–137
Eken S, Sayar A (2015b) Big data frameworks for efficient range queries to extract interested rectangular sub regions. Int J Comput Appl 119(22):36–39
Eldawy A, Mokbel MF (2013) A demonstration of spatialhadoop: an efficient MapReduce framework for spatial data. Proc VLDB Endow 6(12):1230–1233
Ergün U, Eken S, Sayar A (2013) Güncel Dağıtık Dosya Sistemlerinin Karşılaştırmalı Analizi. 6. Mühendislik ve Teknoloji Sempozyumu, Ankara, Turkey, pp. 213–218. (in Turkish)
Fox GC, Aktas MS, Aydin G, Gadgil H, Pallickara S, Pierce ME, Sayar A (2009) Algorithms and the Grid. Comput Vis Sci 12(3):115–124
Khlopotine AB, Jandhyala V, Kirkpatrick D (2013) A variant of parallel plane sweep algorithm for multi-core systems. IEEE Trans Comput Aided Des Integr Circuits Syst 32(6):966–970
Liao H, Han J, Fang J (2010) Multi-dimensional index on Hadoop distributed file system. In: IEEE Fifth international conference on networking, architecture and storage, pp. 240–249
Liu X, Han J, Zhong Y, Han C, He X (2009) Implementing WebGIS on Hadoop: a case study of improving small file I/O performance on HDFS. In: IEEE international conference on cluster computing and work-shops, p. 1–8
Lu W, Shen Y, Chen S, Ooi BC (2012) Efficient processing of K nearest neighbor joins using MapReduce. Proc VLDB Endow 5(10):1016–1027
Lu P, Chen G, Oo BC, Vo HT, Wu S (2014) ScalaGiST: scalable generalized search trees for MapReduce systems. Proc VLDB Endow 7(14):1797–1808
Martınez F, Rueda AJ, Feito FR (2009) A new algorithm for computing Boolean operations on polygons. Comput Geosci 35:1177–1185
McKenney M, McGuire T (2009) A parallel plane sweep algorithm for multi-core systems. In: Proceedings of the 17th ACM SIGSPATIAL international conference on advances in geographic information systems, pp. 392–395
Mohammed Al-Naami K, Seker S, Khan L (2014) GISQF: an efficient spatial query processing system. In: 2014 IEEE 7th international conference on cloud computing, pp. 681–688
Mount DM (2004) Geometric intersection. In: Goodman JE, O'Rourke J (eds) The handbook of discrete and computational geometry, 2nd edn. Chapman & Hall/CRC, Boca Raton, pp 857–876
Official Hadoop Web Site, http://hadoop.apache.org/. Accessed 10 Nov 2015
Puri S, Prasad SK (2014) Output-sensitive parallel algorithm for polygon clipping. In: 43rd international conference on parallel processing, pp. 241–250
Rajaraman A, Ullman JD (2012) Mining of massive datasets. Cambridge University Press, Cambridge
Sayar A, Eken S, Mert U (2013) Registering LandSat-8 mosaic images: a case study on the Marmara Sea. In: IEEE 10th international conference on electronics computer and computation, pp. 375–377
Sayar A, Eken S, Mert U (2014) Tiling of satellite images to capture an island object. Commun Comput Inf Sci 459:195–204
Sayar A, Eken S, Öztürk O (2015) Kd-tree and quad-tree decompositions for declustering of 2-D range queries over uncertain space. Front Inf Technol Electron Eng 16(2):98–108
Schneider BO, van Welzen J (1998) Efficient polygon clipping for an SIMD graphics pipeline. IEEE Trans Vis Comput Graph 4(3):272–285
Shvachko K, Kuang H, Radia S, Chansler R (2010) The Hadoop distributed file system. In: IEEE/NASA goddard conference on mass storage systems and technologies, pp. 1–10

Theoharis T, Page I (1989) Two parallel methods for polygon clipping. In: Computer Graphics Forum, vol 8, no 2. Wiley Online Library, pp. 107–114

Wessler M (2013) Big data analytics for dummies. Wiley, Hoboken

Zaharia M, Chowdhury M, Franklin MJ, Shenker S, Stoica I (2010) Spark: cluster computing with working set. In: Proceedings of the 2nd USENIX conference on hot topics in cloud computing, pp. 1–7

Zhang C, Li F, Jestes J (2012) Efficient parallel kNN joins for large data in MapReduce. In: Proceedings of the 15th international conference on extending database technology, pp. 38–49

The Possibilities of Big GIS Data Processing on the Desktop Computers

Dalibor Bartoněk

Abstract The paper submits the method how to solve big projects in the sphere of geographic information systems (GIS). Our aim is to answer the question whether we can or cannot solve similar projects on commonly used hardware and software. This method is based on making use of parallelism and optimization of individual processes. The whole GIS project is divided according to the territory principle into the individual projects which can be processed concurrently. In the frame of sub-projects data optimization of main theme is performed. After the finishing of the particular phases of the project a manual check of partial results follows. The final step consists in completing the separate results into common database. The project was solved for the GasNet, Ltd. Company which is a part of a RWE group in the Czech Republic. Input data were datasets of orthophoto with a resolution of 25 cm/pixel, layers of communications of ZABAGED CR and vector sets of the route of line of underground engineering networks. Due to the territorial coverage of the CR with the area of 64,350 km^2, these were massive tasks with total data volume more than 500 GB. The data analysis was carried out in the special created application in Python language with the support of ESRI libraries and also in ArcGIS 10.0 environment.

Keywords Big data · GIS · Optimization · Desktop computer

D. Bartoněk (✉)
Institute of Geodesy, Faculty of Civil Engineering, Brno University of Technology, Veveří 330/95, 602 00 Brno, Czech Republic
e-mail: bartonek.d@fce.vutbr.cz

D. Bartoněk
European Polytechnic Institute, Osvobození 899, 686 04 Kunovice, Czech Republic

I. Ivan et al. (eds.), *The Rise of Big Spatial Data*, Lecture Notes in Geoinformation and Cartography, DOI 10.1007/978-3-319-45123-7_20

1 Introduction

The problem dealing with total data volume processing is very topical. Nowadays, modern technologies and their implementation in all possible fields of human activities have caused the generating of great data volume which after the cumulation within a certain period can be marked by the term "Big Data". It concerns data volume orderly terabyte to petabyte. This trend has appeared even in spatial systems including GIS. Navigation systems, Location Based Services etc. are still the sources of continued data growth and need to be effectively processed. This is a highly up-to data topic which deserves a great attention of various levels (academic sphere, state administration, professional applications and other practical usage).

Big data processing is mostly carried out on efficient computers with the special architecture and with the help of sophisticated algorithms. The aim of this paper is to answer if it is possible to solve even so extensive GIS projects on commonly accessible hardware and software within small research teams (up to 5 persons). In conclusion it has been proved that it is possible provided, there is a harmony of factors having decisive influence on the solving of a given problem. These are factors:

- General management of the project (task and applicants management).
- Making use of parallelism in a given task.
- Suitable option and usage of hardware and software.
- Optimization of processing within a chosen software.
- Continuous check of the results quality and time limit.

Realization of extensive GIS project in accessible means in a given term and quality is possible only in maximum harmony of all the above mentioned factors.

The article is involved with the publications (Bartoněk et al. 2014a) and (Bartoněk et al. 2014b), describing the solution of GIS project—classification of the surface above underground gas facilities on the territory of Czech Republic. Within this project it had been proved that it is possible to solve even so extensive GIS projects on commonly accessible hardware and software within small research team (3 persons) in a given term and quality.

2 Related Works

There is a great number of publications dealing with this problem. They can be divided into these categories:

a. data processing from the point of view of inner computer architecture: communication of external memory (disc)—buffer—cache memory L1, L2, CPU registers,

b. access from the point of view of data-driven model e.g. data pre-processing, selection, classification, aggregation with the goal to make use of the results for data mining,
c. access from the point of view of conception and organization (task management) as a complex methodology of solving big data within a given project. Similar methodology is described in this paper.

In the individual categories only typical publications were chosen.

Ad (a) analysis of data volume processing from the point of view of the architecture of computer memory subsystem.

Typical representative of this category is literature overview work (Zhang et al. 2015). The authors are witnessing a revolution in the design of database systems that exploits main memory as its data storage layer. Many of these researches have focused on several dimensions: modern CPU and memory hierarchy utilization, time/space efficiency, parallelism, and concurrency control. In this survey, is provided a thorough review of a wide range of in-memory data management and processing proposals and systems, including both data storage systems and data processing frameworks. This article also gives a comprehensive presentation of important technology in memory management, and some key factors that need to be considered in order to achieve efficient in-memory data management and processing.

Ad (b) approach from the point of view of data stream:

The paper Miller and Goodchild (2014) deals with Big Data and data-driven geography. A data-driven geography may be emerging in response to the wealth of georeferenced data flowing from sensors and people in the environment. Although this may seem revolutionary, in fact it may be better described as evolutionary. Some of the issues raised by data-driven geography have in fact been longstanding issues in geographic research, namely, large data volumes, dealing with populations and messy data, and tensions between idiographic versus nomothetic knowledge. The belief that spatial context matters is a major theme in geographic thought and a major motivation behind approaches such as time geography, disaggregate spatial statistics and GIS-science.

The publication Dahlstrom and Harms (1997) describes the GIS that being implemented at Jackson (Tenn.) Utility Division (which serves residents of Jackson and Madison County mid-way between Nashville and Memphis) illustrates the point. The GIS being implemented at JUD includes core utility GIS functionality and is fully integrated with the Customer Information System (CIS) residing on an IBM AS/400. The presentation will offer an overview of the GIS and related technologies at JUD, and will review the system's functionality as it is configured for four different utilities: electric, gas, water, and wastewater. The presentation will also review the processes behind implementing a project of this scale, including developing executive support, addressing funding considerations, leveraging existing technology investments, and achieving "big system" objectives at a smaller scale.

In paper Lu and Zhang (2014), the distinctive characteristics of generalized geo-spatial information are investigated, and the urgent need for a more generalized concept of GIS is clarified. Then the technical challenges for general GIS are set forward in terms of data collection and cleaning, data management and integration, and data analysis and computing. The progress is summarized and the research issues are discussed in geo-spatial data collection with Internet text mining, moving object database, dynamic and heterogeneous data management, moving trajectory data mining, and complex network analysis. The fusion of geo-computation, urban computing and social computing in the near future is considered at the end of the paper.

The paper Yue and Jiang (2014) gives an overview of recent methods in supporting big data management and analysis in geospatial domain. First, it motivates the necessity to develop and use the Big Data-aware GIS software. By reviewing advanced information technologies and approaches, it can assess what operational system framework and approaches are available and applicable in developing Big Data-enabled next-generation GIS—Big GIS. Key considerations for development of Big GIS are highlighted. The results can help identify critical issues and direct future research agenda for next-generation GIS.

In paper Zhu et al. (2003), the applications of remote sensing and GIS (Geographical Information System) technologies in BJVSIS (planning and development Information System for Beijing Virescence Separator), including three dimensional (313) visualization, integrated databases management and multi-media technologies, were presented. The initial pilot project of BJVSIS was finished successfully in 2001, and up to now this information system runs very well and plays a great role in the construction of the Virescence Separator. Just depended on the applications of so many high-tech like GIS, remote sensing, 3D visualization, and databases integration, BJVSIS can manage all kinds of relevant geographic data, such as vector graphics, DEM (Digital Elevation Model), image and multi-media attributes. And BJVSIS also provides many GIS functions, such as spatial query, 2D/3D display, spatial analysis, statistic calculation, form report and map output.

Publication Kraemer and Senner (2015) presents modular and flexible system, which supports multiple algorithm design paradigms such as MapReduce, in-memory computing or agent-based programming. It contains a web-based user interface where domain experts (e.g. GIS analysts or urban planners) can define high-level processing workflows using a domain-specific language (DSL). The workflows are passed through a number of components including a parser, interpreter, and a service called job manager. These components use declarative and procedural knowledge encoded in rules to generate a processing chain specifying the execution of the workflows on a given cloud infrastructure according to the constraints defined by the user.

Paper Schoier and Borruso (2015) presents an approach to clustering of high dimensional data which allows flexible approach to the statistical modelling of phenomena characterized by unobserved heterogeneity. The approach is compared with classical k-means algorithm. The application concern a synthetic data set and a data set of satellite images.

The publication Lu et al. (2008) deals with the 1100 kV GIS disconnector. This electrical equipment was used in the 1000 kV UHV AC testing demonstration pilot project. Based on two different kinds of structures, three dimensional electric field of 1100 kV SF6 disconnector airchamer are analysed by using the finite element software. For the complex Analysis the extra-high voltage disconnector airchamer, its volume is big and the structure is complex, applies in the different line situation. By analysing the electrical field intensity distribution of disconnector airchamber, the distribution nephogram of the whole area and the maximal electric field intensity area are attained and the comparisons between two distributions are done. Then theoretical basis for construction design of disconnector is supplied according to the analysed results of electrical field.

Ad (c) approach from the point of view of conception, management and task processing:

The architectural solution proposed in Bica et al. (2014) is based on cloud virtualization and aims to provide a flexible and adaptive method to extract and highlight knowledge from the huge data of Earth Observation images. The problem of data processing is solved by virtualizing algorithm objects which are placed in key positions on the hardware level. The users could describe and experiment complex use cases and take advantage of the improvement on execution performance provided by the flexible description and the adaptive processing on high performance computing platforms.

The objective of paper Idrizi et al. (2015) is to introduce the methodology utilised for selecting the most appropriate GIS software for developing a GIS application about 1 year ago. All data in the eight tables represent are real data, which compare eight different software packages. The methodology for the analyses contains eight tables, with eight different topics which are directly related to cost benefit analyses, types of software support by providers and developers, difficulties in developing, using and upgrading GIS applications, system and hardware requirements, and development languages. From the analyses performed, data in three tables (system requirements, hardware specification and development languages) did not show significant differences between different GIS software, so they were excluded from further analysis to select the most appropriate GIS software.

In the work Wang et al. (2015) a special algorithm was developed for time reduction during polygon overlay in big data processing in GIS. The proposed algorithm can reduce times of calling intersection computation by the aid of grid index. Experimental results show that algorithm with spatial grid index consumes less time than the its peer without spatial index. Nevertheless, with the increase of nodes, the upward trend of speed-up ratio slows down.

In paper Frye and McKenney (2015) a special algorithm was developed for big data addressing. The approach includes map/reduce algorithms generating on-demand indexes and maintaining persistent indexes. The paper reviews various approaches, categorizes the spatial queries reported in the testing, summarizes results, and identifies strengths and weakness with the approach.

The publication Boton et al. (2015) characterizes Architecture Engineering and Construction project management data following the conceptual definition of big data and proposes a high level conceptual pipeline aiming at bridging the gap between Building Information Model—based related visualization works and information visualization domain.

In Mondai and Dutta (2015) scalable distributed framework was developed. This framework has two design characteristics: (i) they are using memory scalability in such a manner that the amount of memory required by each process decreases as the number of processes used to solve a given problem instance increases, and (ii) they exploit coarse grain parallelism in the sense that they structure their computations into a sequence of local computation followed by communication phases in which the local computations take a non-trivial amount of time and often involve a non-trivial subset of the process' memory.

Shahrokni et al. (2014) present preliminary findings from a big data analysis and GIS to identify the efficiency of waste management and transportation in the City of Stockholm. The aim of this paper is to identify inefficiencies in waste collection routes in the city of Stockholm, and to suggest potential improvements. Based on a large data set consisting of roughly half a million entries of waste fractions, weights, and locations, a series of new waste generation maps was developed.

All the mentioned publications dealing with the processing of Big Data from different perspectives. None of them, however, submits whether it is possible to process large GIS projects on commercially available hardware and software resources (hardware and software) in a small research team. This is the subject of this article.

3 Theoretical Foundations

Geographic information system (GIS) means five items: *(U, A, V, S, f)*, where *U, A, V, S* are finite sets *f* is a mapping of *U x A* to $V - f$: $U \times A \to V$ in this interpretation: the elements of set *U* are geographical objects (GO), elements of set *A* are attributes of GO, elements of set *V* are values of these attributes; $f(u, a) = v$ indicates, that attribute *a* takes for GO = u value *v*. Set *S* is set of geo-reference coordinate systems. This conception of GIS is based on Pawlak's extension sense of information system (Pawlak 1981). It is differ of the classical structural point of view on information system, where hardware, software, users, data and other components are the basic elements.

Let a non-empty set $U \neq \emptyset$ is the set of GO and *X* is a subset ($X \subseteq U$). The *U* set represents in our case the entire solved area and the *X* subset is part of the solved area. The relation of equivalence *R* divides the *U* set into subsets $U/R = \{X_1, X_2, \ldots, X_n\}$ such that for every *i, j* holds:

(1) $X_i \subseteq U$, $X_i \neq \emptyset$ (all subsets are non-empty),
(2) $X_i \cap X_j \neq \emptyset$ (the intersection of all subsets is empty),
(3) $U_{i=1,2,..,n}, X_i = U$ (the union of all subsets is just the whole *U* set).

The R relation of equivalence can be considered as territorial division of the Czech Republic according to an administrative arrangement that divides this area into lower administrative units (regions, districts, etc.). If the conditions in (1)–(3) are complied, each sub-area in terms of set theory has a character of the X_i class.

Suppose there is another relation $S \subseteq R$ such that defines decomposition of subsets $X_i \subseteq X$ into equivalent X_{ij} classes with same (1)–(3) characteristics:

$$X_i/S = \{X_{i1}, X_{i2}, \ldots, X_{im}\}, \tag{1}$$

where for all $i = 1, 2, \ldots, n$.

Then the system of $\{R, S\}$ relations together with U, X sets represents a hierarchical decomposition of the U set. The S relation in our case defines the further division of territorial units represented by X_i classes into the lower territorial units according to a material principle. The criterion of this division is the fact if relevant geographic objects (GO) are or not are in a given area, i.e. elements that are the main subject of processing in the given GIS project (this is the route of gas line in our case). So the S relation divides areas into X_{ij} subclasses, where for $j = 1, 2, \ldots, m$ according to Eq. (1). The hierarchic decomposition of the set $X \subseteq U$ into equivalent classes is the result of a combination of criteria of subdividing according to:

- the administrative arrangement of territory in the first hierarchical level,
- thematic arrangement, i.e. according to the existence of GO in the second hierarchical level.

Let X is the set of GO in the real world and Y is the image of the X set in digital geo-database. Then the transform $\varphi: X \rightarrow Y$ must have the following characteristics:

- φ is one-unambiguous function,
- φ is continuous function,
- there is an inverse transform $\varphi^{-1}: Y \rightarrow X$, that is also continuous.

Then the φ transform is a homeomorphism, which is fundamental property of topological relations between GO. In our case, the GO (the X set) are the input datasets and the Y set are the relations of these datasets in a digital database of project. These properties are very important for correctness automatic processing of the all tasks within the GIS project.

Next, we define the coefficient of territorial details:

$$\delta = \frac{P_{tot}}{P_{det}}, \tag{2}$$

where P_{tot} is the total area of the modelled territory and P_{det} is the area of the smallest detail, that is contained in the input data. It will be decided on the base of this coefficient according to which sub-territorial units within the administrative structure of the Czech Republic (the R relation, i.e. decomposition in the first hierarchical level) the input datasets of project will be divided. In case of thematic division is the coefficient used individual for each theme.

Let's define the total time of solving of the whole project T_{tot}:

$$T_{tot} = T_{rez} + T_{pa} + T_{pm} \tag{3}$$

where T_{rez} is the overhead time, T_{pa} is the time of automated processing and T_{pm} is manual processing.

Assuming the distribution of the entire project into n partial projects (Eq. 1), then the Eq. (3) has the form:

$$T_{tot} = T_{rez} + \sum\nolimits_{i=1}^{n} (t_{pai} + t_{pmi}) = \min \tag{4}$$

In terms of the efficiency, it is necessary that the overhead time does not exceed the processing time, i.e. the ratio of the processing time $T_{pa} + T_{pm}$ and the overhead time T_{rez} must be maximum (at least greater than 1):

$$\eta = \frac{\sum_{i=1}^{n} (t_{pai} + t_{pmi})}{T_{rez}} = \max \tag{5}$$

The Eqs. (4) and (5) represent the functions to be optimized.

Let's set S be a set of datasets of the project results, the T is dataset of the same type as S and attributes of T have higher quality than attributes of S. Than we can refine the outputs by rewriting values of attributes of dataset S by the values of attributes of dataset T:

$$\text{If } S \cap T \neq \varnothing \text{ then for } a_t \rightarrow a_s \tag{6}$$

where $\cap$ is spatial overlay of datasets, a_s is attribute of $s \in S$ and a_t is attribute of $t \in T$.

4 Method of Solution

The basic approach was the division of the project into component parts and the use of parallelism to speed up processing. For a successful project, it was necessary to competently determine the parameters in Eqs. (1)–(6) as follows:

- optimum processing unit (sub-project) (Eq. 1) in terms of efficiency and reliability,
- δ factor that determines the data requirements of the project (Eq. 2),
- the value of the overhead time periods automated and manual processing, and efficacy (Eqs. 3, 4, 5),
- choice of reference datasets T to improve the quality of results (Eq. 6),
- selection of suitable software,
- selection of appropriate available hardware (number, configuration, etc.).

Mentioned parameters were determined solution to the pilot project of the same type as the main GIS project in a limited area. Parameters for the main project were then determined by extrapolation on the basis of the relative extents of the modelled area.

The proposed methodology for solving large GIS project is shown in Fig. 1. The entire project is divided into sub-projects under the territorial principle. It used a territorial identification register (dial territorial units). To correct automated processing is important that the structure of territorial identification complies homeomorfism mentioned in the previous chapter. Data for sub-project is divided into templates and stored in a directory structure. The names of directories is appropriate to identify the unique identifier in accordance with the code of the territorial unit. Followed by reduction of the volume data based on the main character roles (optimization by topic). It goes e.g. to create a buffer or reduce the extent of the modelled area based on the position of significant elements. Than follows the main processing i.e. the application of spatial analytic functions. The final phase is to improve the results by filtering of the reference data set (Eq. 6). After each of these individual phases performs manual inspection of results and errors correction. In conclusion, all partial results are stored into database. These division is suitable only for this project type. In case of thematic approach we have to divide the whole task according to thematic principle. If we need the whole dataset e.g. in task like the shortest path than we must find independent sub-processes, that can be solved concurrently.

To ensure maximum extent of automation of the entire process, it is necessary to use either the existing software or create custom application. In our case, a set of scripts in Python was used—see the next chapter. It is recommended to use these procedures:

- The procedure for dividing data into individual templates. The control data structure can be territorial identification register with the property homeomorfism.
- The procedure for data reduction (optimization) in terms of the main theme (creation of a buffer or other restrictions).
- The procedure for pre-processing of data in sub-templates (editing of data, creation of auxiliary data sets or structures).
- The procedure for processing in the main sub-projects.
- Procedure to improve the results (the reduction of error rate) using the appropriate reference layer according to Eq. (6).
- The procedure for compiling the results into a common data set or database table.

A prerequisite for success is the optimal composition of the research team. The practice has proved that minimum composition is of three responsible investigators: manager, providing project management as a whole, GIS programmer who creates the above procedure for automated processing and employee for manual and visual inspection of all intermediate results.

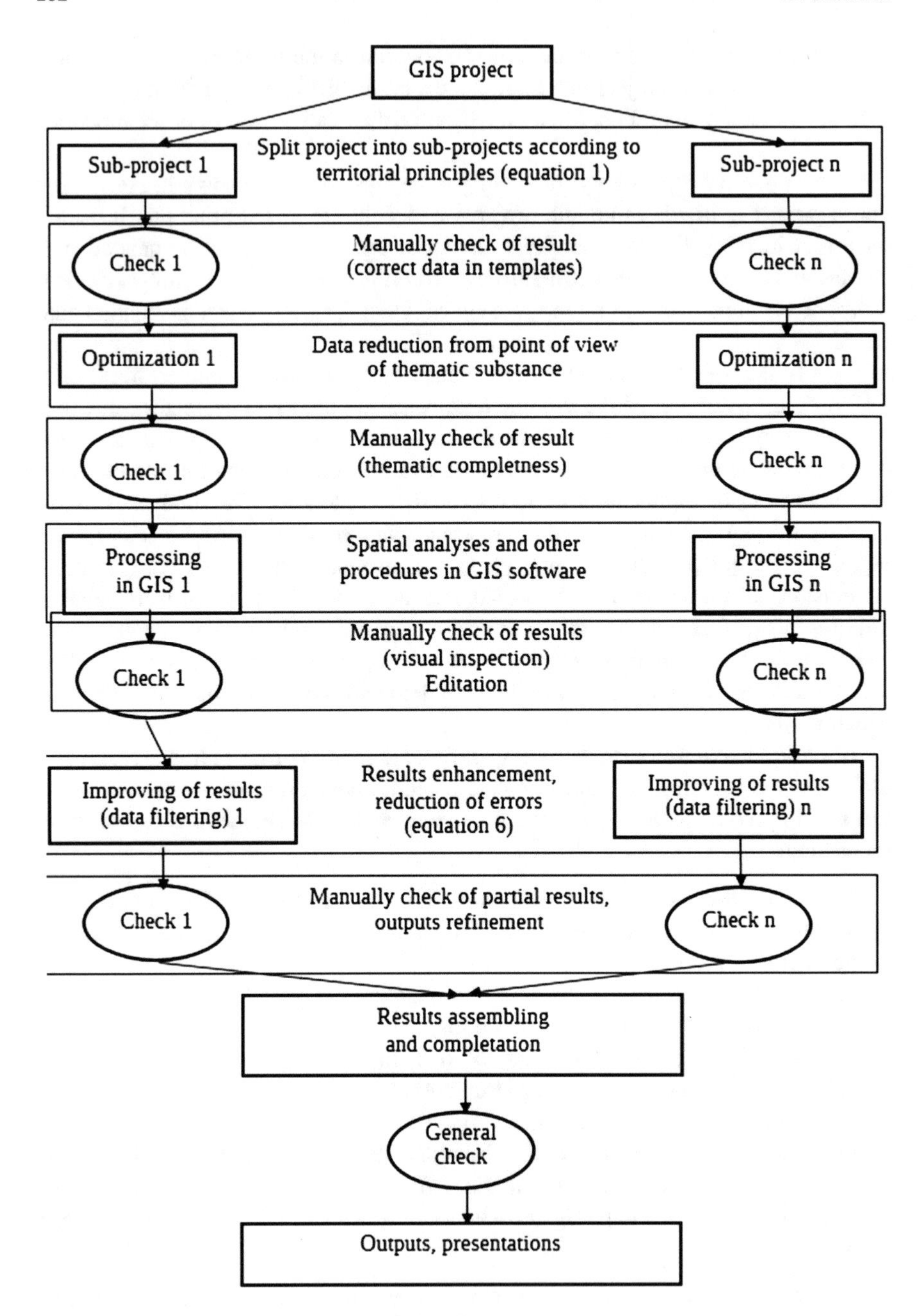

Fig. 1 Workflow of processing of big GIS project

5 Experimental Results

The proposed method was tested in the project of data analysis of storage of gas facilities under certain types of terrain surface in the Czech Republic (CR). The authors solved this project for the GasNet, Ltd. Company which is a part of a RWE group in the Czech Republic. Input data were datasets of orthophoto with a resolution of 25 cm/pixel, layers of roads of ZABAGED CR (fundamental base of geographic data) and vector sets of the route of line of underground engineering networks. Due to the territorial coverage of the CR with the area of 64,350 km^2, these were massive tasks with total data volume of 500 GB.

For quality estimation of parameter values—see relations (1)–(6) was solved the initial project modeled on the city of Brno (CR) with an area of about 250 km^2 (Bartoněk et al. 2014a). In this test project have been set this strategy.

- Optimal territorial unit ORP (municipality with extended power)—see Eq. 1, range ORP is an average of 368 km^2, the average amount of data for ORP is 2.5 GB. These territorial unit was set experimentally in project (Bartoněk et al. 2014b).
- The optimum ratio of η (Eq. 5) for the total of amount of data $x = 500$ GB is based on the following values: $n = 200$, $x_i = 2.5$ gigabytes.
- Optimization of data according to the solved topic was made by cropping orthophoto with buffer 1 m along the pipeline. The buffer value was consulted with RWE company. This data reduction achieved from the original 500 GB to 230 GB.
- After each partial automated processing of ORP were carried out manual checking of partial results.
- The duration of the overhead time, automated and manual processing were determined by the graph in Fig. 2, and after putting the values into a relationship 6 we get relationship 7:
- On the basis of the relation 7 was estimated duration of the entire project for 4 months.
- The ratio of manual and automated processing is 55:45 %.

The implementation of the project was done on machines of type of PC Integra 7025 (parameters: Intel Core i5, 3.8 GHz, 16 GB RAM, NVIDIA GTX650, 2 GB, HD of SDD type and VelociRaptor), which were interconnected to the computer network of 100 Mb/s—see Fig. 3. Main criterion for these decision was value of coefficient $\delta = 10^9$. In the test project, it was found that the higher value of the coefficient δ means the use of more powerful computers.

$$\eta = \frac{\sum_{i=1}^{n} 0,02e^{0,4x_i}}{\sum_{i=1}^{n} 0,002x_i^2 + 0,003x_i + 0,04} \tag{7}$$

$$T_{rez} = 0,002x^2 + 0,003x + 0,04, \quad T_{pa} = 0,02e^{0,4x}$$

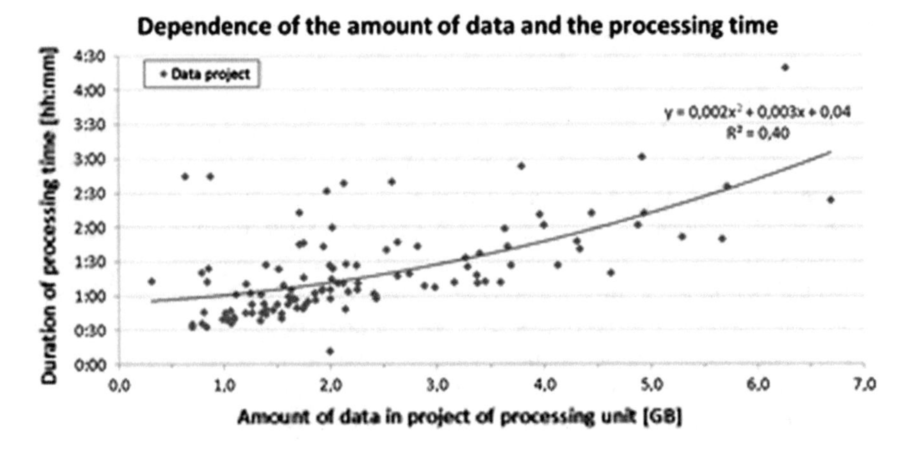

Fig. 2 Dependence of the overhead time T_{rez} on the volume of processed data

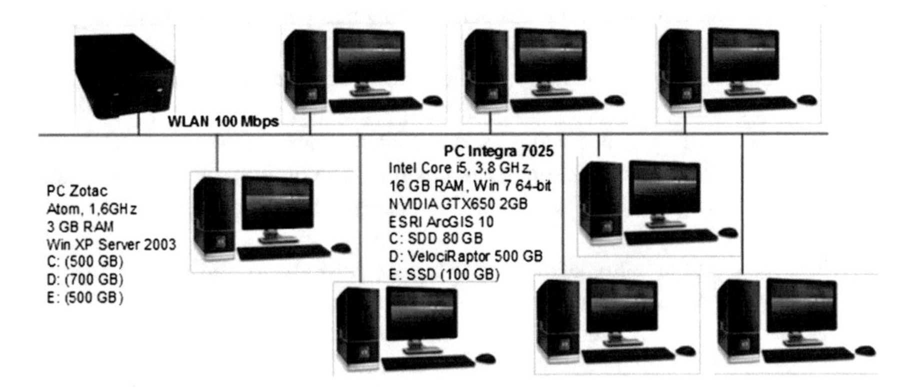

Fig. 3 Hardware for project solution

- The data analysis was carried out in the assigned created application in Python language with the support of ESRI libraries and also in ArcGIS 10.0 environment.
- The main method of solution was supervised classification of raster image (orthophoto) using Maximum Likelihood method in ArcGIS 10.0. However, this classification showed only an average success rate of 70 % and this value was insufficient.
- The input data was supplemented by dataset of 9 layers of communications, which is a part of the ZABAGED (fundamental base of geographical data of the Czech Republic). This resulted in a more accurate results of classification (Eq. 6).

- The results demonstrated the high efficiency of the technology and the low error rate of the classification in the range of 2–3 % was achieved over the whole modelled area (CR).
- The partial results have been written to the Excel sheet, which was open to researchers from the web. This will avoid duplication of processing of individual ORP and improve the overall management of the whole project.

6 Conclusions

It was proposed methodology for solving large GIS projects and verified the possibility of their processing on current desktop hardware and software products. The methodology consists in data-driven procedures realized in Python script with the support of ESRI geoprocessing libraries.

The main features of the proposed methodology are:

- optimum division of the project into sub-sub-projects under the territorial principle,
- parallel processing sub-projects on an appropriate hardware and software,
- optimization of data volume by a thematic principle,
- ensuring the quality of the results by the manual control of output of individual sub-projects,
- increase the quality of results (reduction of error) using the appropriate reference data set—in our case ZABAGED (fundamental base of geographic data in the CR)

Prerequisites for the successful solution of the project are:

- Quality management of the project, coordination of sub-project phases, interoperability and interchangeability of the research team within the individual processing tasks,
- Solution of model pilot project of the same type as the main project in a limited area in order to obtain the necessary values of parameters for determining the above mentioned decision.

The technology described in this paper has general character and it can be used for classification of the surface above linear utilities such as water supply, gas line, electric power distribution, media distribution etc.

Acknowledgments The work was solved within the project marked FAST-S-15-2723 and within research project of MŠMT (Ministry of Education of the Czech Republic) AdMaS ED2.1.00/03.0097 Nr. HS12357021212200 “Data analysis of surfaces above RWE gas lines in the Czech republic”.

References

Bartoněk D, Bureš J, Opatřilová I (2014a) Technology of processing of enormous amounts of geographical data. In: 14th international multidisciplinary scientific geoconference and EXPO, SGEM 2014; Albena; Bulgaria; 17 June 2014 through 26 June 2014; Code 109739, 3(2): 917–924

Bartoněk D, Bureš J, Opatřilová I (2014b) Optimization of pre-processing of extensive projects in geographic information systems. Adv Sci Lett 20(10–12):2026–2029. American Scientific Publishers. doi:10.1166/asl.2014.5664. ISSN:19366612

Bica M, Bacu V, Mihon D, Gorgan D (2014) Architectural solution for virtualized processing of big earth data. In: IEEE international conference on intelligent computer communication and processing (ICCP) 2014, pp 399–404. doi:10.1109/ICCP.2014.6937027. Print ISBN: 978-1-4799-6568-7

Boton C, Halin G, Kubicki S (2015) Challenges of big data in the age of building information modeling: a high-level conceptual pipeline. In: Cooperative design, visualization, and Engineering, pp 205–216

Dahlstrom S, Harms A (1997) Successfully integrating multiple utilities and supporting technologies. In: AM-FM-international conference XX-entering the mainstream. Nashville, TN, pp 691–701

Frye R, McKenney M (2015) Big data storage techniques for spatial databases: implications of big data architecture on spatial query processing. In: Information granularity, big data, and computing. Springer Science, pp 560–586

Idrizi B, Zhaku S, Izeiroski S (2015) Defining methodology for selecting most appropriate GIS software. Surv Rev 46(338):383–389

Kraemer M, Senner I (2015) A modular software architecture for processing of big geospatial data in the cloud. Comput Graph UK 49:69–81

Lu F, Zhang H (2014) Big data and generalized GIS. Geomat Inf Sci Wuhan Univ 39(6):645–654

Lu L, Xin L, Jianyuan X (2008) Numerical analysis of electric field for 1100 kV disconnector in GIS based on two kinds of structures. In: International conference on high voltage engineering and application, Chongqing, Peoples Republic of China, pp 527–530

Miller HJ, Goodchild MF (2014) Data-driven geography. GeoJournal 80(4):449–461

Mondai K, Dutta P (2015) Big data parallelism: challenges in different computational paradigms. In: Proceedings of the 2015 3rd international conference on computer, communication, control and information technology, C3IT 2015, 12 March 2015, Article number 7060186

Pawlak Z (1981) Information system theoretical foundations. Inf Syst 6:205–218

Schoier G, Borruso G (2015) On the problem of clustering spatial big data. In: Osvaldo G, Beniamino M, Sanjay M, Gavrilova ML, Ana Maria Alves Coutinho R, Carmelo T, David T, Apduhan BO (eds) Computational science and its applications. Proceedings, Part I: 15th international conference, Banff, AB, Canada, 22–25 June 2015, pp 325–338. ISBN 978-3-319-21403-0 (Print) 978-3-319-21404-7 (Online)

Shahrokni H, Van Der Heijde B, Lazarevic D, Brandt N (2014) Big data GIS analytics towards efficient waste management in Stockholm. In: 2nd international conference on ICT for sustainability, ICT4S 2014; Stockholm; Sweden; 24 August 2014 through 27 August 2014; Pages 140–147, Code 111677

Wang Y, Liu Z, Liao H, Li C, (2015) Improving the performance of GIS polygon overlay computation with MapReduce for spatial big data processing. In: Salim H (ed) Cluster computing, Springer Science + Business Media, New York, pp 506–516. ISSN: 1386-7857 (Print version), ISSN: 1573-7543 (Electronic version)

Yue P, Jiang L (2014) BigGIS: How big data can shape next-generation GIS. In: The 3rd international conference on agro-geoinformatics, agro-geoinformatics 2014, Beijing, China, 11 August 2014 through 14 August 2014. Category number CFP1448T-ART, Article number 6910649, Code 114697

Zhang H, Chen G, Ooi BC, Tan K-L, Zhang M (2015) In-memory big data management and processing: a survey. IEEE Trans Knowl Data Eng 27(7): 1920–1948. ISSN:1041-4347

Zhu XY, Zhu Q, Zhang YH (2003) Applications of remote sensing and GIS technologies in the planning and development information system for Beijing virescence separator. In: 2nd Conference on remote sensing for environmental monitoring, GIS applications and geology. Book series: proceedings of the society of photo-optical instrumentation engineers (SPIE), vol 4886, pp 107–114

Utilization of the Geoinfomatics and Mathematical Modelling Tools for the Analyses of Importance and Risks of the Historic Water Works

Lucie Augustinková, Vladimír Fárek, Jiří Klepek, Aneta Krakovská, Martin Neruda, Iva Ponížilová, Marek Strachota, Jan Šrejber, Jan Unucka, Vít Voženílek, Ivo Winkler and Dušan Židek

Abstract Old and historic water works are the subject of interdisciplinary interest, which covers the spectrum from hydrology and environmental protection to heritage preservation. The reason is the very meaning of these works in terms of nature and landscape protection and technical landmarks conservation, but also their impact on the water dynamics in the landscape during the normal and extreme runoff situations. Because the documentation of these objects is often conforming to their age and changes in ownership together with the issues about the standardization of these data, and often not even in the digital form. However, Czech Hydrometeorological Institute is responsible for the hydrometric and expert

L. Augustinková · I. Ponížilová
Faculty of Civil Engineering, VSB-Technical University of Ostrava,
Ludvíka Podéště 1875/17, 708 33 Ostrava-Poruba, Czech Republic
e-mail: lucie.augustinkova@vsb.cz

I. Ponížilová
e-mail: iva.ponizilova@vsb.cz

V. Fárek · J. Šrejber
Czech Hydrometorological Institute Ústí Nad Labem,
Kočkovská 18/2699, 400 11 Ústí nad Labem, Czech Republic
e-mail: vladimir.farek@chmi.cz

J. Šrejber
e-mail: jan.srejber@chmi.cz

J. Klepek · A. Krakovská · J. Unucka (✉)
Faculty of Mining and Geology, VSB-Technical University of Ostrava,
17. Listopadu 15/2172, 708 33 Ostrava-Poruba, Czech Republic
e-mail: jan.unucka@vsb.cz

A. Krakovská
e-mail: aneta.krakovska@vsb.cz

J. Klepek
e-mail: jiri.klepek@vsb.cz

I. Ivan et al. (eds.), *The Rise of Big Spatial Data*, Lecture Notes
in Geoinformation and Cartography, DOI 10.1007/978-3-319-45123-7_21

surveying on these objects. Above mentioned aspects formed the main motivation for the progressing cooperation with partners on the issues such is the methodical data collection and evaluation of these objects in pilot areas using the tools of the hydrometry, geoinformatics and mathematical modelling. Existing results of the hydrometrical measurements, GIS analyses and hydraulic modelling within the pilot areas together with the other tasks and issues to be solved are presented in this paper.

Keywords Flume · Small water structures · Hydrometry · GIS analyses · Hydraulic modelling

1 Introduction and the Aim of Study

Water races or flumes, as a human's creations used for redistribution of water in the countryside, is a phenomenon which worth an attention in several respects. Its origin is proven and documented from the very first civilization, including the Indus Valley Civilization (see e.g. Tainter 2009) through the Roman period and the Middle Ages (see Squatriti 2000) to the onset of the Industrial Revolution. Even Karl Wittfogel emphasizes the meaning of the redistribution of water, through the artificial constructions at the earliest civilizations, with the term of "hydraulic civilization" (in Bárta and Kovář 2013). An interesting point of view about the importance of a small water structures in the mountain, forest or agricultural watersheds can be found in Solnický (2007), Štěpán et al. (2008), Augustinková et al. (2011) or Belisová (2014). Fascinating fact on these hydraulic buildings and landscape sightings, with no exaggeration, is that many medieval water structures of this type still serves its purpose, such as gutters in Třeboňsko or Poodří nature

M. Neruda
Faculty of Environment, Jan Evangelista Purkyně University,
Králova Výšina 3132/7, 400 96 Ústí nad Labem, Czech Republic
e-mail: martin.neruda@ujep.cz

M. Strachota · J. Unucka · I. Winkler · D. Židek
Czech Hydrometorological Institute Ostrava, K Myslivně 3/2182,
708 00 Ostrava-Poruba, Czech Republic
e-mail: marek.strachota@chmi.cz

I. Winkler
e-mail: ivo.winkler@chmi.cz

D. Židek
e-mail: dusan.zidek@chmi.cz

V. Voženílek
Faculty of Science, Palacký University Olomouc, 17. Listopadu 50,
771 46 Olomouc, Czech Republic
e-mail: vit.vozenilek@upol.cz

reserve. Also nowadays slope or super elevation flume paremeters, which the former builders coped without using any modern geodetic methods, are incredible. As well as the technical and architectural design of buildings themselves. This is also a reason, why it is completely understandable, that some of them are subjects of the nature protection (Act no. 114/92 Coll.) and care of the historical monuments (Act no. 20/1987 Coll.). However, there is still no methodical approach to their records and subsequent classification of hydrological, water management, historical or other significance. Perhaps, because approach to the appreciation of their importance would vary in opinion of the water managers or historian. The studies and its evidence is necessary in this case, as their basic function and attribute is to transfer a water within the basin or between the basins, what presents an indispensable problem in drainage characteristics of the territory and creation of essential data for zoning during the low water period. Another aspect in some cases is inconvenient technical conditions of buildings after the change of owners, so that the negative impacts can include for example seepages or alluviation race by the sediments, what often leads to their dysfunction without actual possibility of rehabilitation or liquidation (unclear ownership, etc.). There is no doubt that these structures are often located on plots of forest land resources, because their task was to redistribute water for rafting of wood, operation saw mills or small textile factories. These races can be found in relatively large amounts on the northern part of the Sudetenland, where team of authors concentrates and searches for parallels e.g. between Bohemian Switzerland National Park, Lužické hory nature reserve or Jeseníky nature reserve at the moment. So we can tell, that this regard states are important in terms of forest and small watershed hydrology as well. This article focuses on a creating, if possible, a consistent methodical approach within the mapping and documentation of races as a phenomenon, which is affecting the local drainage conditions and also water structures like an important and interesting aspects.

2 Used Methods

Authors focused on the pilot area and several objects, where have been taken identical steps in their inventory and analysis functionality, hydrological significance and potential risks of individual objects. These following steps can be divided into several areas of activities:

1. Preliminary analysis of the territory in GIS.
2. Field mapping and localizing of these constructions (geodetic GNSS, total stations etc.).
3. Hydrometric measurements using various types of the hydrometric propellers, ADCP, ADV flow tracker and inductive flow meter (for measurement principles see Boiten 2008).

4. Creation of mathematical models for analysis of hydrological aspects (transfers of water in the regime of large and small water contents, seepages, etc.).
5. Calibrating of the models according to the measured data.
6. Creation of the GIS geodatabase, register of the layer races with an uniform structure of the attribute table.

The above steps lead to the final thus registration these constructions, the structure of the attribute table allows queries and filtering so that it is possible to select constructions according to selected criteria. At this stage, it is about the design and testing of the suitability of geodatabase structure, so the structure may be customized by requirements of the user. Apart from the register and territorial attributes were (due to the prevailing professional focus of authors) emphasized selected hydrologic and hydraulic characteristics of the objects. Introductory GIS analysis of the pilot areas were focused on selected morphometric parameters DTM (vertical alignment, depth and width of the valley, density of the valley networks etc.) using the DTM 4G ČÚZK (Czech State Administration of Land Surveying and Cadastre). From the GIS modules for these DTM analyzes were used platforms ESRI ArcGIS and extensions Spatial Analyst, 3D Analyst, Network Analyst and ArcHydro, and also GRASS GIS and SAGA GIS. Further, through the extension of HEC-GeoHMS and GeoHMS-AddIn conducted schematization for rainfall-runoff model HEC-HMS, which is cares about the balance calculations and scenarios of affecting runoff conditions.

Geodetic localization of construction was carried out with the use of GNSS technologies. There has been used dual frequency GNSS equipment, which allows to localize objects by high positional accuracy. This accuracy is influenced by the quality of received signal available by satellites and the signal of mobile operators for receiving correction data from the reference stations CZEPOS. CZEPOS is a network of permanent stations GNSS, which is managed and operated by State Administration of Land Surveying and Cadastre. Most of the objects of interest is situated on forested land that is way measurement carried out in extra-vegetative period. This ensured the highest possible signal and thus it achieved the highest possible accuracy. Current geometric/topographic and elevation data for individual races have been completed and specified with field measurement data. Also, there have been localized sites with visible seepages. Further there were cross sections of races focused on using GNSS technologies, especially at the Ploskovický potok, which unlike races in Žimrovice and Hanušovice do not have regular channel. Race for Hübel weaving mill has not been aimed, as it is dry, clogged with sediment and dysfunctional. The resulting data were used as a basis for schematization hydraulic model HEC-RAS and MIKE 11.

Hydrometric measurements (excluding flume for Hübel weaving mill) proceeded on several stages in use of conventional hydrometric propeller (OTT 31 and OTT C-2), ADV SonTek FlowTracker, OTT Nautilus 2000 and OTT MF-Pro inductive device and ADCP RDI Teledyne StreamPro and SonTek M9. Measuring the profile velocity was used for subsequent parameterization and calibration of hydraulic models and created flows rating curve. Measuring programs using ADCP has

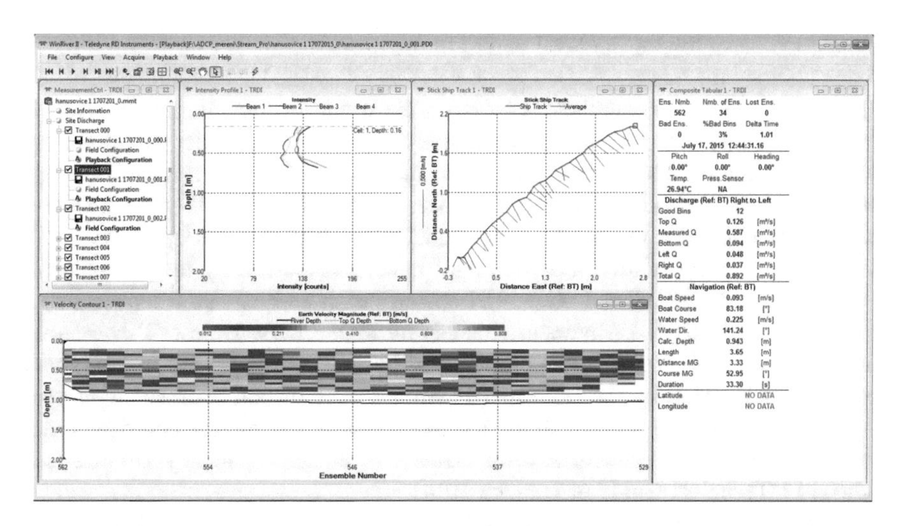

Fig. 1 Result of the discharge measurement using ADCP RDI StreamPro on the Hanušovice pilot area (software WinRiver II)

analyzed gradually decrease of Q, due to seepages or other influencing factors. StreamPro ADCP measurement result is shown as an example in Fig. 1. Measurement using ADCP in these types of construction is specific, because the signal at extreme positions at the banks is reflected not only from the bottom but also from the perpendicular walls of the race. For this reason it is very important to accurate adjustment device, especially type parameters measurement and extrapolation, of velocity field in these peripheral segments.

Next phase of processing was the schematization of the mathematical models (rainfall-runoff, hydraulic, hydrogeological). Unambiguous criterion was a use of tools, which allow simulation of the impact of water divider, hydraulic of technical channel and subsequent return of the water to the main channel. Another criterion was the use of industry standards FEMA/NFIP or software, which is flexibly used by Flood Forecasting Service of the Czech Republic (hereinafter FFS CR). It was utilized rainfall-runoff model HEC-HMS 4.0, which is currently the most widely used tool for rainfall-runoff modeling, and it is validated on a global scale. Another argument for choosing this tool is a good connectivity to GIS systems routinely used in practice within the Czech Hydrometeorological Institute (advisory activities, support for the hydrologic forecasts) and the actual support for schematic and simulation water divider and race. For hydraulic modeling there have been used HEC-RAS and MIKE 11 as industry standards FEMA/NFIP and essentially 1D hydraulic modeling of global significance. They are also used in practice within Czech Hydrometeorological Institute for hydraulic studies (assessment of flood levels or flows rating curves derivation). As well as HEC-HMS they allow schematization and simulation of technical objects on watercourses, including water divider and simulation of operational failures of these objects (e.g. improper manipulation, dam bursts, etc.). Among the methods of simulation there belong a simulation according to the level and Q in the main stream or using a time series manipulations. In this case of complex solution of hydraulic conditions was used HEC-RAS, which are complicated by races and water dividers, Bernoulli equation in 1D. MIKE 11 uses dynamic wave approximation of higher order, which allows to simulate the effects of backflow. Kinematic wave approximation, which is the standard for rainfall-runoff modeling, has been used to simulate a 1D flow in natural and artificial channels in rainfall-runoff model HEC-HMS and it is by the way used in operation model of FFS CR. Seepages have been solved by lossy coefficient "Groundwater Leakage" and "Gaining/Loosing Reach", which allow connection of HEC-RAS and MIKE 11 hydraulic models on MODFLOW/FEFLOW hydrogeological models. Similarly, it has been solved by HEC-HMS rainfall-runoff model parametrically using by the "Channel Loss/Gain" percolation parameters, respectively using constants. Functionality and flow in the water divider and dam structures have been solved with several variants of equations, mostly Honma (see MIKE 11 Technical Guide or Unucka 2014).

Diffusion wave equation in 1D can be described by following relation:

$$\left(\frac{1}{A}\right)\frac{\partial Q}{\partial t}+\left(\frac{1}{A}\right)\frac{\partial\left(\frac{Q^2}{A}\right)}{\partial x}+g\frac{\partial y}{\partial x}-g(S_0-S_f)=0, \tag{1}$$

where:

Q flow rate ($m^3\ s^{-1}$)
A flow area (m^2)
q lateral inflow per unit length ($m^3\ s^{-1}/m$)
x distance at the direction of longitudinal profile and flow (m)
y depth (m)
g gravitational acceleration (9.81 $m\ s^{-2}$)
S_0 bottom slope ($m\ m^{-1}$)
S_f water level slope ($m\ m^{-1}$)

Honma 1 formula, used for the simulation of the flow on weirs and water dividers, has following form:

$$Q=\begin{cases}C_1W(H_{us}-H_w)\sqrt{(H_{us}-H_w)}\\ C_2W(H_{ds}-H_w)\sqrt{(H_{us}-H_{ds})}\end{cases} \tag{2}$$

W is the width of the crest (m)
C_1 and C_2 weir coefficients
H_{us} upstream water level (m)
H_{ds} downstream water level (m)
H_w weir level (m)

MIKE 11 allows extended Honma equation for three variants of the ratio levels above and below the weir or water divider, respectively Villemont's relation. HEC-RAS uses basic equation Q = C.W.H3/2 [3] (Dyhouse et al. 2003). If a bridge or other structure billows, then HEC-RAS and MIKE 11 will use proven numeric-analytical solver WSPRO. Other aspects of mathematical solutions can be found in the references of the manual for the software HEC-RAS and MIKE 11, resp. Dyhouse et al. (2003).

3 Description of the Pilot Areas

There have been chosen four pilot areas. Two of them are on the territorial scope of branch ČHMÚ Ústí nad Labem (flume for Hübel weaving on Chřibská Kamenice and Ploskovický potok) and two of them on the territorial scope of branch ČHMÚ Ostrava (flumes in Žimrovice and Hanušovice). We can also mention localities such as Malá Morávka, Karlov and Ždárský potok in Jeseníky. The system of Holanské rybníky in the catchment basin is undoubtly an interesting area in the region of Ústí.

Ploskovický potok is located in the western part of the basin Třebušín, which is the geographical scope of catchment basin of Ploskovický potok. Catchment basin of Luční potok to the estuary of the Elbe is a water form of OHL_0020. It belongs to the sub-basin Ohře, sub-basin Dolní Labe and other tributaries of the Elbe. According to the Ministry of Agriculture Decree no. 292/2002 Coll., about catchment areas, it belongs to the number of hydrological order 1-12-03: from the river Elbe to Ohře. Ploskovický potok belongs to the hydrogeological region 4520—Chalk of the right-hand tributaries of the Elbe river in catchments 1-12-03, 1-13-05 and 1-14-03. The area is located on the southern edge of the basal Cretaceous collector 4730—basal collector Benešov syncline, the southern part of the area of interest interferes in 4720—basal Cretaceous collector from Hamr along the Elbe. Ploskovický potok is not measured creek. Ploskovický potok has elongated basin in the NW–SE direction, spring is above the Staňkovice village and it drains eastern slopes of Dlouhý vrch. There was original flow W–E orientation at the locality of Staňkovický mlýn. It was diverted to the direction N–S where continues along contour line. On the right side there is a relatively well-preserved Chudoslavice mill with a race and pond. The flow behind the road Chudoslavice–Myštice is cutting through the deep ravin, which has been formed by the fluvial erosion. The flow meanders and creates vast stretches of the mighty lateral erosion on the botttom of the ravin. Flow receives a tributary from the right side in between of Myštice and Chudoslavice, and then the creek continues as the Těchobuzický potok. Water divider is composed on the left bank, whose settlement in the terrain can be considered as a disrepair. This duct drains some of the water to the Těchobuzický potok. Ploskovický potok (The Ploskovicky creek) continues by the right bank mounted channel, whose left bank is initially defined by a dam and it flows in the length of 1.1 km in the SE direction. On the right side there are the remains of a race and a pond for increasing the flow of the mill in Ploskovice. In the last section of the creek in front of Ploskovice there is considerably clogged channel, sometimes the water pipes are disrepair and their permeability decreases. This is the cause of water shortage in the castle grounds and the potential risk during floods. It is certain that the current state of the whole Ploskovice creek is the result of extensive water management measures from the period of building the water systems of Ploskovice´s castle after 1720. The artificial character of the flow is still evident especially in the about of the Staňkovice mill—the road from Chudoslavice to Myštice. In the locality of the Staňkovice mill (where the original flow was diverted) there is a concrete water divider. Some of the water is converted by an underground converter into the Chudoslavice creek. It is approximately in the direction of the original flow. This fact (in relation to the further course of the Ploskovice creek) should be considered for the values of flow rates obtained by an extension of the ArcGIS-AG posudek and presented in Table 1.

The flume in Chřibská (a flume of Hübel's weaving mill) is situated in the middle of the stream Chřinská Kamenice (Chřibský potok) in the area between Dolní Chřibská and Na Potocích. The Chřibská Kamenice has a spring on Jelenní vrch (552 m above sea level). It flows into the Kamenice in the area of Srbská Kamenice. The hydrological order code is 1-14-05-0140. The headwater and upstream (approximately to the Chřibská) are parts of the hydrogeological zone 4650—the Cretaceous Dolní Ploučnice and Horní Kamenice. Another part of the river catchment belongs to the hydrogeological zone 4660—the Cretaceous Dolní Kamenice and Křinice. The length of the river is about 22.2 km (the value of the stationing profile in Všemily is often incorrectly stated as the length of the flow), the catchment area is 62.2 km^2. On the lower course of the stream there is a profile of category C Všemily (stationing 21.8 km), whose main characteristics are shown in Table 1. The river basin of the Chřibská Kamenice can be divided into two geomorphologically different areas: upstream is a part of Lužické hory in the west. The relief is determined by metamorphic rocks of Lužice pluton. The lower part of the river basin is a part of sandstone landscape of Českosaské Švýcarsko. Both areas are affected by the tertiary neovulcanism. The race of Hübel's textile factory was built in 1884–1894. The race was in operation until the Great Depression of the early thirties for the needs of the textile factory. In the literature, there are different figures about the length of individual sections of the race. The data presented here are based on preliminary measurements in situ and they will be specified after studying the technical documentation (stored in SOkA Děčín) and after detailed measurements. The character of the race is influenced by the complicated morphology of floodplain of the Chřibská Kamenice consisting of distinctive bends and a narrow, sandstone valley (Belisová 2014). The total length of the race, from its defunct water divider at the dam structures in Dolní Chřibská, is about 750 m. The first well-preserved tunnel in the rock is situated about 250 m from the water divider and has about 35 m. The race continues to the sandstone aqueduct above the Chřibská Kamenice. The height of 8 m is mentioned but, according to our preliminary measurement, it is only 5 m. The race leads trough a second tunnel before entering into the former textile factory. The length of the tunnel is about 140 m and its mouth is in the original technical building (now completely defunct). The race continues from the building through the former factory courtyard by an underground passage and then continues to its mouth into the Chřibská Kamenice. Nowadays, the race is completely dry and its impact on drainage conditions during floods will be subject to modelling.

Table 1 Basic hydrologic characteristics of the pilot areas derived from GIS extension ArcGIS/AGPosudek (CHMI), hydrometric measurements and rainfall-runoff models

Locality name	Hydrological order	Q_A	Q_1	Q_2	Q_5	Q_{10}	Q_{20}	Q_{50}	Q_{100}
Moravice nad Mlýnskou strouhou	2-02-02-0710	4.93	57.9	82.5	120	151	185	233	273
All hydrological characteristics are in $m^3\ s^{-1}$			Q_{30d}	Q_{60d}	Q_{90d}	Q_{120d}	Q_{150d}	Q_{180d}	Q_{210d}
			13.3	8.23	5.99	5.03	3.93	2.34	1.84
		A (km^2)	Q_{240d}	Q_{270d}	Q_{300d}	Q_{330d}	Q_{355d}	Q_{364d}	–
		653.18	1.50	1.25	1.00	0.736	0.437	0.183	–
Locality name	Hydrological order	Q_A	Q_1	Q_2	Q_5	Q_{10}	Q_{20}	Q_{50}	Q_{100}
Morava pod Krupou	4-10-01-0270	4.20	25.7	39.2	58.6	74.1	90.3	113	131
All hydrological characteristics are in $m^3\ s^{-1}$			Q_{30d}	Q_{60d}	Q_{90d}	Q_{120d}	Q_{150d}	Q_{180d}	Q_{210d}
			9.20	6.60	5.06	4.10	3.41	2.93	2.54
		A (km^2)	Q_{240d}	Q_{270d}	Q_{300d}	Q_{330d}	Q_{355d}	Q_{364d}	–
		216.9	2.22	1.94	1.64	1.36	1.04	0.764	–
Locality name	Hydrological order	Q_A	Q_1	Q_2	Q_5	Q_{10}	Q_{20}	Q_{50}	Q_{100}
Chřibská Kamenice -Dolní Chřibská	1-14-05-0160	0.407	5.32	8.87	15.3	21.4	28.9	40.6	50.7
All hydrological characteristics are in $m^3\ s^{-1}$			Q_{30d}	Q_{60d}	Q_{90d}	Q_{120d}	Q_{150d}	Q_{180d}	Q_{210d}
			0.870	0.610	0.480	0.400	0.340	0.290	0.250
		A (km^2)	Q_{240d}	Q_{270d}	Q_{300d}	Q_{330d}	Q_{355d}	Q_{364d}	–
		43.3	0.220	0.190	0.160	0.130	0.110	0.090	–
Locality name	Hydrological order	Q_A	Q_1	Q_2	Q_5	Q_{10}	Q_{20}	Q_{50}	Q_{100}
Všemily - měrný profil	1-14-05-0180	0.576	6.42	10.7	18.4	25.8	34.6	48.6	61.0

(continued)

Table 1 (continued)

Locality name	Hydrological order	Q_A	Q_1	Q_2	Q_5	Q_{10}	Q_{20}	Q_{50}	Q_{100}
All hydrological characteristics are in $m^3 s^{-1}$			Q_{30d}	Q_{60d}	Q_{90d}	Q_{120d}	Q_{150d}	Q_{180d}	Q_{210d}
			1.23	0.860	0.680	0.560	0.480	0.410	0.350
		A (km^2)	Q_{240d}	Q_{270d}	Q_{300d}	Q_{330d}	Q_{355d}	Q_{364d}	–
		61.79	0.310	0.270	0.230	0.190	0.150	0.130	–
Locality name	Hydrological order	Q_A	Q_1	Q_2	Q_5	Q_{10}	Q_{20}	Q_{50}	Q_{100}
Ploskovický potok—Staňkovice	1-12-03-0810	4.21	0.650	0.970	1.62	2.27	2.67	3.64	4.45
Daily-return period discharge values are in v $l.s^{-1}$, year-return period values are in $m^3 s^{-1}$			Q_{30d}	Q_{60d}	Q_{90d}	Q_{120d}	Q_{150d}	Q_{180d}	Q_{210d}
			8.29	6.40	5.35	4.59	3.98	3.49	3.07
		A (km^2)	Q_{240d}	Q_{270d}	Q_{300d}	Q_{330d}	Q_{355d}	Q_{364d}	–
		0.78	2.67	2.29	1.92	1.54	1.00	0.590	–

Žimrovice (Weisshuhn's) flume or Papírenský flume near Žimrovice and Hradec nad Moravicí belongs to Moravice basin. Number of the hydrological order is 2-02-02-0710. Considering hydrologic division, this area belongs to 6611—Kulm of low Jeseníky in Odra basin and in the north it borders with area subdivision 1520—Quaternary of Opava. Surrounding terrain reaches 400–450 m (e.g. Kozí ridge at Papírenský weir, Bradlo, etc.). Channel bottom of Moravice at confluence is 287 m above the sea level. Building of the flume was initiated and financed in 1890 by local businessman of German origin Carl Weisshuhn, grandfather to the famous Joy Adamson. Purpose of this building was to provide sufficient amount of water and electricity for paper factory, which was to be built in Žimrovice. Flume was built by Italian workers from Baraba company, who had plenty of experience in the field of punching tunnels in rocky terrain of their homeland and Alpine regions, so the flume was finished, incredibly, in 1 year. Although there were some complications, like choking up of freshly punched tunnel beneath Kozí rigde by sediments and rocks during floods, the flume was finished and put to use in 1891. Disjunctive object is a part of Papírenský weir on the left bank of Moravice (see Fig. 2) (Brosch 2005; Křivánek et al. 2014). The flume is separated from Moravice before the river reaches Žimrovice. It starts at Papírenský weir, it's length is 3.5 km, and it involves two aqueducts and three tunnels (the longest is lies right next to Papírenský weir and is 45 m long, originally there were four tunnels, but one caved in). Besides using electricity generated by Francis's horizontal turbine (which is still functional) and supplying paper factory with operational water, the

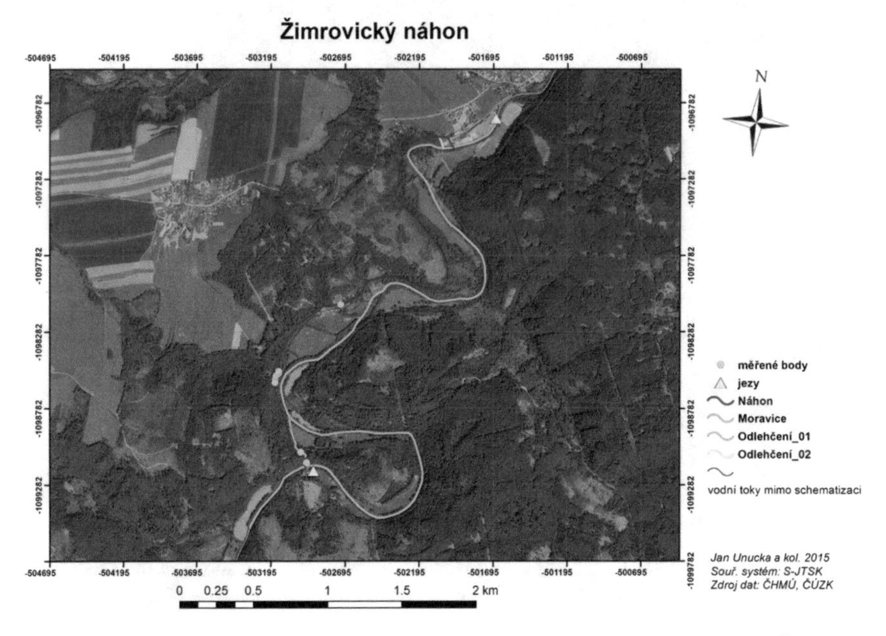

Fig. 2 Map of the MIKE 11 a HEC-RAS hydraulic models schematizations of the Žimrovice flume and adjacent part of the Moravice river (map derived using the ESRI ArcGIS)

flume was from the start used for barging wood until the year of 1966. At the mouth into Žimrovice paper factory the flume is 22 m higher compared to the river bed of the Moravice watercouse, and this only magnifies significance of this construction, which furthermore, was built in quite hilly and hardly accessible terrain. The flume outlet is provided by Žimrovice fall, which is attractive for tourists and is one of the biggest artificial falls in Czech republic. It starts by side outlet and then short cascade back to Moravice river. It's height is approximately 22 m and flow rate reaches hundreds of liters per second.

Hanušovice flume is situated in NW part of Hanušovice cadastral area on the left bank of Morava river. Disjunctive object is located at confluence of the rivers Morava and Krupá near quarry and railway-bridge across the Morava river. The flume is 2.7 km long. At the diversion object, flume's channel bed is 404 m above sea level and at the end it's altitude is approximately 391 m above the sea level. In regard to hydrologic division, area belongs to regions 6432—crystalline metamorphic rocks of the south part of East Sudetenland (Demek et al. 1992). At small hydropower plant situated in the old textile factory the flume's altitude is 22 m above Morava river bed level, which is almost identical to value and slope of Žimrovice flume. Hanušovice flume was a part of textile factory in Hanušovice and at the time autonomous settlement Holba. Building was initiated by textile factory

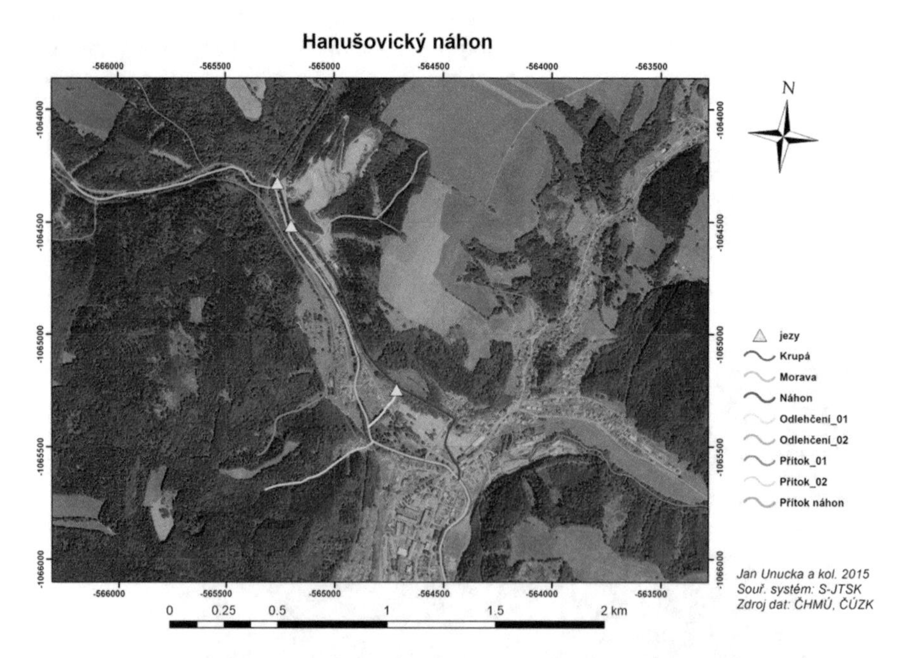

Fig. 3 Map of the MIKE 11 a HEC-RAS hydraulic models schematizations of the Hanušovice flume and adjacent part of the Morava river (map derived using the ESRI ArcGIS)

owner from Šumperk of German origin, Eduard Oberleithner and in 1857 flax weaving factory had started producing yarn for weaving factory in Šumperk. Oberleithner family also initiated building of famous beer-brewery. Left-bank mill flume outflows back to Morava river using fall, right behind railway-bridge. Situation overview and hydraulic scheme for models HEC-RAS and MIKE 11 can be seen in Fig. 3.

4 Selected Results

In this chapter there will be demonstrated results of hydrometric measurements together with coupled hydraulic modelling, which took place up until now and results of hydrologic-hydraulic analysis of pilot areas compared to schemed hydraulic models DHI MIKE 11 and HEC-RAS with linkage to hydrogeological model MODFLOW (numerically solving the groundwater leakage, percolation and water losses). Extent of this paper don't allow demonstration results of all hydrometric measurements, GIS analyses and simulation results of the hydraulic models, therefore authors focused on what's important in their opinion—demonstration of possibilities granted by modern hydrometric and hydroinformatics methods and tools in regard to outlined research and inventory (Fig. 4). One of the side goals was also testing if hydraulic models HEC-RAS and MIKE 11 are able to simulate such complicated hydraulic circumstances. This presumption was confirmed within

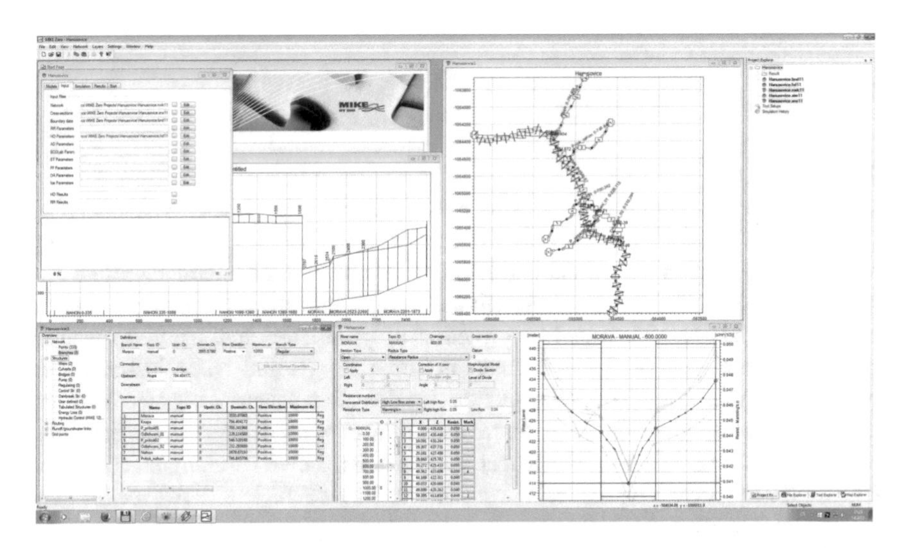

Fig. 4 Schematization of the Morava river a Hanušovice flume in the hydraulic model DHI MIKE 11

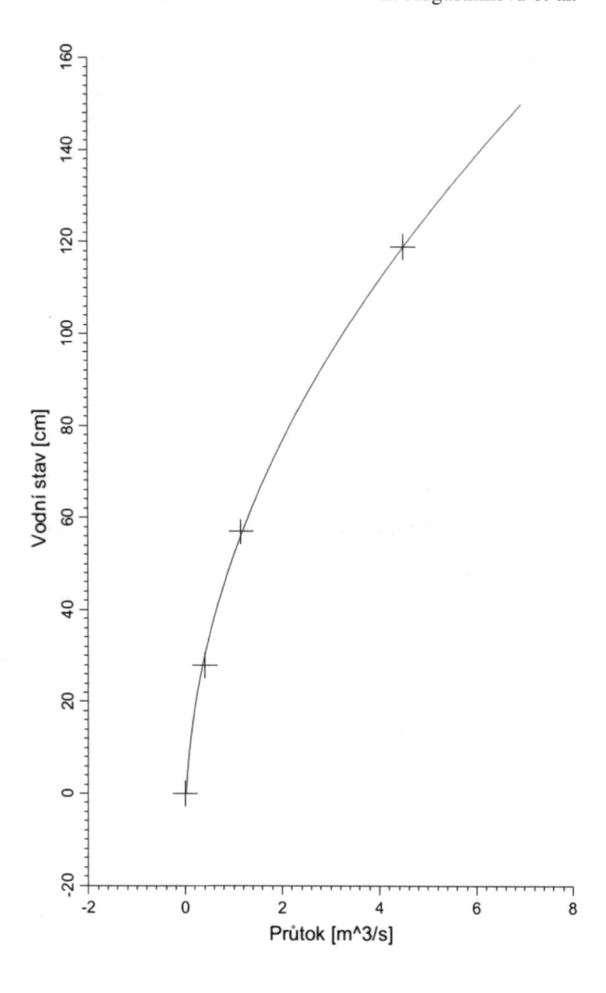

Fig. 5 Consumption curve for the Žimrovice flume derived from the measurements (software WinZPV) and approximated by hydraulic modelling approaches (software MIKE 11)

finished analysis, and further in situ measurements and simulations, including numeric solution of saturated zone will further test abilities of chosen GIS program resources and mathematic modelling. Figure 5 demonstrates the particular flow-rate curve constructed for the Žimrovice flume using conventional methods based on hydrometric measurements and construction of parametric and analytic shape of specific flow-rate curve. Table 2 contains flow-rate (Q) and water losses within flume course due to leakage or disjunction and diversion canals, compared to in situ measurement using ADCP RDI StreamPro. It's obvious from results, that after proper calibrating it is possible to get valid results. Level of compliance can be determined by regression analysis or Nash–Sutcliffe coefficient, however for generalization it is necessary to do more measurements in pilot areas and it is very vital to record process of flood with manipulation for the opportunity of setting complex boundary conditions to simulate unsteady flow.

Table 2 Comparison of the measured discharges using ADCP RDI StreamPro with the simulation results of the hydraulic models HEC-RAS a MIKE 11

	Žimrovice flume		
Length (km)	0.107	0.892	1.59
Date and time	2.7.2015 15:50–16:12	2.7.2015 16:28–16:55	2.7.2015 17:13–17:38
	Q (m^3 s^{-1})		
ADCP RDI StramPro	0.617	0.525	0.457
HEC-RAS	0.628	0.541	0.463
MIKE 11	0.631	0.529	0.442
	Hanušovice flume		
Length (km)	0.101	0.69	1.41
Datum a čas měření	17.7.2015 12:43–12:52	17.7.2015 13:29–13:49	17.7.2015 14:27–14:51
	Q (m^3 s^{-1})		
ADCP RDI StramPro	0.904	0.749	0.693
HEC-RAS	0.911	0.823	0.717
MIKE 11	0.917	0.835	0.725

5 Discussion

It's important to let be known, that collective authors had to deal with shortage of studies concerning this kind of cases, which affected literature choice for pilot areas (Figs. 6, 7). Objects of interest are very often also historical objects, so there's much more information in terms of historical monuments conservation, historical analysis of the technical objects rather than information about hydrological aspects of aforementioned areas and objects. Brosch (2005) and Křivánek et al. (2014) dedicate in their publications to certain hydrological parameters of these objects. In terms of anthropogenic forms of relief and geomorphology, Janků (2009) partially dedicates his thesis to areas of interest (Hanušovice, Jesenicko). Buryšková (2015) partially dedicates her bachelor paper to problematics of flumes and small water managing objects. It is obvious, that for measurements of flumes and pilot areas to be systematic, it is desirable to do as much in situ measurements and hydrologic simulations as possible. Suitability of each program resources for particular purposes can be verified on FEMA/NFIP website or in publications of authors Beven (2009), Bedient et al. (2013), Dyhouse et al. (2003), Chaterjee et al. (2008), Mujumdar and Kumar (2012) or Di Baldassarre (2012). Partial suitability analysis of hydrologic or hydraulic models was done by collective of authors in publications Fárek and Unucka (2015), Fárek et al. (2014), Kožaná et al. (2014) or Jancíková and Unucka (2015). Belisová (2014) quite systematically analyses problematic of small water objects in area of NP Czech Switzerland and Karel (2008) do so for Rýmařovsko region. Interesting addition and in certain aspect, still unequalled, to

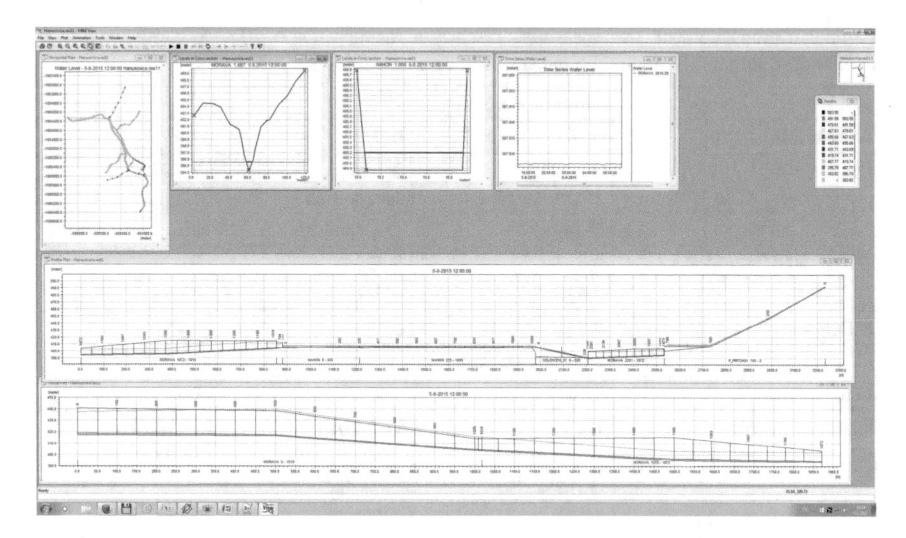

Fig. 6 Simulation result of the MIKE 11 hydraulic model viewed in MIKE View for the Hanušovice flume together with the confluence of Morava and Krupá rivers

Fig. 7 ADCP RDI StreamPro during measurement in Hanušovice flume

general analyses of small basins, is publication of Haan et al. (1994). Large datasets were processed during the treatment of the ADCP measurement results and their importing into the GIS and hydraulic models schematizations. Another large datasets were processed during the DMR data transfer into the rainfall-runoff, hydraulic and hydrogeological models schematizations to achieve the integrated distributed hydrologic models for the simulating and predicting dynamics of surface and groundwater during various hydrosynoptical situations and their impact on the runoff from the catchment involving particular hydraulic conditions in flumes during these situations.

6 Conclusion

As was said in the article's introduction, flumes represent interesting phenomena deserving further multidisciplinary research. If we look at them in terms of fields of study like hydrology, water management, forestry, historical research of the technical landmarks or geomorphology of anthropogenic forms of relief, it is evident, that without multidisciplinary approach, they cannot be systematically studied. Authors are trying to analyze these objects primarily from hydrological viewpoint using modern hydrological methods, which nowadays allows quality research of hydrologic objects—both artificial and natural. In the opinion of collective of authors inventory of flumes isn't self-serving. Firstly it is desirable to understand impact of the flumes on small and big water systems, in regard to manipulation and transfers of water and secondly by using technological hydrometric equipment, geoinformation technologies and hydroinformatics, analyze their hydrological importance and partial parameters of their technical state and risk. Whether we talk about still functioning flumes or objects which are now just reminders of our ancestor's architectonical and building craftsmanship, their systematic inventory is meaningful in order to understand their hydrologic role in the landscape, which especially concerning small water systems, can be crucial. In this regard, simulation capabilities of geoinformatics and hydroinformatics can be seen as a tool for reconstruction of not only past times, but future as well, including extreme outflow situations, such as droughts and floods. And in this case there is a simple, but crucial principle—the more relevant data we have at our disposal, the better and the more precise simulation and predictive tool we are able to create.

References

Augustinková L et al (2011) Moravská brána do Evropy. Soupis technických a zemědělských památek regionu Poodří. MAS regionu Poodří, o.s. ISBN:978-80-87427-21-7

Bárta M, Kovář M (2013) Civilizace a dějiny: historie světa pohledem dvaceti českých vědců. Vyd. 1. Praha, Academia. 557 p. ISBN:978-80-200-2301-8

Bedient PB, Huber WC, Vieux BC (2013) Hydrology and floodplain analysis, 4th edn. Prentice Hall, London. ISBN:978-0-273-77427-3

Belisová N (2014) Tulákům Jetřichovicka. Jetřichovice, Občanské sdružení pro záchranu a konzervaci Dolského mlýna, 2014, 672s. ISBN:978-80-260-7010-8

Beven KJ (2009) Environmental modelling: an uncertain future?. Routledge, London. ISBN:978-0-415-46302-7

Boiten W (2008) Hydrometry: a comprehensive introduction to the measurement of flow in open channels, 3rd edn. CRC Press, Boca Raton. ISBN:978-0415-46763-6

Brosch O (2005) Povodí Odry. Ostrava, Anagram. ISBN:80-7342-048-1

Buryšková L (2015) Vodní mlýny v povodí Bíleho potoka. Historie a dopady na fluviální systém. Bakalářská práce PřF MU Brno

Chaterjee C, Förster S, Bronstert A (2008) Comparison of hydrodynamic models of different complexities to model floods with emergency storage areas. Hydrol Process 22(24):4695–4709

Demek J, Novák V et al (1992) Neživá příroda. 1. vyd. Brno, Muzejní a vlastivědná společnost. Vlastivěda moravská, sv. 1. ISBN:80-85048-30-2

Di Baldassarre G (2012) Floods in a changing climate. Inundation modelling. Cambridge University Press, Cambridge. ISBN:11-070-1875-7

Dyhouse GR et al (2003) Floodplain modelling using HEC-RAS. Bentley Institute Press, Watertown. ISBN:978-1-934493-02-1

Fárek V, Unucka J (2015) Results comparison of the flow direction and accumulation algorithms together with distributed rainfall-runoff models in Czech Switzerland National Park. In: Lecture notes in geoinformation and cartography. Springer, Dordrecht, pp. 87–98. ISBN:978-3-319-18406-7

Fárek V, Unucka J et al (2014) Využití GIS a distribuovaných srážko-odtokových modelů v odhadu hydrologických parametrů malých povodí. Případová studie z NP České Švýcarsko a Jetřichovického potoka. 5 p. In Vodní hospodářství 1/2014. ISSN:1211-0760

Haan CT, Barfield BJ, Hayes JC (1994) Design Hydrology an Sedimentology for Small Catchments. Academic, London. ISBN:978-0123123404

Jančíková A, Unucka J (2015) DTM impact on the results of dam break simulation in 1D hydraulic models. In: Lecture notes in geoinformation and cartography. Dordrecht, Springer, p. 125–136. ISBN:978-3-319-18406-7

Janků P (2009) Vodohospodářské tvary reliéfu v Šumperské kotlině. Diplomová práce, PřF UPOL, Olomouc 76 p

Karel J (2008) Příběh lesů a lidí Rýmařovska. Dějiny lesů a jejich užívání. 316 p. Moravská Expedice, Rýmařov. ISBN:80-86511-33-2

Kožaná B, Štěrba O, Unucka J et al (2014) Příspěvek k možnostem stanovení vlivu lužního lesa na tlumení povodňových vln s využitím 1D a 2D hydraulických modelů a GIS. 7 p. In Zprávy lesnického výzkumu vol. 59, 2/2014. ISSN:0322-9688

Křivánek J, Němec J et al (2014) Drobné vodní toky v ČR. Praha, Consult. 295 p. ISBN:978-80-905159-0-1

Mujumdar P, Kumar N (2012) Floods in a changing climate. Hydrologic modeling. Cambridge University Press, New York. ISBN:978-110-701876-1

Solnický P (2007) Vodní mlýny na Moravě a ve Slezsku. 1. vyd. Praha, Libri. 221 p. ISBN:978-80-7277-244-5

Squatriti P (ed) (2000) Working with water in medieval Europe: technology and resource-use. Brill, Boston. ISBN:9004106804

Štěpán L, Urbánek R, Klimešová H et al (2008): Dílo mlynářů a sekerníků v Čechách II. Vyd. 1. Praha, Argo. 316 p. ISBN:978-80-257-0015-0

Tainter JA (2009) Kolapsy složitých společností. Praha, Dokořán, 2009, 319 s. ISBN:978-80-7363-248-9

Unucka J (2014) Environmentální modelování 1. Skriptum PřF OU. 209 p. Ostrava, SVZZ CZ.1.07/2.3.00/35.0053 & PřF OU

Creating Large Size of Data with Apache Hadoop

Jan Růžička, David Kocich, Lukáš Orčík and Vladislav Svozilík

Abstract The paper is focused on research in the area of building large datasets using Apache Hadoop. Our team is managing an information system that is able to calculate probability of existence of different objects in space and time. The system works with a lot of different data sources, including large datasets. The workflow of data processing is quite complicated and time consuming, so we were looking for some framework that could help with system management and, if possible, to speed up data processing as well. Apache Hadoop was selected as a platform for enhance our information system. Apache Hadoop is usually used for processing large datasets, but in a case of our information scystem is necessary to perform other types of tasks as well. The systems computes spatio-temporal relations between different types of objects. This means that from relatively small amount of records (thousands) are built relatively large datasets (millions of records). For this purposes is usually used PostgreSQL/PostGIS database or tools written in Java or other language. Our research was focused to determination if we could simply move some of this tasks to Apache Hadoop platform using simple SQL editor like Hive. We have selected two types of common tasks and tested them on PostgreSQL and Apache Hadoop (Hive) platform to be able compare time necessary to complete these tasks. The paper presents results of our research.

J. Růžička (✉) · D. Kocich · V. Svozilík
Institute of Geoinformatics, Faculty of Mining and Geology,
VŠB-Technical University of Ostrava, 17. listopadu 15/2172,
708 33 Ostrava-Poruba, Czech Republic
e-mail: jan.ruzicka@vsb.cz

D. Kocich
e-mail: david.kocich@vsb.cz

V. Svozilík
e-mail: vladislav.svozilik@vsb.cz

L. Orčík
Department of Telecommunication, Faculty of Electrical Engineering
and Computer Science, VSB-Technical University of Ostrava, 17.
listopadu 15, 708 33 Ostrava, Czech Republic
e-mail: lukas.orcik@vsb.cz

I. Ivan et al. (eds.), *The Rise of Big Spatial Data*, Lecture Notes
in Geoinformation and Cartography, DOI 10.1007/978-3-319-45123-7_22

Keywords Data creating · Apache hadoop · Cloudera · Hive · PostgreSQL

1 Introduction

We are building information system that works with a lot of different sources of data, including spatial data for example from RÚIAN (Registry of Territorial Identification, Addresses and Real Estate) (COSMC 2015) or Czech Office for Surveying, Mapping and Cadastre (COSMC) download or view services (COSMC 2016). The systems computes spatio-temporal relations between different types of objects. The system is mainly focused on calculating probability of existence of different objects in space and time.

We are working with large datasets and some computations take more than 1 week on a single computer. We have decided to use Apache Hadoop to distribute computation among several computers in a cluster.

Apache Hadoop technology is a leading technology in the area of large data processing. The technology is used by Google, Facebook, Microsoft or IBM. The technology is distributed according to open source licence and is published by Apache Software Foundation. Apache Hadoop is popular not only in commercial area but open source area as well. The main feature that makes Apache Hadoop so popular is possibility to create cluster built of computers with less power that can compete with specialised powerful hardware. That hardware configuration is named distributed storage with distributed computation. The main idea is based on distribution of data according to predefined conditions. That distribution is done by file system named HDFS (Hadoop Distributed File System). That file system is used for computation based on MapReduce architecture, where Map distributes tasks and Reduce joins results. There are other tools that together build complex platform for large data processing such as YARN, HBase, Hive, Spark or Solr.

We use Cloudera distribution for Hadoop (CDH) (Cloudera 2015), which contains tools like Hue, Hive, Impala, Workflow or Pig editors. We are now slowly moving our system from traditional RDBMS or file-based information system to Hadoop. Our original system is based on PostgreSQL/PostGIS, GDAL/OGR or GRASS GIS tools.

Each process in our computation workflow usually contains following steps:

- downloading data from open data source (like RÚIAN or Open Street Map),
- importing data to PostGIS and/or GRASS GIS,
- perform specific SQL queries,
- perform further analyses with GRASS GIS.

We have decided to use workflow scheduler and management in Hadoop to have a better control over processes. This is the first step of our migration to Apache Hadoop. Second step is to move as much tasks as possible to MapReduce architecture. If we can move all tasks to MapReduce architecture, the management of all processes is going to be very simple.

Another advantage of Hadooop is to speed up data processing just simply add another computer to our cluster. Most of the tasks in Hive will be done faster when the number of nodes in cluster increases.

We know that there is project named GIS tools for Hadoop (ESRI 2016), geoprocessing tools and spatial framework for Hadoop by ESRI (2016). There is also Spatial Hadoop maintained by Minnesota's research university (Eldawy and Mokbel 2013). We would like to test these tools in the future as well.

2 Methodology

Calculation of several matrices of relations were done in the past, for example distance between several places based on aerial distance or distance using transport network. These matrices are used for several purposes and are usually calculated on demand, so the speed of calculation is quite important. Another important requirement for this task is to make it customizable and simple enough for general user of table calculator or relational database.

We expected that running these kind of queries on Hive will be slower than on PostgreSQL, but we were interested how big is the difference. If the difference is not enorm than we could use Hive instead of PostgreSQL to perform such queries and store them as tasks to be able reuse them in a workflow.

2.1 Hardware and Software Configuration

The PostgreSQL and cluster for Hadoop uses virtual computers with the same hardware configuration.

Performance of the PostgreSQL 9.3 was tuned according to optimization tips from official wiki sources (The PostgreSQL Global Development Group 2015).

Apache Hadoop is configured as a cluster with one master and 3 slaves. Performance of Hadoop cluster was not tuned because we do not have enough information how to tune Hadoop yet.

2.2 Aerial Distances Between Basic Settlement Units in the Czech Republic

The first tested task was calculation of aerial distances between basic settlement units in the Czech Republic.

For calculating such distance matrix we can use three possible ways:

- Using SQL on PostgreSQL/PostGIS.
- Using simple programs written for example in Java or C++.
- Using SQL on Hive (or Impala) in Apache Hadoop.

We were interested in comparison of both SQL approaches. Our colleagues are usually able to construct simple SQL queries, so if they could use Hive/Impala (which are very similar to general SQL editors) to do the task it would be very useful for us in the future. Such task could be stored in Hadoop and used in a workflow.

We have imported set of basic settlement units to PostgreSQL and to Hadoop databases. The data were based only on simple data types, so we could use simple functions to compute distances between basic settlement units.

We have constructed identical SQL queries and run them in Hive and PostgreSQL. The SQL was constructed as follows:

$$\begin{aligned} &\text{CREATE TABLE bsu\_distances AS SELECT concat(a.id,}\,'\_',\,\text{b.id) ids,}\\ &\text{sqrt}(\text{pow}(\text{a.x} - \text{b.x}, 2) + \text{pow}(\text{a.y} - \text{b.y}, 2))/1000\ \text{distance FROM bsu a, bsu b}\\ &\text{WHERE a.id} < > \text{b.id;} \end{aligned} \tag{1}$$

where:

- *bsu* is table with records for basic settlement units
- *id* is identifier of basic settlement unit
- *x* is X coordinate of centroid of basic settlement unit (Czech CRS Krovak, EPSG: 5514)
- *y* is Y coordinate of centroid of basic settlement unit (Czech CRS Krovak, EPSG: 5514)

We have counted time that was required to perform the tasks.

2.3 Aerial Distances Between Selected Facilities and Basic Settlement Units in the Czech Republic Within Defined Distance

The second tested task was calculating aerial distances between selected facilities and basic settlement units in the Czech Republic within defined distance. This kind of task works with two data sources and uses condition to reduce results.

We have imported set of basic settlement units to PostgreSQL and to Hadoop databases. We have imported set of selected facilities to PostgreSQL and to Hadoop databases. The data were based only on simple data types, so we could use simple functions to compute distances between basic settlement units. We have imported RÚIAN address points (COSMC 2015) to locate selected facilities.

We have constructed identical SQL queries and run them in Hive and PostgreSQL.

There were two SQLs:

The first one was used to locate selected facilities:

$$\begin{aligned}&\text{CREATE TABLE facility AS SELECT f.id, f.capacity capa, a.xjtsk x, a.yjtsk y}\\&\text{FROM ruian.facility\_cap f, ruian.adresnimista a WHERE f.ruian\_kod} = \text{a.kodadresy;}\end{aligned} \tag{2}$$

where:

- *facility_cap* is table with selected facilities
- *adresnimista* is table with address points
- *id* is identifier of facility
- *capacity* is possible number of persons who can use a facility
- *xjtsk* is X coordinate of address point (Czech CRS Krovak, EPSG: 5514)
- *yjtsk* is Y coordinate of address point (Czech CRS Krovak, EPSG: 5514)

The second one was used to calculate distances between basic settlement units and facilities within defined radius:

$$\begin{aligned}&\text{CREATE TABLE bsu\_facility\_distances AS SELECT concat(a.id,}'\_'\text{, b.id) ids,}\\&\text{a.capa, sqrt(pow(a.x}-\text{b.x, 2)} + \text{pow(a.y}-\text{b.y, 2))}/\text{1000 distance FROM facility a,}\\&\text{bsu b WHERE a.id} < > \text{CAST(b.id AS VARCHAR) AND (sqrt(pow(a.x}-\text{b.x, 2)} +\\&\text{pow(a.y}-\text{b.y, 2))}/\text{1000)} < \text{radius;}\end{aligned} \tag{3}$$

where:

- *bsu* is table with records for basic settlement units
- *facility* is table with records for facilities
- *id* is identifier of basic settlement unit or identifier of facility
- *capa* is number of persons possible to use a facility
- *x* is X coordinate of centroid of basic settlement unit or X coordinate of location of facility
- *y* is Y coordinate of centroid of basic settlement unit or Y coordinate of location of facility

We have counted time that was used to perform the tasks.

3 Results

3.1 *Aerial Distances Between Basic Settlement Units in the Czech Republic*

The tested input tables with basic settlement units have from 10,000 to 22,000 records. We were not able to run queries on Impala, queries ends with an unexpected error (Table 1; Fig. 1).

3.2 *Aerial Distances Between Selected Facilities and Basic Settlement Units in the Czech Republic Within Defined Distance*

The tested input table has about 22,000 records of basic settlement units and about 6000 records of selected facilities. We were not able to run queries on Impala, queries ends with an unexpected error (Table 2; Fig. 2).

Table 1 Aerial distances between basic settlement units in the Czech Republic

Records count (in millions cca)	Time on PostgreSQL (s)	Time on Hive (s)
100	137	247
150	201	368
250	340	606
500	721	1205

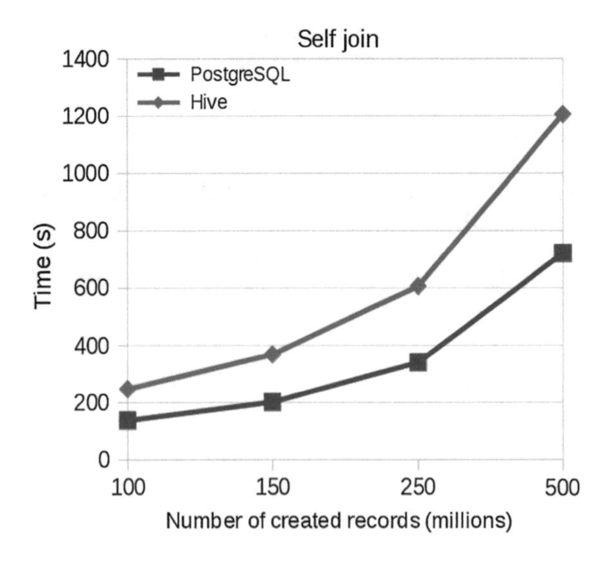

Fig. 1 Aerial distances between basic settlement units in the Czech Republic

Table 2 Aerial distances between selected facilities and basic settlement units in the Czech Republic within defined distance

Records count (in millions cca)	Time on PostgreSQL (s)	Time on Hive (s)	Limits
2	66	114	20 km
10	79	140	50 km
35	115	216	100 km
115	187	305	No limits

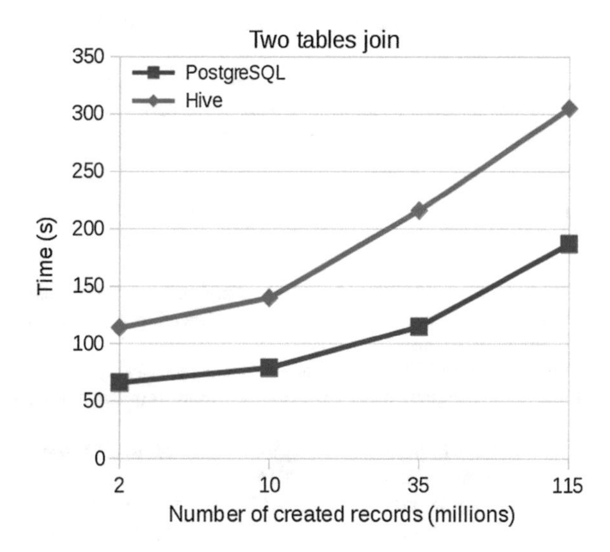

Fig. 2 Aerial distances between selected facilities and basic settlement units in the Czech Republic within defined distance

4 Conclusion

We can conclude that running both types of queries (self join query for calculating aerial distances between objects or join of two tables query for calculating aerial distances between objects) are slower on Hive than on PostgreSQL. The queries on PostgreSQL were close to twice faster than on Hive. This is acceptable for our purposes. If we run query on PostgreSQL than later the data must be transferred to Hive and it will take some time as well.

We can conclude that both types of queries in a case of changing number of generated records are still close to two twice faster on PostgreSQL than on Hive.

We can conclude that we were not able to run selected queries on Impala. Queries ends with an unexpected error.

5 Discussion

We did not tested queries when running Apache Hadoop on different configuration than was mentioned in Methodology section. According to Wang et al. (2010) the queries in Hive could be speed up by adding more nodes to the infrastructure. The question is if more nodes will speed up self join queries as well. We would like to test this in the future.

We did not tested time necessary for transferring results from PostgreSQL to Hive, it should be done in future to prove that our conclusion is correct.

Acknowledgments Supported by grant from Student Grant Competition, FMG, VSB-TUO. We would like to thank to all open source developers.

References

Cloudera (2015) http://www.cloudera.com/content/www/en-us/downloads.html

COSMC (2015) Registry of territorial identification, addresses and real estate. http://www.cuzk.cz/ruian/RUIAN.aspx

COSMC (2016) COSMC download or view services. http://geoportal.cuzk.cz

Eldawy A, Mokbel M (2013) SpatialHadoop. http://spatialhadoop.cs.umn.edu/. Accessed 5 Jan 2016

ESRI (2016) Esri/geoprocessing-tools-for-hadoop. https://github.com/Esri/geoprocessing-tools-for-hadoop. Accessed 5 Jan 2016

The Postgresql Global Development Group (2015) Performance optimization—PostgreSQL wiki. https://wiki.postgresql.org/wiki/Performance_Optimization. Accessed 5 Jan 2016

Wang K, Han J, Tu B, Dai J, Zhou W, Song X (2010) Accelerating Spatial data processing with mapreduce, parallel and distributed systems (ICPADS). http://ieeexplore.ieee.org/xpl/articleDetails.jsp?arnumber=5695607&tag=1

Datasets of Basic Spatial Data in Chosen Countries of the European Union

Václav Čada, Jindra Marvalová and Barbora Musilová

Abstract The fundaments for creating high-quality European data infrastructure, which is based on different national data infrastructures, are good basic spatial data of individual European countries. The purpose of using this infrastructure is especially European environmental policy. Expected quality calls for appropriate methods of data collection, its content and management. On national level the need of actual data with guaranteed quality applied to decision making in public administration, is much more important. Geographic data fulfilling such requirements should distinguish itself by high level of details, information about ownership or higher accuracy. Activities in passive infrastructure, transport infrastructure, spatial planning and other services explicitly require such type of geodata. The aim of this paper is to compare the content of national datasets, the level of detail and quality, collection method and data management in the Netherlands, the United Kingdom and the Czech Republic according to chosen criteria.

Keywords European data infrastructure · National data infrastructures · Geographic data · Geographic information · National datasets · Vector database · State map series

V. Čada (✉) · J. Marvalová · B. Musilová
Geomatics Section of Department of Mathematics, Faculty of Applied Sciences, University of West Bohemia, Technická 8, 306 14 Pilsen, Czech Republic
e-mail: cada@kma.zcu.cz

J. Marvalová
e-mail: jindram@kma.zcu.cz

B. Musilová
e-mail: barbora2@ntis.zcu.cz

I. Ivan et al. (eds.), *The Rise of Big Spatial Data*, Lecture Notes in Geoinformation and Cartography, DOI 10.1007/978-3-319-45123-7_23

1 Introduction

The complexity of processes for society management and administration is influenced by biding legislation, which generates many obligations for public administration, institutions and citizens. Given that these obligations are required under certain sanctions, it is necessary to create in advance appropriate conditions and environment for their fulfilling including measures to evaluate the outputs of these obligations. From this point of view, the role of each country is essential also in the area of collection, management and provision of relevant and guaranteed reference geographic data. It is necessary to stress that the quality of decision making processes is in many cases influenced by the quality and accessibility of geographic data.

In the past, the required functionality was ensured by the national map series. This was a cartographic product, created and published by state administration authority in public interest including reference and generally exploitable content that seamlessly portrays the territory in a unified form. Currently, this role is taken over by data sets consisting of selected geographic features which are uniformly and clearly described. A map series represents a visualisation and combination of these data sets.

The generalised conception of geographic data as multi-dimensional vectors stored in optimised data structures, enables data generalisation (selection, reclassification, spatial reduction, aggregation, simplification, amalgamation) and creation of secondary data models including generalised objects from primary data (primary data models). It is important to start from unified primary data and to avoid duplicate data collection. It is well known that the costs for primary data collection and their update can be up to 80 % of all expenditures for operational information systems. This fact contributes to an intensive rationalisation of efforts spent to maximise the effect. The optimisation of processes of data collection, processing, management and publication is one of the main priorities of all infrastructures for spatial information that are being built in abroad. These optimisation processes are currently supported also by many EU projects.

The aim of this study is to compare the contents of national datasets, the level of detail and quality, methods of collection, implementation and management of basic spatial data sets in the Netherlands, the United Kingdom and the Czech Republic according to selected criteria. These countries were chosen because of different situation that is determined by the different historical development in different cultural conditions as well as the current approach to the collection and management of the basic geodata. Moreover, both of these countries are in the forefront among European countries in terms of the access to basic geodata management.

The issue of some comparisons of different legislative environment and with non-unified terminology is shown on concrete examples. Furthermore there is mentioned the last year approved ‘Strategy of the Development of the National Infrastructure for Spatial Information in the Czech Republic up to 2020’ (GeoInfoStrategy), especially in the area of National Set of Spatial Objects.

2 Definition of Basic Geodata in Compared Countries

Guaranteed reference geodata (spatial data with the relevant quality certificate, over which public authorities conduct transparent decision-making processes with a subsequent legal responsibility), including particularly top-level detailed geodata with defined quality and set of properties for the selected objects of real world are called basic geodata. It is a set of abstractions of objects with a guarantee of identification and spatial position throughout the national territory, which is generally useful for decision-making processes of public administration, the needs of the private sector, educational institutions and contributes to solving everyday life situations. These basic geodata are acquired and updated with clearly defined standards of data quality and data flow, data content, management processes and providing data services.

In the past, this principle was partly applied in the Czech Republic in the development of large-scale maps according to Czech technical standard ČSN 013410 "Large-scale maps. Base and thematic maps". The base large-scale map was a part of state map series that was created for the needs of the economy (Article 70). It was a technical basis for real estate registration, a basis for the renewal of base medium-scale maps and for the thematic maps (thematic maps were used for detailed phenomena and objects localization, under the ground surface and above the ground surface) creation (Article 71). There was a need for maximal using of base maps and results of previous geodetic and cartographic works in the creation of thematic maps.

If the definition of basic spatial data complies requirements of current level of knowledge and needs of society, there is no possibility to relate this definition to the state map series. Conversely the state map series may use the resources of basic spatial data to update its defined content or visualize its primary data model objects (Čada 2014).

3 Specification of Basic Geodata

3.1 The Czech Republic

The Czech Republic Act no. 200/1994 Coll. (Zákon č. 200/1994 Sb.) Section 4, para. 3 specifies that surveying and cadastral institutions, that create and maintain spatial data from the Czech Republic, provide the Infrastructure for spatial information (set of principles, knowledge, arrangements, technologies, data and human resources that enable sharing and effective use of spatial information and services) in the European Community (INSPIRE) with following basic spatial data sets:

(a) base state map series for public use,
(b) geodetic data of survey control points,
(c) database,

(d) orthoimagery throughout the country,
(e) database of geographic names.

The phrase "basic spatial data" is used only once and incidentally in the Act no. 200/1994 Coll. (Zákon č. 200/1994 Sb.), without any deeper meaning and other related issues. Moreover, this establishment relates to the provision of information in the European Community and it is clear that this is not a top-level detailed data. Similarly, the phrase "public use" is not defined or specified.

Therefore, there was given a significant attention in the 'Strategy of the Development of the National Infrastructure for Spatial Information in the Czech Republic up to 2020' (GeoInfoStrategy) to the area of the improvement and further development of the data resources for its use in public administration. It was emphasized that it is necessary to define the spatial data (resp. objects and phenomena clearly described by spatial data), which requires public authorities for their activities. It should be defined also the processes of acquisition, management and updating, the level of required quality, including specifying the managers and coordinators for such data. Gradually, processes to validate and guarantee such data must be set up. It is also necessary to ensure data standardization and interoperability.

3.2 The Netherlands

The crucial role in the Dutch National Spatial Data Infrastructure (NSDI) play the organization GI-Council, created in 2006 as an advisory council of The Ministry of Infrastructure and the Environment, and Geonovum, that was created in 2007 and is mandated by GI-Council and public geo-sector. Geonovum's key tasks are: representing the Dutch geo-information community, standardization, creating and maintaining the national geoportal, the operational INSPIRE coordination, and international networking. Furthermore, there was RGI (Dutch innovation and program to improve and innovate the Dutch National Geographic Information Infrastructure) in 2005–2009. Pursuant to advice from the GI Council, the Geonovum and RGI foundations have devised this approach and strategy in consultation with parties in the Geo-meeting. As a result in 2008 GIDEON was created. GIDEON is an approach and implementation strategy for the national facility for location-specific information. Some of GIDEON strategies are: to encourage the use of existing key registers, to embed INSPIRE directive into Dutch legislation, to encourage the use of geoinformation in government policy and to form a government-wide geoinformation facility that include geodata standardization, new infrastructure, and collaborative maintenance (Spatial Data Infrastructures in The Netherlands 2011).

The key element of the Dutch NSDI is the system of key registers (basisregistraties). The key registrations are registrations with a uniquely defined core dataset that governmental agencies are obligated to use. The collection and maintenance of the data is regulated in legislation, and the data consistent, adhering to the data specifications, across the registration. The users are obliged to report incorrect information to producers, and there is a stringent policy on quality assurance (Spatial Data Infrastructures in The Netherlands 2011).

In terms of the Netherlands, which is active in the creation of European Data Infrastructure and data opening, the mentioned definition of basic data should be extended by one additional feature—data openness. According to this criteria the basic geodata sets in the Netherlands include:

(a) key register cadastre (Basisregistratie Kadaster, BRK),
(b) key register large-scale topography (Basisregistratie Grootschalige Topografie, BGT),
(c) key register of addresses and buildings (Basisregistraties adressen en Gebouwen, BAG),
(d) Top10NL.

3.3 The United Kingdom

There are no legislative constitutions in the United Kingdom that would define the "national basic geodata set", the more the content of this data set. In 90th was formulated "British Standard 7666 for Geographical Referencing of Spatial Datasets", however it does not deal with the definition, but it tries to set conditions for cooperation and interaction of particular datasets that contain information about address places. The standard focuses on four sections—Street Gazetteers, Land and Property Gazetteers, Addresses, and Rights of Way (Morad 2002). It suggests a unique addressing scheme, based on Unique Property Reference Numbers (UPRNs; UPRNs are unique twelve digit numbers assigned to every addressable location in the UK. Every address place is thus identifiable by UPRN that is consistent throughout a property's life.) (Evans 2006). There exists also a statutory instrument Land Registration Rules 2003 that regulates the role of Her Majesty's Land Registry and that deals with legal acts regarding registration of land property rights and duties and with evidence of this (Land Registration Rules 2003). Its equivalent in the Scottish legislative is Land Registration etc. (Scotland) Act 2012. The main purpose of the act is to reform the law on the registration of rights to land. It connected the law of land registration with property law (Land Registration etc. 2012).

In the UK there exists also a set of recommendations—Location Strategy. This strategy was proposed in 2008 for national policy in the field of geospatial information (Vandenbroucke et al. 2011). After interaction with requirements of INSPIRE, the Spatial Data Infrastructure in The United Kingdom was developed in 2011. This set of recommendations (however, they are not normative and moreover they do not deal with the definition of basic geodata) is the general strategic document on which the implementation of the UK NSDI is based (Vandenbroucke et al. 2011). Its main objectives are data exchange, data sharing and data re-use. Before the development of the UK Location Strategy, no national strategy on the coordination of the providing and sharing of geographic information existed. A legal framework for the National Spatial Data Infrastructure in the UK has not been developed (Vandenbroucke et al. 2011).

Scotland has its own strategy "One Scotland, One Geography". It provides a strategic vision and leadership to ensure a coordinated and pragmatic approach to geographic information in Scotland and it also sustains the technical standards for efficient and effective use of the information (The Scottish Government 2005).

4 National Mapping Agencies as Providers of Basic Geodata

To ensure the implementation of the activities of creation, renewal and issuance of base and thematic state map series, building geodetic control and to ensure the documentation of results of survey activities, the Act no. 359/1992 Coll. [40] set up the Czech Office for Surveying, Mapping and Cadastre (CUZK) as the central administrative office of surveying and cadastre in the Czech Republic.

In accordance with §3a of the Act 359/1992 Coll. (Zákon č. 359/1992 Sb.) the Land Survey Office with headquarters in Prague enforces the management of geodetic bases and management of base state map series and thematic state map series. Furthermore, the Land Survey Office also administrates the Fundamental Base of Geographic Data of the Czech Republic (ZABAGED). The base state map series is in Act no. 200/1994 Coll. (Zákon č. 200/1994 Sb.) defined as a cartographic work with a basic and generally usable content, consistently showing the territory according to unified principles and that is created and issued by the government authority in the public interest. Thematic state map series is defined as the cartographic work depicting thematic reality usually on the basis of the base state map series and issued by the government authority in the public interest.

The Dutch institution, that provides data collection, field survey and mapping, is The Netherlands' Cadastre, Land Registry and Mapping Agency, simply called Kadaster.

The crucial legal regulation in the field of mapping is the Cadastral Act (Wet van 1989). According to this Act and its amendment (Wet van 2003) the task of Kadaster is keeping public registers, keeping cadastral registers, management of the geodetic control and registration of the ships and aircrafts. Kadaster also ensures uniform and consistent collection of geographic data, it collects and records also the underground infrastructure data (pipelines and cables). It is also obliged to collect and update geographic data for the armed forces in behalf of the Ministry of Defence. It is obliged to provide recorded data and promote its availability.

Before 2004 there was another institution that provided mapping—Topographical Service (Topografishe Dienst). It was a part of the Ministry of

Defence. Its outputs were for example 1:10,000, 1:25,000 or 1:50,000 scale topographic maps.

In the United Kingdom there exist several agencies that collect, process or provide geographic data (resp. data related to land property) on national level. The main geospatial data provider is Ordnance Survey (resp. Ordnance Survey Northern Ireland) (Vandenbroucke et al. 2011). Ordnance Survey (OS), as a national mapping agency, maintains large-scale mapping for England, Scotland, and Wales. In Northern Ireland this is the responsibility of Ordnance Survey Northern Ireland. Ordnance Survey was established by Ordnance Survey Act 1841, however its history dates back to the mid of 18th century, to the beginnings of its antecedent, the Board of Ordnance, a government's quasi-military office. In 1973 the OS released its first large-scale digital map, and the entire coverage of 230,000 current maps was digitized by 1995 (Evans 2006). Ordnance Survey was originally established by British government, but in nowadays it is based on business principles, with a governmental commission. OS provides (among others) British national topographic base map—OS Master Map. Data for OS maps are collected by employees of OS, and also are provided by several national offices (such as Hydrographic Office, British Geological Surveys, etc.) (Ordnance Survey 2015).

Ordnance Survey is also the main producer of geographic names datasets, addresses maps [Complex information about addresses is collected, managed and provided by GeoPlace, which is a Limited Liability Partnership owned by the Local Government Association and Ordnance Survey (GeoPlace 2015)] and they also provide other important datasets including transport networks. Geodata sets with maritime features (coastal zone and offshore) are produced by the UK Hydrographic Office. Land and property data are held by Her Majesty's Land Registry, Registers of Scotland and the Land Registers of Northern Ireland. For thematic data (such as socio-economic, geological, environmental including protected sites or land cover), many other agencies are involved (Vandenbroucke et al. 2011).

5 National Cadastres in Compared Countries

The central administrative authority of surveying and cadastre of the Czech Republic is the Czech Office for Surveying, Mapping and Cadastre [§1 (Zákon č. 359/1992 Sb.)] It ensures the uniform implementation of the land registry management, creates and manages the information system of cadastre of real estates (ISKN) and manages the central database of the cadastre. Furthermore it manages the Register of territorial identification, addresses and real estate [§3 (Zákon č. 359/1992 Sb.)].

Cadastre is defined in the Czech Republic by the Act No. 256/2013 Coll. as a public register, which contains a set of data about real estate defined in this Act, their list, description, their geometric and positional determination and registration of rights to this property. The cadastre keeps (a) land in the form of parcels, (b) buildings that have a house number or registration number unless they are not a part of the plot or part of the right to build, (c) buildings, which do not have a house number or registration number unless they are not a part of the plot or part of the right to build, they are a main building on the plot and they are not a small building, (d) units defined by the Civil Code, (e) units defined by the Act no. 72/1994 Coll., on ownership of flats (f) the right to build.

The content of the cadastre is arranged in cadastral documentation by cadastral units. The cadastral documentation consists primarily of a set of geodetic information (SGI), which includes the cadastral map and its numerical representation. Furthermore, a set of descriptive information, documentation of survey results and measurement for a management and recovery of the set of geodetic information, including local names, a collection of documents (decisions of public authorities, contracts and other documents) and protocols on entries, records, notes and additional records into the cadastre.

Cadastral map is included in the national large-scale map series. It contains a planimetry and map lettering. Planimetry of the cadastral map includes the boundaries of cadastral units, units of territorial administration, state borders, plots boundaries, building's perimeters and perimeters of waterworks registered in cadastre, boundaries of protected areas and protective zones and points of geodetic control. Planimetry of the digital cadastral map includes boundaries of the easement scope to the part of the plot. Another elements of planimetry are bridges, culverts and tunnels in the road if a watercourse or a road goes through it.

Digital cadastral map is kept electronically in the national coordinate system S-JTSK. The deadline for completion of digitization of cadastre was originally set for 2006. It was postponed several times and it should be completed in 2017. Remaining analog cadastral maps are still kept graphically on the plastic foil.

In the Netherlands both land registration and cadastre are under the responsibility of the same institution as mapping, therefore Kadaster. By the law (Wet van 1994) it is independent public body since 1994 and it is subordinated to the Ministry of Infrastructure and the Environment. The competence of Kadaster includes land registration, cadastre and maps and for this reason it is called 'unified cadastre' (Dijkstra and Booji 2003). The competence of Kadaster in the area of land registration is legally determined by cadastral act (Wet van 1989).

For understanding the concept and the role of the land property registration and the associated mapping system in the UK, the specific approach to evidence of land property boundaries should be mentioned. In the UK, the land property boundaries are considered as "general", i.e. their position does not have to be exact (see the difference between general and fixed boundaries, as distinguished e.g. in Bogaerts and Zevenbergen (2001). Namely, general boundaries are based on the visible feature on the ground, but their position is considered to show the boundary approximately (even when it has measured coordinates). These visible boundaries do not have indicate the exact location of the legal boundary. Whereas for fixed boundaries there must be specified exact position of each boundary point, determined usually by coordinates, and these coordinates define the boundary.), as follows from legal anchorage (Section 60 of Land Registration Act 2002). Every land property (or its change) is registered with "title plans" (Based on OS

MasterMap, see further). However, the position of boundaries on the plan do not indicate the exact position of the property, unless the owners had worked out the exact boundaries (A landowner can apply to achieve more precise boundaries. The setting the exact boundaries is based on neighbours' agreement, when the neighbours should agree on the position of the exact boundary between their properties (Gov.uk. Your property boundaries 2015; Gov.uk. Practice guide 40 2015)).

In the United Kingdom, there does not exist a complex cadastre known from European concept, i.e. as a register of legal title and as a map showing the legal title. Particular tasks of collecting and maintaining of land property information are divided into two main responsible subjects—Land Registries, which hold land property data, and Ordnance Survey, the national mapping agency. Moreover, these two tasks are ensured by several agencies in UK, depending on the part of the country (see Table 1). Basically, in the UK there is no single organization responsible for the cadastre.

As it is seen from the Table 1, the responsibility for the recording of land rights in the UK is divided between Her Majesty's Land Registry in England and Wales, Registers of Scotland in Scotland, and Land Registers of Northern Ireland. The basic information about the land property is in Land Registries (both Her Majesty's Land Registry and Registers of Scotland) recorded by title register along with title plan. The title register contains title number, address, owner(s) and price, the price and the location of property are initial data for land value tax. The plan shows the land owned. Title plans are prepared on the latest Ordnance Survey map (Land Registry associates their boundary information with OS topographic maps, specifically with boundary objects as fences, building walls, roads, etc. (Gov.uk. National Polygon Service—Detailed guidance 2015) available at the time of registration and are not updated as a matter of course (Land Registry 2015). However, because of the concept of general boundaries, the title plan shows the general position of the boundaries: it does not show the exact line of the boundaries (Land Registration Act 2002). Ordnance Survey itself does not provide (and shall not, according to Section 12 of the Ordnance Survey Act 1841) any map with property boundaries, nor information about ownership of physical features, even that some property boundaries may be coincident with surveyed map features (Ordnance Survey. Property boundaries: the roles of Land Registry and OS 2015; Ordnance Survey Act 1841).

Table 1 Agenda of British offices. Data source: (Manthorpe and John 2004; Registers of Scottland, 2014

	England a Wales	Scotland	Northern Ireland
National mapping	Ordnance Survey	Ordnance Survey	Ordnance Survey for Northern Ireland
Land Property Registration	Her Majesty's Land Registry	Registers of Scotland, namely Land Register	Land Registry of Northern Ireland
Land use and other thematic layers	Environment and Agricultural Departments and County and Local Authorities (private, public)		

Her Majesty's Land Registry provides The National Polygon Service, which serve as auxiliary tool for working with land property data. It contains three datasets. The National Polygon dataset (The dataset is a published dataset of Land Registry Index Polygons. Index Polygons are a representation of a registered land properties showing their location and extent. Its purpose is to provide an index to show the indicative location of a registered title. Land Registry Index Polygons are mapped against Ordnance Survey large-scale data, MasterMap (Gov.uk. National Polygon dataset specification 2015)) that shows the indicative shape and position of each boundary of a registered title for land and property in England and Wales (In the provided map, the property boundaries are either based on real-world objects from OS MasterMap (fences, building walls, streams etc.), then a solid red line is used, or the boundary is not defined by a physical feature and Land Registry indicates it on the plan by a dotted line.) (Gov.uk. National Polygon Service—Detailed guidance 2015). The second dataset is The Title Descriptor dataset that describes the legal interest, it provides a freehold or leasehold label for each title [13]. The third dataset is the "code key"—The Title Number and Unique Property Reference Numbers (UPRN) Look Up dataset, which contains Land Registry title numbers and associated UPRNs for the land property (Gov.uk. National Polygon Service—Detailed guidance 2015).

Her Majesty's' Land Registry provides also the INSPIRE Index Polygons, an open source dataset, developed to comply with the INSPIRE Directive. It contains the locations of freehold registered property in England and Wales and a sub-set of Index Polygons for all freehold land and property (Gov.uk. INSPIRE Index Polygons spatial data—Detailed guidance 2015). The INSPIRE Index Polygons are a sub-set of the National Polygon dataset (In compare to National Index Polygons, INSPIRE Index Polygons have a much simpler set of attributes (Gov.uk. National Polygon dataset specification 2015)), made up of polygons related to freehold titles (Gov.uk. National Polygon dataset specification 2015).

Registers of Scotland produce their own parcel map The Cadastral Map, which takes over information from local title lists. The Land Registration etc. (Scotland) Act (2012) supports this and puts the creation of the cadastral map on a statutory basis, introducing it as a part of the Land Register (Registers of Scotland 2014).

Even that in UK exist maps showing parcels and land property, those map representations do not represent the legal boundaries exactly, and therefore they do not determine a boundary in the sense of resolving a disagreement where the exact boundary is located (Ordnance Survey. Property boundaries: the roles of Land Registry and OS 2015).

6 Content of National Basic Spatial Datasets

6.1 The Czech Republic

In the areas of the Czech Republic, where a digitalization of dataset of objects of the Cadastre of Real Estate of the Czech Republic was already processed, this digitized dataset is the source of basic geodata with the highest detail level. However, because of the historical development of the land cadastre documentation lasting more than 180 years, the Set of Geodetic Information (SGI; SGI is a summary name for cadastral maps, both analog and digital.) is inhomogeneous and with different quality. Due to the fact that in the last 25 years it was resigned from any revision of cadastral data and the real situation in the field, today's cadastre map is considered only as a reflection of the legal state registered in the national cadastre. Because of this approach the map cannot fulfil the functions of the large-scale base map, nor (in case of the digitalized form) the set of basic geodata.

The poor condition of the national geodata with the highest detail level should have been compensated by project Digital Map of Public Administration (DMVS). The project was included in the program of electronization of public administration agendas (eGovernment). The aim of the program was to increase the efficiency of the public administration and to simplify the contact between public and governmental offices. Through the project it was reflected the demand for geodata with the detail level of the land parcel data model.

For the DMVS project, as well as for the following projects Purpose-built Cadastral Map (UKM), Digital Technical Map (DTM) and Tools for Creating and Maintaining Spatial Planning Analytical Data (UAP), the responsibility hold administrative regions under supervision of the Ministry of the Interior of the Czech Republic. From the technical point of view, the DMVS is composed from digital ortophoto of the Czech Republic, digital and digitized cadastre maps, digital purpose-built cadastre maps, and technical map of town or village if maintained (§36 of the Act No111/2009 Coll.). Obviously, this combination of non-harmonized datasets without legal continuity can guarantee the decision-making processes of public administration including legal relevance only with difficulties.

Next geodata sets are covered by basic registers of the Czech Republic. In the system of basic registers, as it is stated in the Act No 111/2009 Coll., is included among others (Next registers defined in the Act are for example: base register of legal entities, business persons and public authorities (held by Czech Statistical Office), base register of inhabitants (held by Ministry of the Interior of the Czech Republic), and others. Even that the Act states the system of basic registers, the system does not have to be definitive, as it is obvious from the example of submitted bill on measures to reduce the cost of deploying high-speed electronic communications networks, which indirectly presuppose creation of a register of passive infrastructure.) also the Registry of Territorial Identification, Addresses and

Real Estate (RUIAN), administered by Czech Office for Surveying, Mapping and Cadastre (CUZK).

Since the digitalization of cadastre map is not completed yet, the data for RUIAN were taken over from DMVS dataset, because the territorial features from RUIAN are shown on maps of State Map Series, or on Digital Map of Public Administration. Datasets of Purpose-built Cadastral Map are continuously being replaced as the digitalization of Cadastre maps held by CUZK proceeds. However, it leads to changes in topography in maps of State Map Series, which cause considerable difficulties for everyone, who uses State Map Series as a base for deriving thematic maps (e.g. features for land planning analytical data, ground plans, and easements).

Next dataset held by CUZK is the Fundamental Base of Geographic Data (ZABAGED) (Since 2013 ZABAGED ®.), which has been built since 1994, based on Government Resolution No. 492 from 8. 9. 1993. ZABAGED was created by digitalization of updated printing masters (topography, hypsography, hydrography, selected land covers and lettering) of 4555 map sheets of Base map of the Czech Republic 1:10 000 into raster. Since 1998, the ZABAGED has become a topological-vector model. Geographic data being led in ZABAGED are spatial data that represent location of features including topological relations and attributes for describing characteristics of features or for the identification. The collection, processing and keeping the data is held by CUZK, according Act No. 200/1994 Coll. on land surveying. Descriptive data should be provided by offices responsible for relevant features (roads and highways, motorways, water bodies, etc.). A condition of successful management of ZABAGED is close interdepartmental cooperation.

The definition of geographic features and the data model (and its changes) of ZABAGED is published in Catalogue of feature types ZABAGED®. Nowadays ZABAGED® contains 8 categories and 116 types of geographic objects with more than 350 attributes. Between 2009 and 2011, based on INSPIRE requirements, a harmonization of geodata in border areas of the Czech Republic and Free State of Saxony was processed (With a support of Program Ziel3), followed by bilateral cooperation with surveying offices of Bavaria, of Slovakia and of Poland.

Based on current data from ZABAGED, all medium-scale state map series are maintained and produced. This makes ZABAGED® the main data source for cartographic models of all medium-scale state map series (as specify Government Resolutions No. 116/1995 Coll. and No. 430/2006 Coll.). By the realization of topological-vector topographic base and by using cartographic models for computerized production of all medium-scale state map series, the Czech Republic ranks among the technical most advanced European countries, as for the collection and processing of geodata representing the land reality (Šíma 2016), however with no connection to parcel data model and land property estates. Though, the level of detail in the model (e.g. for buildings) is far short of fulfilling requirements of range of agencies of public administration bodies (building permission proceeding, land use planning, etc.) and it is not realistic to expect widespread improvement of positioning and increasing the level of detail of all objects from the topographic model, which was primarily designed for different purpose.

Concurrently with ZABAGED, another digital topological-vector model of territory of the Czech Republic (DMÚ 25) was created for the need of army. It was based on vectorization of military topographic map in the scale of 1:25,000, due to duality of military and civil national map products since 1969. The difference is in the topographic objects defined in the Catalogue of the objects (7 categories, 112 catalogue sheets). The database is created and maintained in ArcInfo ESRI software, the codes are strictly following NATO DIGEST and its catalogue FACC. The original coordinate system was S-42, after admission in NATO it was changed to WGS84.

6.2 The Netherlands

Dutch key cadastral register was included in the law by the act (Wet van 1989). The previous analogue map was unsatisfactory and due to economic growth and growth in the real estate market there was a need for quality and available cadastral data. So in 1990 a seamless digital cadastral map and digital cadastral evidence began to develop. It is managed in the key register cadastre (BRK). BRK database content is similar to the content of the Czech Information System of the Cadastre of Real Estate (ISKN); it contains land registrations and cadastral maps. The register has links to many other registries, for example key register persons (BRP), national trade register (NHR) and BAG.

The key register large-scale topography BGT was included in the law on 25. 8. 2013 by the act (Wet van 2013). BGT is the most detailed large-scale digital dataset in the Netherlands. It replaced the previous Large Scale Base Map of the Netherlands (GBKN). GBKN has been created since 1975 as an analogue 1:500 to 1:5000 map. After a long period of development, in 1992 the LSV-GBKN (Landelijk Samenwerkingverband—National Cooperation GBKN) was established which finished the production of the GBKN in 1999. The LSV-GBKN was a national joint venture with 11 regional partners. It was a public private partnership of the municipalities, utility companies, water boards, the Dutch cadastre and the Dutch administration. GBKN was established according to user requirements and funding opportunities. For this reason, the map content was inconsistent and there were differences in its quality and level of detail. For the needs of state and government such a map was insufficient and it was decided to switch from GBKN on BGT, uniform and consistent map on the whole territory of the country (Peersman et al. 2015; Goorman 2010).

BGT is an object-based topographical database that covers the entire Netherlands. This full coverage is achieved by having a multitude of source holders who provide the data, including not only the various level of government (national, provincial, water boards, and municipalities), but also some private parties like utility providers, who provide datasets covering their own networks.

The BGT has links to the other key registers. The major links are with BAG, BRT and the final major link from the BGT is with the BRK that derives building geometry from the BGT register (Goorman 2010).

The key registration for addresses and buildings was included in the law by the act [38] on 24 January 2008. The law entered into force on 1 July 2009, and the local authorities are since then under the obligation to deliver their addresses and building data to the national registry. From 1 July 2011, all public bodies have to use the register (Spatial Data Infrastructures in The Netherlands 2011).

The BAG actually consists of two separate key registers: the Key Register for Addresses (BRA) and the Key Register for Buildings (BRG). However, the two are intricately linked and are usually considered to be a single register. BAG data originate from municipalities that are the source holders of all BAG data. Like all key registers, the BAG register has links to the other key registers. For example the assessment of immovable objects registration (WOZ) derives addresses from the BAG database. WOZ objects are also linked to BAG objects where possible. Secondly, the BAG database references the municipal personal records database (GBA) key register. Thirdly, the BAG register has links with the two topographic key registers, BGT and BRT, which store large-scale and small-scale topography (Goorman 2010).

The law making Top 10NL the key registration for topography was adopted in January 2008 by amendments to the Law on the Cadastre. From 1 January 2009, all public bodies in the Netherlands are under an obligation to use it. Local authorities that had their own 1:10,000 topographic maps have to use it since 2010 (Spatial Data Infrastructures in The Netherlands 2011).

Top 10NL is the most detailed output from the key register topography (BRT) that is object-oriented digital topographic dataset founded in 1997. Top10NL data are uniform, consistent, nationwide, that follows the previous data set Top-10 Vector. In addition, by the automatic generalization derived 1:50,000, 1:100,000, 1:250,000, 1:500,000 and 1:1,000,000 maps are also part of the authentic registration for topography. A covenant about the maintenance of the registration was signed by the Ministry of Infrastructure and Environment and by the Cadastre (Bakker and Kolk 2003).

Top 10NL data are based on a scale of 1:10,000, but can be used in 1:5000 to 1:25,000. The individual ob-jects of this data set are divided into ten classes: road section, railway section, portion of water, building, field, design element, relief, administrative area, geographic area, functional area.

There are two ways of linking key registers with each other: administrative or georeferenced. Administrative linking is done with the help of unique identifiers. Linking an address to a person is done by combining the identifier of the premise which has an address with the corresponding person identifier. This relation is main-tained until some change has occurred like the person moved to another address. Georeferenced linking is done by using the location in maps. If information in a key register shares the same location with information in another key register there is a georeferenced link between them (Ellenkamp and Maessen 2009). The relationship of selected Dutch da-tasets is shown in Fig. 1.

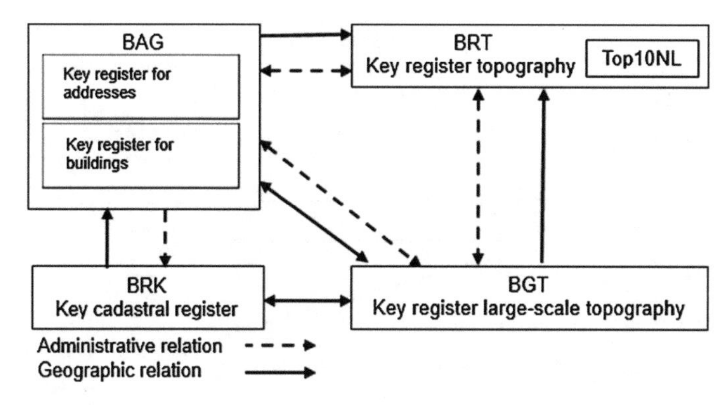

Fig. 1 The relationship of selected Dutch data sets

6.3 The United Kingdom

In the UK there is not any national organization responsible for the whole concept of collecting, processing and maintaining the geospatial data, and thus the datasets which could be considered as basic geodata sets are provided by several subjects. However, because of the importance and role of Ordnance Survey and its products, the OS MasterMap can be considered as a basic national map in UK (namely its Topographic Layer). For urban areas the data has been captured and maintained at a recommended scale of 1:1250, for rural areas at scale of 1:2500 and for remote (mountain and moorland) areas the scale of 1:10,000. However, for particular purposes can be appropriate any scale from 1:1250 to 1:10,000 (Wet van 2003). The Topography Layer is updated with frequency of 6 weeks. Because the term of basic national geodata is not in UK put on a statutory base, and because Ordnance Survey is the national mapping agency with a special governmental commission, the data from OS MasterMap could be from a specific perspective considered as a basic geodata in UK.

The OS MasterMap Topographic Layer contains 7 themes: administrative boundaries, buildings, heritage and antiquities, land, rail, roads, tracks and paths, structures, terrain and height, and water (Wet van 1991).

Also a dataset of address places could be considered as one of basic geodata sets. Addresses, their database and administration, are held by GeoPlace (GeoPlace is a Limited Liability Partnership jointly owned by the Local Government Association and Ordnance Survey. It brings together local and national data into unified dataset to build a single, definitive address database. Their main products are the National Street Gazetteer and the National Address Gazetteer (Ordnance Survey 2001), a map of address places is produced by Ordnance Survey (the AddressBase map) based on GeoPlace data. Postal addresses are in responsibility of Royal Mail (Katalysis Limited 2014). Registries of land property titles, as it is held by Land Registries should be also involved into basic geodata of land property titles, as it is

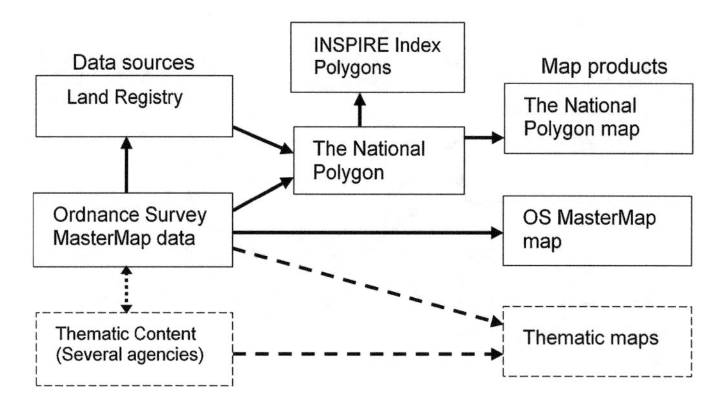

Fig. 2 The relationships of main agencies and map products in the UK

held by Land Registries (See description of the National Cadastre above.). The relation of main agencies and map products in the UK is shown in Fig. 2.

7 Summary of Knowledge Gained from Analysis

The main characteristics of dealing with geospatial data in the United Kingdom is that there exists a private national mapping agency, Ordnance Survey, with a special commission for managing and providing geospatial data for the government, but also providing its data for public. Data for OS maps are collected by employees of OS in the field, or are provided by several national offices (e.g. Hydrographic Office). In fact they create the base national map, even that the term is not specified in any legislative. Ordnance Survey collects topographical data, maintains this data and besides the topographic map it produces also several thematic maps. However, it does not provide any map with property boundaries. There is not only one agency responsible for cadastre in UK, but its functionalities and responsibilities are divided into more offices. In general, particular tasks of collecting and maintaining the information about land property are divided into two main subjects—Land Registries, dealing with land property data, and Ordnance Survey, the national mapping agency.

In the Netherlands, there is only one institution, which is responsible for land registration, cadastre and mapping services. This connection has a positive effect on the development of the Dutch NSDI, which is based on the system of key registers. Key registers are regulated and established under Dutch law and governmental organisations are obligated to use key register's data. Key registers try to meet the objective of creating the Dutch NSDI, thus increasing data quality and reducing the cost of the acquisition, effective management of data and making data available to the public.

The system of key registers tries to meet the objective of creating the Dutch NSDI, thus increasing data quality and reducing the cost of the acquisition, effective management of data and making data available to the public. There are links between registers to ensure its relations. This ensures efficient data management and easy access to different registers. However, there is some data redundancy apparent in the system. For example, the theme buildings is kept in BAG register, BRK register and in topographic registers, too. The term ‘building’ is not defined and specified for use in the cadastre and its specification according to specific catalogues is different for individual registers. Therefore, building data can be inconsistent in various registers.

In the Czech Republic there does not exist an adequate, consistent and homogenous legal framework in the field of geospatial information. Within the activities of development of geospatial data one faces the existing complicacy, resp. inconsistencies of legislation for pro obtaining, gathering, transfer, management, guarantee, publishing and conditions of providing the geospatial information. Due to the situation, it will be necessary to assemble a comprehensive systematic analysis of legal framework that deals with spatial information or proceedings of the information. As a result it should be also indicated which legislative is eventually lacking or does not govern the problems sufficiently.

It is also necessary to define the framework of regulation and its links to services of public administration in the field of geospatial information. It should be made changes in legislative and regulatory standards for the geospatial information. An adoption of the law on national infrastructure in the field of geospatial information should be also a part of the changes. The law should unify the requirements for managing, updating, providing and using geospatial information and through this it will create conditions for their effective sharing and for the development of public services dealing with geospatial data.

It is necessary to set up and ensure coordination of activities in the field of geospatial information and set up roles and responsibilities for all participants by jurisdictional defining of public authorities. From this reason, an establishment of the coordinating and control system on the governmental level should be institutionally ensured. The institution should coordinate and manage the development of national infrastructure, including coordination of legislative preparations or funding.

In general: effective use of geospatial information is characteristic for the contemporary society. From this point of view creating particular information systems with fragmented data background cannot be considered as perspective. The more, when the background is specific only for the certain department, with consideration of duplicity or only slight modifications of range of object. Not only that such data ensuring implies disproportionately high financial demands, but first of all it does not lead to the data interoperability even to interoperability of geospatial information generated for those data.

The main advantage of the maintaining the reference geographic data on the national level is, that the data are obtained only once, but with a multiple use. The crucial benefit of reference data are savings when upgrading data and mutual consistency of data sets and users’ thematic data.

It is obvious that foreign functional systems and experience in the field of basic geodata are transferable only with difficulty due to a different legal environment and conditions of the different historical development. It is possible to accept and use proven and generally valid principles, on which the INSPIRE Directive (Zákon č. 89/2012 Sb.) international standards such as ISO (International Organization for Standardization) series 19100, the standards W3C (World Wide Web Consortium), or OGC (Open Geospatial Consortium, Inc) are based. It is always necessary to take into account national specification.

8 Next Development of Infrastructure for Spatial Information in the Czech Republic

In the 'Strategy of the Development of the National Infrastructure for Spatial Information in the Czech Republic up to 2020' (GeoInfoStrategy), approved by the Czech government in 2015, the main emphasis is placed on providing the services of the public administration. It was declared, that it is important to specify the essential information society services in the field of geospatial information. However, such services for information society involves continuous improvement and subsequent development of data resources for geospatial data and their use by public administration or the whole society. This is solved in the GeoInfoStrategy by creating the National Set of Spatial Objects (NaSaPO).

The concept of NaSaPO is based on general definition of NaSaPO objects including their reference attributes. It defines principles of unique identification of geospatial features for the NaSaPO and also terminological and factual definition of the features. The Plan for implementation and launch of NaSaPO Information System (NaSaPO IS) and ensuring the flow of spatial data assumes creating the NaSaPO IS in several phases (Fig. 3).

NaSaPO IS will be built and maintained by an administrator of NaSaPO. The key task for the administrator will be to define current flows of geodata, which arise

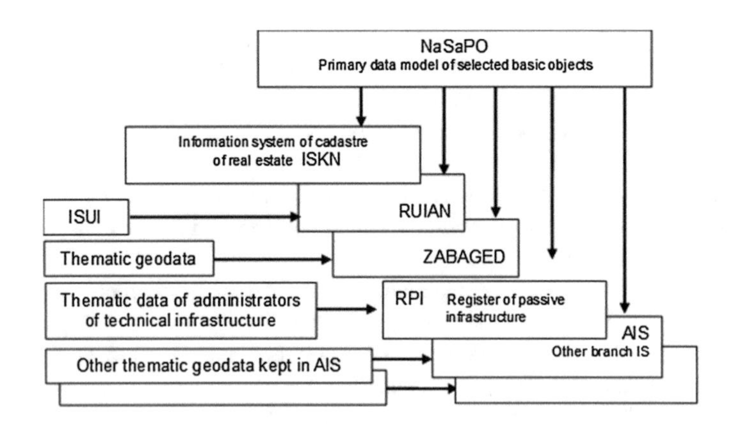

Fig. 3 NaSaPO Information system and links to other information systems

during operations of the public administration in the territory and which are related with execution of surveying activities. The next important task is to provide such an implementation of these flows in NaSaPO, so that it will create quality and certified source for NaSaPO objects. This database of spatial features, maintained and updated in the primary data model and sufficiently detailed, will allow creating several derived NaSaPO objects by data generalization. The data model of objects, with clear identification of spatial position, should also contain some other clearly defined characteristics of this object. Through this requirements new approaches to collecting geodata are defined, not only in terms of content and accuracy, but also and above all in term of recency in conformity with the real situation in the field.

By the Czech government approved 'Strategy of the Development of the National Infrastructure for Spatial Information in the Czech Republic up to 2020' accepts general principles of the development of the infrastructure for spatial information, it brings new initiatives to guarantee public administration services in the field of spatial information. However, it requires a fundamental rethinking mainly of the data flows generated by the public administration, the commercial sector, research sphere and academia sphere to achieve legal, organizational, semantic and technical interoperability.

References

Bakker N, Kolk B (2003) Top10 nl, a new object-oriented topographical database in gml. 21st, pp 841–851

Besluit van 29 augustus 2011 houdende vaststelling van voorschriften met betrekking tot het bouwen, gebruiken en slopen van bouwwerken (Bouwbesluit 2012)

Bogaerts T, Zevenbergen J (2001) Cadastral systems—alternatives. Comput Environ Urban Syst 25(4–5):325–337. doi:10.1016/S0198-9715(00)00051-X

Čada V (2014) Improvement and Further Development of Reference Geographic Data Within The Strategy for the Development of the Infrastructure for Spatial Information in the Czech Republic until 2020. GEODÉZIA, KARTOGRAFIA A GEOGRAFICKÉ INFORMAČNÉ SYSTÉMY 2014, Tatranské Matliare, 17–19 September 2014

Dijkstra T, Booji AS (2003) Renewal of automated information systems at the Netherlands cadastre and land registry agency. In: Strategies for renewal of information systems and information technology for land registry and cadastre, Symposium held by FIG Commission

Ellenkamp Y, Maessen B (2009) Napoleon's registration principles in present times: The Dutch System of Key Registers. In: 11th International conference for spatial data infrastructure. Rotterdam, Netherlands, pp 15–19

European Parliament and Council Directive 2007/2/ EC of 14 March 2007 establishing an Infrastructure for Spatial Information in Europe (INSPIRE). http://eur-lex.europa.eu/LexUriServ/LexUriServ.do?uri=OJ:L:2007:108:0001:0014:CS:PDF, cit. 20. 12. 2015

Evans AJ (2006) The National Geographical Information Policy of Britain. In: Proceedings of the GIS international conference/GIS Korea 2006, Seoul, Korea, 17–19 May 2006, pp 275–299

GeoPlace (2015) GeoPlace Annual Statement 2015–16. London. https://www.geoplace.co.uk/documents/10181/40244/GeoPlace+annual+statement+2015-16/817b6dcc-3278-4980-945b-2fce49240841, cit. 2. 12. 2015

Gooman N (2010) BAG & BGT: Spatial Key Registers: Compatibility and municipal use in Zwolle

Gov.uk. Your property boundaries. https://www.gov.uk/your-property-boundaries. cit 30. 11. 2015

Gov.uk. National Polygon Service—Detailed guidance. https://www.gov.uk/guidance/national-polygon-service. cit. 28. 11. 2015

Gov.uk. National Polygon and Title Descriptor datasets published—News stories. https://www.gov.uk/government/news/national-polygon-and-title-descriptor-datasets-published. cit. 28. 11. 2015

Gov.uk. INSPIRE Index Polygons spatial data—Detailed guidance. https://www.gov.uk/guidance/inspire-index-polygons-spatial-data. cit. 1. 12. 2015

Gov.uk. National Polygon dataset specification. https://www.gov.uk/government/publications/national-polygon-dataset-specification/national-polygon-dataset-specification. cit. 28. 11. 2015

Gov.uk. Practice guide 40: Land Registry plans, supplement 4, boundary agreements and determined boundaries—GOV.UK. https://www.gov.uk/government/publications/boundary-agreements-and-determined-boundaries/practice-guide-40-land-registry-plans-supplement-4-boundary-agreements-and-determined-boundaries. cit. 28. 11. 2015

Katalysis Limited (2014). An Open National Address Gazetteer. Ref: BIS/14/513

Land Registration Act 2002

Land Registration Rules 2003. SI 2003/1417

Land Registration etc. (Scotland) Act 2012

Land Registry. Example title plan. https://eservices.landregistry.gov.uk/www/wps/QDMPS-Portlet/resources/example_title_plan.pdf. cit. 30. 11. 2015

Manthorpe, John (2004) Comparative analysis on the cadastral systems in the European Union: United Kingdom. http://www.eurocadastre.org/pdf/john_manthorpe_comparative.pdf

Morad M (2002) British Standard 7666 as a framework for geocoding land and property information the UK. Computers, Environment and Urban Systems, 26(5), September 2002, pp 483–492. ISSN 0198-9715

Ordnance Survey Act 1841

Ordnance Survey (2001) OS MasterMap™ real-world object catalogue. v1.0 Nov 2001

Ordnance Survey. Data collection, management and analysis. https://www.ordnancesurvey.co.uk/international/knowledge/data-collection-management.html. cit. 25. 11. 2015

Ordnance Survey. Property boundaries: the roles of Land Registry and OS. http://www.ordnancesurvey.co.uk/resources/property-boundaries-owners.html. cit. 4. 12. 2015

Peersman, Martin, Hans van Eekelen, Sieb Dijkstra a Jenny Lisapzly., http://www.cuzk.cz/O-resortu/Nemoforum/Akce-Nemofora/Seminare/Promeny-mapy-VM/GBKN-transition-paper_CZ_ver5.aspx. Transition process of the Large Scale Base Topography of the Netherlands in historical perspective. cit. 24.11.2015

Registers of Scottland (2014). General Guidance: The Cadastral Map, v.01—08/10/14. https://www.ros.gov.uk/__data/assets/pdf_file/0018/11385/General_Guidance_CM.pdf. cit. 1. 12. 2015

Spatial Data Infrastructures in The Netherlands: State of play 2011, http://inspire.ec.europa.eu/reports/stateofplay2011/rcr11NLv123.pdf. cit. 14. 12. 2015

Šíma J (2016) Základní báze geografických dat (ZABAGED®)—dílo jedné generace českých zeměměřičů. Geodetický a kartografický obzor, ČÚZK, Praha

The Scottish Government (2005) One Scotland—one geography; a geographic information strategy for Scotland. ISBN 0755947843

Vandenbroucke, Danny and Dimitrios Biliouris, 2011. Spatial Data Infrastructures in The United Kingdom: State of play 2011. K.U.Leuven (Spatial applications division). rcr11UKv122

Wet van 3 mei 1989, houdende regelen met betrekking tot de openbare registers voor registergoederen, alsmede met betrekking tot het kadaster

Wet van 29 augustus 1991 tot herziening van de Woningwet

Wet van 14 februari 1994, houdende verzelfstandiging van de Rijksdienst van het Kadaster en de Openbare Registers

Wet van 9 oktober 2003 tot wijziging van de Kadasterwet en de Organisatiewet Kadaster (aanpassing van doeleinden en taken van de Dienst voor het kadaster en de openbare registers alsmede enkele andere wijzigingen)
Wet van 24 januari 2008, houdende regels omtrent de basisregistraties adressen en gebouwen (Wet basisregistraties adressen en gebouwen)
Wet van 25 september 2013, houdende regels omtrent de basisregistratie grootschalige topografie (Wet basisregistratie grootschalige topografie)
Zákon č. 359/1992 Sb. o zeměměřických a katastrálních orgánech, ve znění zákona č. 107/1994 Sb., zákona č. 200/1994 Sb., zákona č. 62/1997 Sb., zákona č. 132/2000 Sb., zákona č. 186/2001 Sb., zákona č. 175/2003 Sb., zákona č. 499/2004 Sb. a zákona č. 227/2009 Sb
Zákon č. 200/1994 Sb., o zeměměřictví a o změně a doplnění některých zákonů souvisejících s jeho zavedením, ve znění zákona č. 120/2000 Sb., zákona č. 186/2001 Sb. a zákona č. 319/2004 Sb., zákona č. 413/2005 Sb., zákona č. 444/2005 Sb., zákona č. 124/2008 Sb., zákona č. 189/2008 Sb., zákona č. 223/2009 Sb., zákona č. 281/2009 Sb., zákona č. 380/2009 Sb., zákona č. 350/2012 Sb. a zákona č. 257/2013 Sb
Zákon č. 89/2012 Sb., občanský zákoník

Spatial Data Analysis with the Use of ArcGIS and Tableau Systems

Szymon Szewrański, Jan Kazak, Marta Sylla and Małgorzata Świąder

Abstract Nowadays there are different technologies enabling visualisation of spatial data. The combination of two different systems may enhance the visualisations and therefore better communication of the results to decision makers and the wider public. The aim of our contribution is to assess the possibility of combining the functionality of Geographic Information Systems (GIS) and Business Intelligence (BI) systems for spatial data visualisation. We assess the analytical and visualisation features of combined ESRI ArcGIS and BI Tableau systems with the use of the visual data exploration approach. For the purpose of this study, Geographic Information System is used as a data manager and a data blender. The geoprocessed feature class was stored in the personal geodatabase and then loaded into the Tableau environment. We present the selected functionality of visual data discovery on the example of land change flows in the Czech Republic, Poland and Slovakia. In order to highlight the possibility to conduct analyses on different spatial levels, we ran the simulation at the local level and aggregated it to the regional level. The use of computational capabilities of GIS and BI enhance the geovisualisation on the map by quantitative analysis of tabular data, facilitate the visualisation of the results, and improve communication.

Keywords Big data visual analytics · Visual data discovery · Tableau · GIS and business intelligence integration · Visualisation of spatial data

S. Szewrański · J. Kazak (✉) · M. Sylla · M. Świąder
Department of Spatial Economy, Faculty of Environmental Engineering and Geodesy, Wrocław University of Environmental and Life Sciences, ul. Grunwaldzka 55, 50-357 Wrocław, Poland
e-mail: jan.kazak@up.wroc.pl

I. Ivan et al. (eds.), *The Rise of Big Spatial Data*, Lecture Notes in Geoinformation and Cartography, DOI 10.1007/978-3-319-45123-7_24

1 Introduction

Well-designed visualisation of spatial data facilitates communication of the results. Communicating is a creative process that involves many loops: goal definition, data preparation, suitable visualizations selection, designing for aesthetics, using effective medium and channel, and checking the results. Communication can be distorted by a large information flow. Therefore, the object visualization requires limiting the amount of information that is processed. Sometimes visualisation is only used to determine the location of the concentration of the studied phenomenon (Yuan et al. 2012) or the spatial distribution of the certain feature (Grassi 2014) at different spatial levels (Dolega et al. 2016). In many cases, in order to facilitate the interpretation of the results, classical spatial visualisation on the map must be supported by other graphic elements, such as simple bar charts or diagrams representing the more advanced non-spatial features (Dykes et al. 2010). The use of advanced methods of data visualization may improve the communication of big data (Benenson 2011).

The big data analysis is often preceded by the preparation of data, data blending and visualisation. Data preparation requires aggregation and generalisation of information included in the big data sets. Aggregation is extremely useful in summarizing big data to find trends and patterns in collected data (Silver 2012). Data blending allows gathering, formatting and cleaning data from various sources. Real life data comes from everywhere, in all different formats, models, ranges and sizes (Baesens 2014). This kind of data could be structured, semi-structured or unstructured at all. Also, it could be incomplete, duplicated and inconsistent. All these issues cause merging and unification problems (Wessler 2015). Therefore, when starting analyses, data types and formats shall be well recognized and evaluated (Morton et al. 2012).

1.1 GIS and BI Tools

The aim of this paper is to assess the possibility of combining the functionality of Geographic Information Systems (GIS) and Business Intelligence (BI) systems. We would like to prove that the dynamic connection of the systems is possible and they are complementary. The complementarity of the systems stems from their characteristics and functionalities. ArcGIS Desktop is the software for mapping and managing spatial data. Extensive toolboxes allow a user to make countless types of spatial analysis and advanced geoprocessing. GIS software allows automating processes using the Model Builder. ArcGIS is also commonly recognized as a powerful tool for data visualization (Law and Collins 2015). However, in this study we use the GIS system to a limited extent as the data manager mainly for data blending. Tableau Desktop is originally used as the business intelligence tool. According to Negash (2004), BI are systems which "combine data gathering, data storage, and knowledge management with analytical tools to present complex (…) information to planners and decision makers". Currently, BI is widely applied in big data analysis in the environmental, health, education, telecommunications and sport studies (Murray 2013). Tableau facilitates the data discovery (finding knowledge in big data) and data communication process by creating explanatory graphic, dynamic and interactive visualisations (Jones 2014). Tableau is a read-only table application. It has native connectors to many common databases systems such as Microsoft SQL Server, MySQL, Teradata, Oracle, EMC Greenplum, HP Vertica, and more. Also, it connects data sources such as Hadoop technologies (Hortonworks Hadoop Hive and Cloudera Hadoop) and cloud sources. Finally there's also ODBC (Open DataBase Connectivity) protocol. Tableau provides an optimized, live connector to many data sources so you can work directly with your data and it allows to bring data in-memory to work faster, and also work offline. Apart from the database server connection, Tableau allows direct connection to files such as MS Excel, MS Access.csv—comma separated values text files as well as: sav—IBM SPSS file; .sas7bdat—SAS data file; .rdata—R data file; .cub—Local Cube file; .tsv—a tab-separated file. Tableau offers easy (in comparison to GIS systems) features such as geovisualisation, connection to WMS, and custom geocoding (Peck 2014). The key feature of Tableau technology is the new language for data exploration. VizQL is a formal language for describing tables, charts, graphs, maps, time series and tables of visualizations (Hanrahan 2006).

2 Materials and Methods

For the purpose of this research, we applied the visual data exploration approach. The approach is based on the assumption that the user does not define the preliminary hypotheses, but discovers trends and interconnections while designing the graphs and visualizations. Data exploration was conducted according to the following procedure (Fig. 1). The spatial data has been blended in the GIS environment using geoprocessing tools. Also, it has been stored in the personal geodatabase and connected to the BI system. Visual analytics was conducted by searching for appropriate combination of dimensions and measures of the attribute table and presenting them on graphs and maps. The final results presentation took the form of an interactive dashboard on which the results of visual analytics are placed. Dashboard may also include the developed base maps as well as those available through web services (e.g. WMS, Mapbox).

For the presentation of the analytical and visualization feature of the proposed solution, we used the latest database of land cover changes: Corine Land Cover—Change Layer. The Corine Land Cover (CLC) inventory was initiated in 1990; next updates were produced in 2000 and 2006. The latest 2012 update is under production. It consists of an inventory of 44 land cover classes. CLC database is widely used in research and spatial analysis (Büttner 2014; Büttner et al. 2000; Feranec et al. 2007, 2010; Gutry-Korycka et al. 2015; Pekkarinen et al. 2009). Data for the

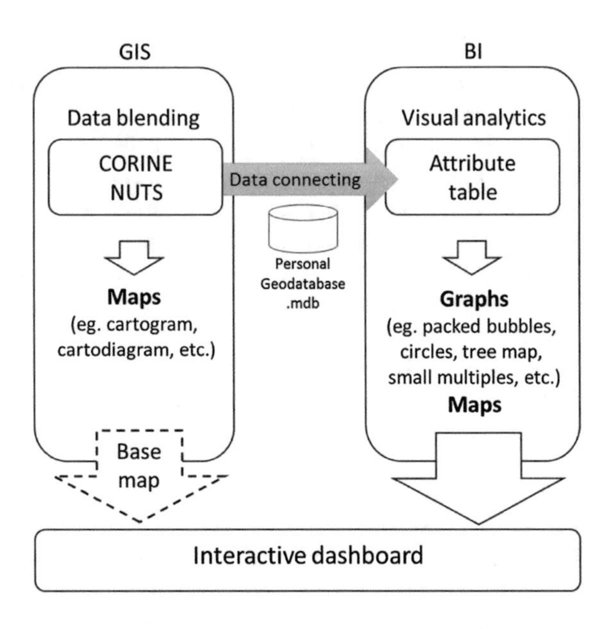

Fig. 1 Research conceptual framework

presentation was downloaded from the Copernicus Land Monitoring Services repository. The ESRI Geodatabase contains spatial information concerning the unique ID code, CLC classes in 2006 and 2012, and the type of land cover change or shape area. Data projection is EPSG: 3035 (ETRS89, LAEA). CORINE Land Cover Changes (LCC) 2006–2012 is yet the non-validated data and any interpretations and conclusions concerning land cover changes are unauthorised. The quantitative analysis will be carried out upon the completion and publication of the final products of the CLC 2012. That is why the aim of this paper is limited to presenting features of the data discovery and visual analytics of the integration of Tableau and ArcGIS.

For the presentation, we used the geodatabase containing information about the borders of the European NUTS 2013 (the copyrights to EuroGeographics for the administrative boundaries). The geographic database was extracted from the Eurostat repository—GEOSTAT. It contains information about the borders of NUTS 1, 2, 3 units in the European Union. Data projection is also ETRS89. For the initial preparation of the data, we used ArcGis for Desktop 10.2.2.

3 Results

3.1 Data Blending

Combining the functionality of GIS ArcGIS and Tableau spatial data visualisation requires data blending, data connecting and visual analytics. After adding the

NUTS data of the Czech Republic, Poland and Slovakia to map file, the selected objects have been exported as feature class to the new personal geodatabase. NUTS codes were assigned to the names of regions. Then the LCC data was added to the data frame. With the use of two geoprocessing tools clip and intersect the CLC changes in the three countries been extracted. That is how we received the new feature class. The original data contains only a CLC code. In order to facilitate subsequent interpretation of the results, we added additional tables containing descriptive legend to NUTS level 1, 2 and 3 classes which were joined to the CLC feature class. The new data table contained ca. 27,000 records × 12 attribute fields. Furthermore, geometric calculations were made and surface area for each object was designated. The next step was to export data to the personal geodatabase. It is the non-aggregated table containing spatial data that is the data source for the Tableau system. At this stage ArcGIS is used only as the data manager. In this regard, one can use its full technological potential: the geoprocessing tools: clip, union, intersect, merge as well as the toolbox Data Management Tools, and the toolset Fields.

3.2 Data Connecting

The first step of the analysis is to create a personal geodatabase containing the data feature class. In the research work, ArcGIS uses.mdb files for personal geodatabases, which can be imported natively into Tableau. Attribute table of the feature class can now be opened in Tableau Desktop software (version 9.1.2 was used). For this purpose, we start the application and select Connect to File > Access Server Connection. Tableau recognizes the geodatabase as the Access type database. After connecting to the database file, we get access to the panel for the tables' management in the database. The next step is to load the attribute table. The user may just drag it and drop in the window (Fig. 2).

The result table is the same that was created in ArcGIS. The user can choose whether to remain in the live mode (the changes made in the GIS are seen in Tableau) or extract (Tableau create its own native table—this allows for faster and safer analysis but the outside connection will be broken). Tableau automatically identifies the type of data stored in columns: numbers (floating-point and integer), date and time, date and string. These types can be set manually. The important feature is the ability to assign specific Geographic Role. When Geographic Roles are marked, the globe icon appears. That means that Tableau has automatically geocoded the information in that field and associated each value with the latitude and longitude value. If it contains geographic data that is not supported; it is possible to import custom geocoding. The data table can now be processed and visually analysed in the worksheet panel.

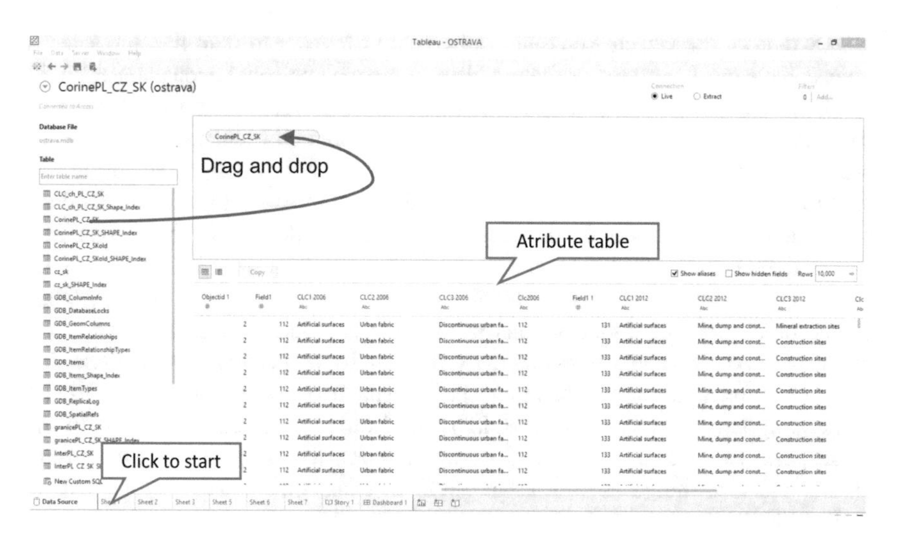

Fig. 2 Managing data sources in Tableau Desktop

3.3 *Visual Analytics*

The basic analytical panel in Tableau is called worksheet (Fig. 3). The worksheet was divided into panels and shelves. It contains information about the data: the database sources, dimensions and measures. Dimensions can be used to group or categorize data. Measures are fields which can be calculated by counting, summing or averaging. Measures can be either discrete or continuous. The drag and drop technology allows to click on dimensions and measures and move them onto other shelves and cards to create the views of the data. The Show Me card suggests applicable visualization types. These may include the text tables, heat maps, highlight tables, pie charts, horizontal bars, stacked bars, side-by-side bars, tree maps, circle views, side-by-side circles, continuous and discrete lines; continuous and discrete area charts; scatter plots, histogram; box-and-whisker plots, Gantt chart, bullet graph, packed bubbles as well as symbol and filled maps. Each analysis can contain multiple sheets, each with a different view of the data (Jones 2014).

Land cover changes can be analysed as land cover flows (LCF) (Feranec et al. 2007, 2010). The analysis is based on the combination of two land cover layers aggregated at the third level of CLC classification. In the case of CORINE land cover (44 classes), the intersection between two layers produces 1936 possible combinations. All types of possible changes in land cover are illustrated in the packed bubbles graph. For this purpose, we used two dimensions: CLC change (different colours) and the size of the circles which illustrates the surface changes. Figure 4 shows example of database visualisation.

Our next step was to aggregate information contained in the NUTS field. Tableau allows you to group attribute values and assign aliases. All PL, CZ and SK fields are grouped separately and are assigned aliases of the name of the countries.

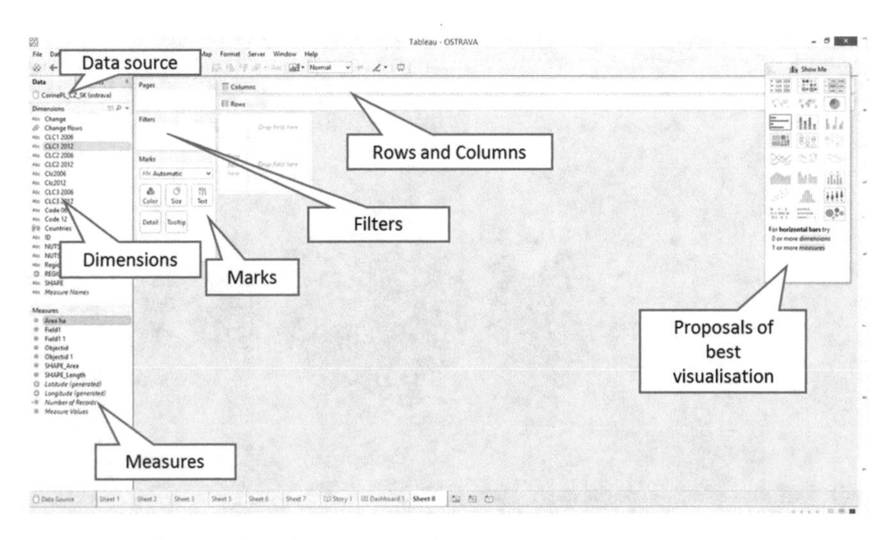

Fig. 3 Working in Tableau Desktop

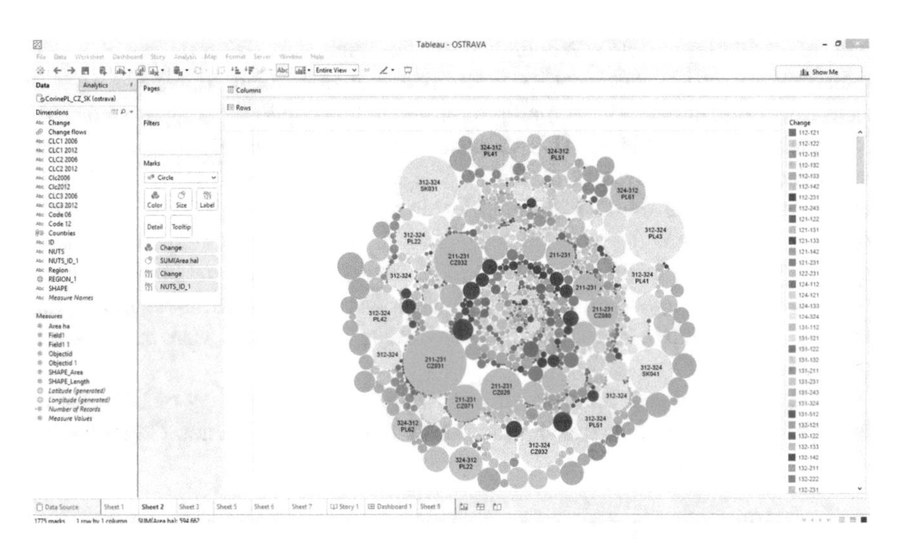

Fig. 4 Packed bubbles

Modified dimension was later dragged to the shelf columns. In a split second we obtained graphs for individual countries (Fig. 5).

The changes can be aggregated and grouped to land cover flows and classified according to the major land use processes. In the European Environmental Agency study (2008), 9 classification groups are identified: (lcf1) Urban land management; (lcf2) Urban residential sprawl; (lcf3) Sprawl of economic sites and infrastructures; (lcf4) Agriculture internal conversions; (lcf5) Conversion from forested and natural

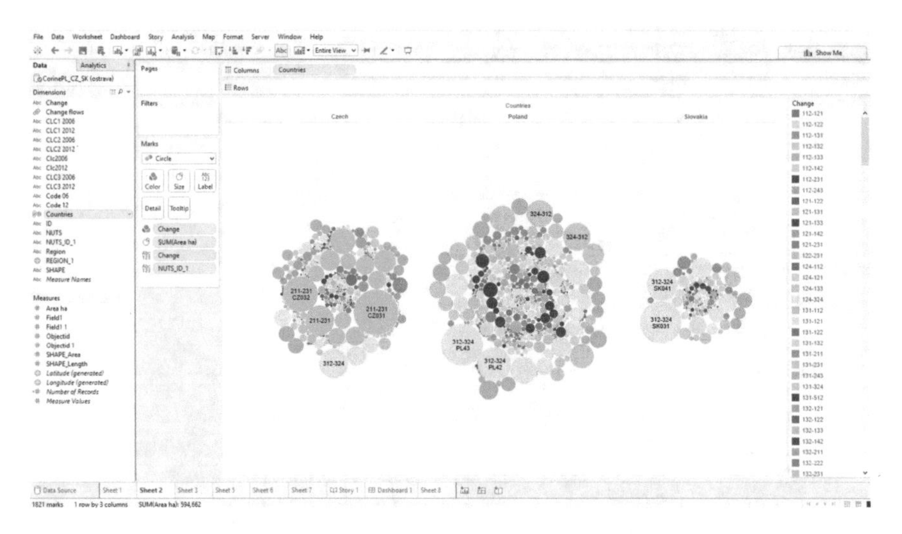

Fig. 5 Separated packed bubbles

land to agriculture; (lcf6) Withdrawal of farming; (lcf7) Forests creation and management; (lcf8) Water bodies creation and management; (lcf9) Changes of Land Cover due to natural and multiple causes. In this paper, we suggest land cover flow analysis at CLC2 level. This simplistic approach has already been used in the Romanian study (Popovici et al. 2013). Our next step was to use the fields containing the legend of the CLC 1 and 2 levels for the visualisation. The aggregated values of the field change were attributed with the new aliases. Grouped dimension is marked with a paperclip icon and as a new re-created data can be used for further analysis. At any stage each result of analysis can be exported to an image file, database, or crosstab to Excel. In our case, the data change flows were exported to a graphic file (Fig. 6).

All operations, which are necessary to illustrate a new typology of LCF, took a few minutes. The same data using previously developed dimension Countries can

CLC1 2006	CLC1 2012	CLC2 2012	Change flows
Agricultural areas	Agricultural areas	Arable land	Agriculture intensification
		Heterogeneous agricultural ar..	Agriculture intensification
		Pastures	Agriculture intensification
		Permanent crops	Agriculture intensification
	Artificial surfaces	Artificial, non-agricultural vege..	Agricultural land abandonment
		Industrial, commercial and tra..	Industrialisation
		Mine, dump and construction ..	Agricultural land abandonment
		Urban fabric	Urbanisation
	Forest and semi natural areas	Forests	Afforestation
		Scrub and/or herbaceous veg..	Agricultural land abandonment
	Water bodies	Inland waters	Water bodies construction
Artificial surfaces	Agricultural areas	Arable land	Agriculture intensification
		Heterogeneous agricultural ar..	Agriculture intensification
		Pastures	Agriculture intensification
		Permanent crops	Agriculture intensification

Change flows
- Afforestation
- Agricultural land abandonment
- Agriculture intensification
- Deforestation
- Industrialisation
- Other changes /abandonment, land development/
- Urbanisation
- Water bodies construction

Fig. 6 Fragment of the exported visualisation table

be illustrated in a different way. Obtaining the side-by-side circles requires selection of the appropriate data in the panel Data and selecting an appropriate chart in the Show Me tab (Fig. 7).

We can get the information what types of LCF changes occur in any of the three countries between 2006 and 2012. To evaluate the surface changes you only need to select the tab ShowMe, for instance the Treemap option. In the Marks field you must type the appropriate dimensions and measures; in this case these were change flows, countries and area (Fig. 8). The size of the field and the intensity of colour is proportional to the recorded changes. All chart properties can be modified: colour bar, palette styles, fonts, etc.

Finally, the data can also be presented on maps. Tableau allows automatic geocoding of the objects which have assigned the geographical roles. For the Czech Republic, Poland and Slovakia Tableau, regions have been defined. It was only important to indicate them because the feature class prepared for analysis contained the field with geographical regions. Tableau automatically detects the geographic location of the object. Symbology can be freely defined. One may choose visualized attributes by the use of filters and parameters (Fig. 9). It should be recalled that all the visualizations and data products have been made at one table, originally prepared in ArcGIS environment.

The above examples are just a small representation of analytical possibilities of Tableau. Multiple sheets can be brought together onto a dashboard, in which interactions with data elements on one sheet can filter or highlight data on other sheets (Fig. 10).

Dashboard can be made public and presented on-line. Tableau allows you to make calculations on data, logical operations and ranking—in a similar way as the

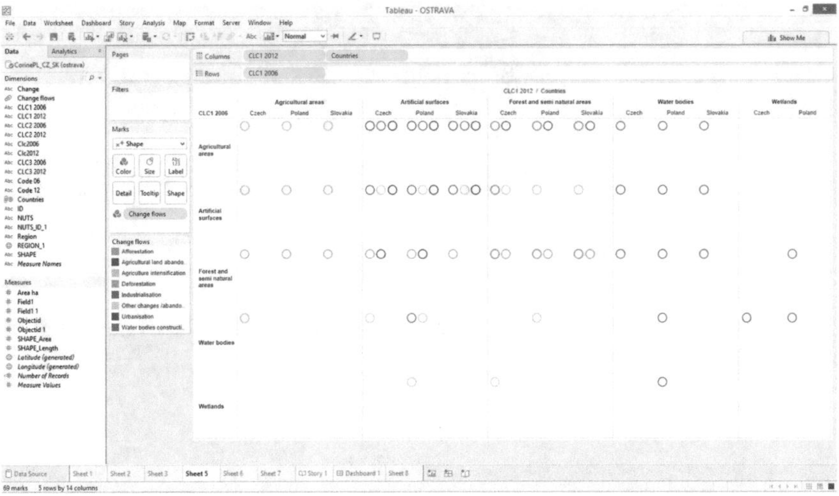

Fig. 7 Side-by-side circles

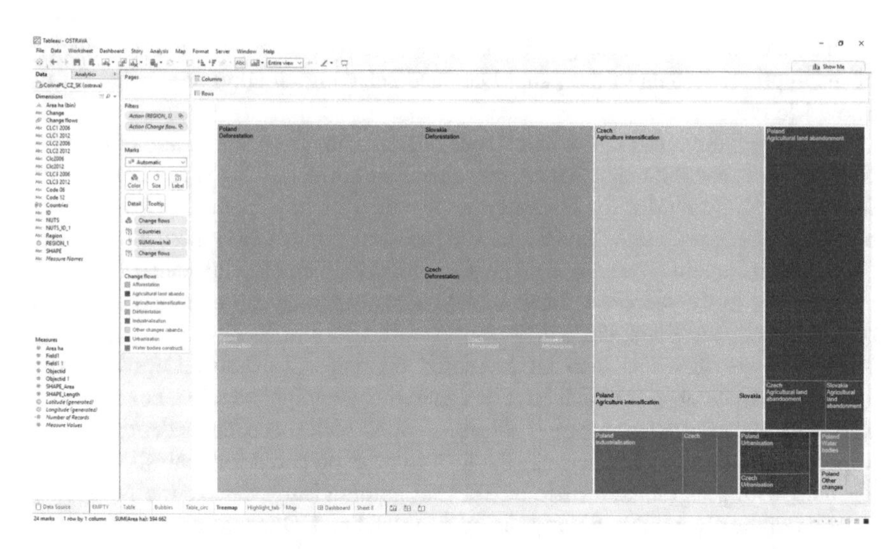

Fig. 8 Treemap of land change flows

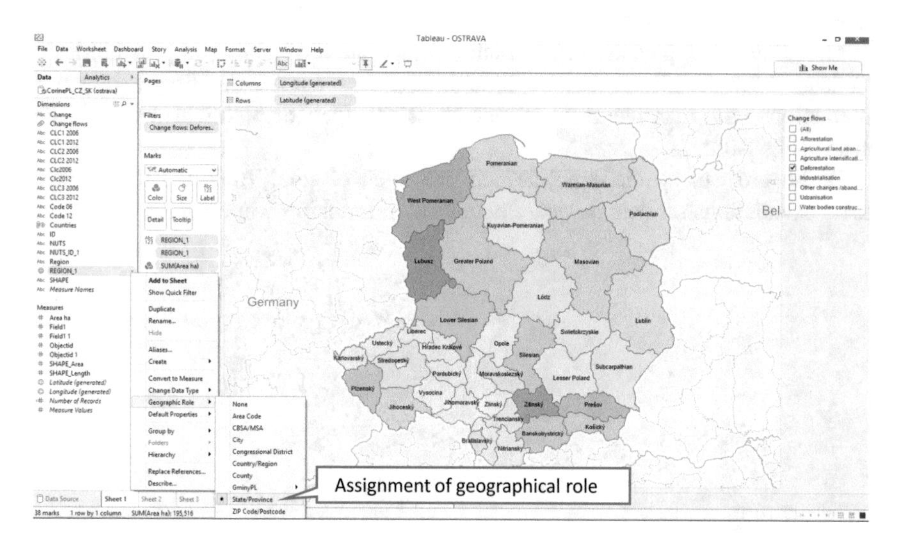

Fig. 9 Cartogram of land change flows

Field Calculator works in a GIS software. Analytics module includes models of the trend line and forecast.

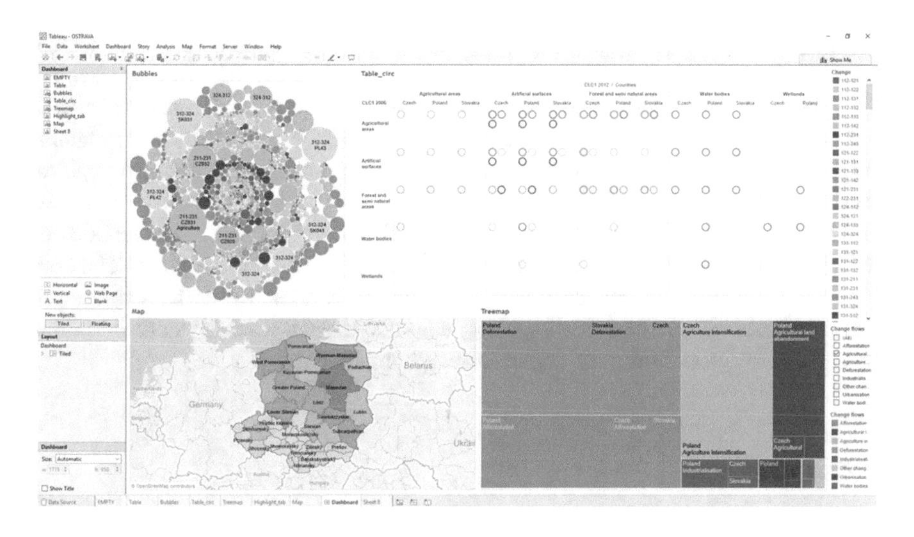

Fig. 10 Tableau dashboard

4 Conclusions

This article presents selected possibilities of visual analysis and visualization of tabular data contained in a geodatabase. According to the main goal we combined functionality of GIS and BI systems. The both systems are complementary to each other. Tableau in no way replaces the possibilities of spatial analysis and modelling in ArcGIS. It, however, allows exploring, illustrating, and computing the content of the tables. Any table operations are many times faster than in ArcGIS. ArcGIS also does not help to create complex and aesthetic graphs. ArcGIS-Tableau is a combination of two independently developed platforms. Together they constitute the data discovery and data visualisation tool, which is very fast, user friendly and offers enhanced possibilities of communication. The ArcGIS-Tableau integration model establishes a clear division between the analytical part of data processing (ArcGIS), and the part of geovisualisation (Tableau). All spatial operations and geoprocessing made in ArcGIS are automatically visualized in the Tableau environment. Any change in the content of the ESRI geodatabase requires only to refresh a live connection to the data source in the Tableau panel. However, it is not possible to modify the geodatabase from the Tableau because it is a read-only table program. The only limitation is 2 GB file size of single geodatabase (Walker 2013). This problem can be overcome by sharing and storing data in several geodatabases. In the Tableau application, divided data can be easily combined by establishing relation rules between the tables.

As the analytical part of the big data analysis is a time consuming process, it appears advisable to test other system than ArcGIS in the future. Other solution presented in this paper could be Feature Manipulation Engine (FME) (Wang et al. 2015). This software enables processing of spatial data, carrying out computational operations only on tabular values, without losing the spatial information about objects. FME, however, was not equipped with spatial data visualization component, which is a big inconvenience in geographical research. The search for alternatives for computing and geovisual programs integration can provide the answer to many research problems posed in geographical research in an era of increasing access to big data.

The described solution contributes to the discussion on the characteristics of such integrated models and the quality of results expected by the users. The suggested combination of GIS and Tableau is currently used by us for other environmental research on carrying capacity, land use change as well as spatial policy assessment and monitoring of regional development.

References

Baesens B (2014) Analytics in a big data world. The essential guide to data science and its applications. Wiley, Hoboken

Benenson I (2011) Geospatial analysis and visualization: keeping pace with the data explosion. Comput Environ Urban Syst 35:91–92

Büttner G (2014) CORINE land cover and land cover change products. In: Manakos I, Braun M (eds) Land use and land cover mapping in Europe, remote sensing and digital image processing. Springer, Dordrecht, pp 55–74

Büttner G, Steenmans C, Bossard M, Feranec J, Kolár J (2000) Land cover—land use mapping within the European Corine Programme. In: Buchroithner M (ed) Remote sensing for environmental data in Albania: A strategy for integrated management, NATO Science Series. Springer, Dordrecht, pp 89–100

Dolega L, Pavlis M, Singleton A (2016) Estimating attractiveness, hierarchy and catchment area extents for a national set of retail centre agglomerations. J Retail Consum Serv 28:78–90

Dykes J, Andrienko G, Andrienko N, Paelke V, Schiewe J (2010) Editorial—geovisualization and the digital city. Comput Environ Urban Syst 34:443–451

European Environmental Agency (2008) Land cover flows based on Corine land cover changes database (1990–2000)

Feranec J, Hazeu G, Christensen S, Jaffrain G (2007) Corine land cover change detection in Europe (case studies of the Netherlands and Slovakia). Land Use Policy 24:234–247

Feranec J, Jaffrain G, Soukup T, Hazeu G (2010) Determining changes and flows in European landscapes 1990–2000 using CORINE land cover data. Appl Geogr 30:19–35. doi:10.1016/j.apgeog.2009.07.003

Grassi S, Junghans S, Raubal M (2014) Assessment of the wake effect on the energy production of onshore wind farms using GIS. Appl Energy 136:827–837

Gutry-Korycka M, Mirończuk A, Hościło A (2015) Land cover change in the Middle River Vistula catchment. In: Romanowicz RJ, Osuch M (eds) Stochastic flood forecasting system. GeoPlanet: Earth and Planetary Sciences. Springer, Dordrecht, pp 3–16

Hanrahan P (2006) VizQL: a language for query, analysis and visualization. In: Proceedings of the 2006 ACM SIGMOD international conference on management of data. ACM, Chicago, pp 721–721

Jones B (2014) Communicating data with Tableau. O'Reilly Media, Sebastopol

Law M, Collins A (2015) Getting to know ArcGIS for desktop, 4th edn. ESRI Press, Redlands

Morton K, Bunker R, Mackinlay J, Morton R, Stolte C (2012) Dynamic workload driven data integration in tableau. In: Proceedings of the 2012 ACM SIGMOD International Conference on Management of Data (SIGMOD '12). ACM, New York, pp 807–816

Murray D (2013) Tableau your data!: fast and easy visual analysis with tableau software. Wiley, Indianapolis

Negash S (2004) Business intelligence. Commun Assoc Inf Syst 13:177–195

Peck G (2014) Tableau 8 the official guide. McGraw-Hill Education, New York

Pekkarinen A, Reithmaier L, Strobl P (2009) Pan-European forest/non-forest mapping with Landsat ETM+ and CORINE land cover 2000 data. ISPRS J Photogramm Remote Sens 64:171–183

Popovici E-A, Bălteanu D, Kucsicsa G (2013) Assessment of changes in land-use and land-cover pattern in Romania using corine land cover database. Carpath J Earth Environ Sci 8:195–208

Silver N (2012) The signal and the noise: why so many predictions fail—but some don't. Penguin Press, New York

Walker A (2013) ESRI to tableau connection. Web forum post. Tableau Support Communities. Tableau, Published: 10 May 2013. Accessed 15 Nov 2015

Wang T, Kazak J, Qi H, de Vries B (2015) Semi-automatic rule extraction for land use change simulation—an application in industrial area redevelopment process. In: Proceedings of ILUS 2015, Dresden, 11–13 Nov 2015

Wessler M (2015) Data blending for dummies. Wiley, Hoboken

Yuan Y, Raubal M, Liu Y (2012) Correlating mobile phone usage and travel behavior—a case study of Harbin, China. Comput Environ Urban Syst 36:118–130

Processing LIDAR Data with Apache Hadoop

Jan Růžička, Lukáš Orčík, Kateřina Růžičková and Juraj Kisztner

Abstract The paper is focused on research in the area of processing LIDAR data with Apache Hadoop. Our team is managing an information system that is able to calculate probability of existence of different objects in space and time. The system works with a lot of different data sources, including large datasets. We may process LIDAR data in the future as well, so we were interested how to process LIDAR data with Apache Hadoop. Our colleagues from the institute of geology are using LIDAR data and have a lot of problems with their processing. The main problem is time that is necessary to process the data to build simple DEM in GRID format that is commonly used in the geology (geomorphology) area. We were interested if we could help them with processing such kind of data with MapReduce architecture on Apache Hadoop to produce GRID data suitable to their needs. The paper presents results of our research.

Keywords LIDAR · MapReduce · GRID · Apache Hadoop · Cloudera

J. Růžička (✉) · K. Růžičková
Institute of Geoinformatics, Faculty of Mining and Geology,
VŠB-Technical University of Ostrava, 17. listopadu 15/2172,
708 33 Ostrava-Poruba, Czech Republic
e-mail: jan.ruzicka@vsb.cz

K. Růžičková
e-mail: katerina.ruzickova@vsb.cz

L. Orčík
Department of Telecommunication, Faculty of Electrical Engineering
and Computer Science, VSB-Technical University of Ostrava,
17. listopadu 15, 708 33 Ostrava, Czech Republic
e-mail: lukas.orcik@vsb.cz

J. Kisztner
Institute of Geological Engineering, VSB-Technical University of Ostrava,
17. listopadu 15, 708 33 Ostrava, Czech Republic
e-mail: juraj.kisztner@vsb.cz

I. Ivan et al. (eds.), *The Rise of Big Spatial Data*, Lecture Notes
in Geoinformation and Cartography, DOI 10.1007/978-3-319-45123-7_25

1 Introduction

The amount of data produced in last years is enormous and it rises every day, the data should be processed to handle useful information. We could expect that the speed of amount data growing will be faster every day in the future. At the other hand hardware performance is not rising as fast as data amount. Organisations have to invest into new hardware because upgrade of old one is not efficient. That is why exists demand for technologies that could use potential of less powerful hardware.

One of these technologies is Apache Hadoop. It is an open source technology, that stores data to several single computers (general or specialised). It utilises distributed file system named HDFS (Hadoop Distributed File System). Processing of the data is distributed and it takes place in parallel on single nodes (computers). Main architecture in Apache Hadoop is based on MapReduce concept originally developed by Google, where Map distributes tasks and Reduce joins results.

There are several tools available for Apache Hadoop and they together build complex platform for data processing. There are several distributions that covers Apache Hadoop and its tools to build complex system. One of the most well known is Cloudera (Cloudera distribution) (Cloudera 2015).

We are building quite complex information system that uses Cloudera technology to manage sources and distribute tasks in our cluster. We would like to place as much as possible tasks to distributed computing environment. Cloudera is a tool that could be used to manage cluster of computers.

Not all of our tasks could be simply re-factored to be able use advantages of Apache Hadoop, but many of them could be simply re-factored.

LIDAR (Light Detection And Ranging) is a remote sensing technology that measures distance by illuminating a target with a laser and analysing the reflected light. LIDAR data are very popular in last years. The user can quite simply obtain millions of points in the scanned area. The point clouds could be used to build 3D models of the scanned objects. There are several types of LIDAR instruments. For purposes of our research we used data from mobile LIDAR instrument.

Our colleagues from the Institute of geology are using LIDAR data and have a lot of problems with their processing. The main problem is time that is necessary to process the data to build simple DEM in GRID format that is commonly used in the geology area. When the amount of the data is not huge, they can import the whole dataset and build GRID using for example IDW interpolation method. When the amount of the data is higher, they have to divide studied area into sub-areas or reduce number of input points during import.

Following image (Fig. 1) shows part of scanned rock massif in the tested area.

We were interested if we could help them with processing such kind of data with MapReduce architecture on Apache Hadoop to produce GRID data suitable to their needs.

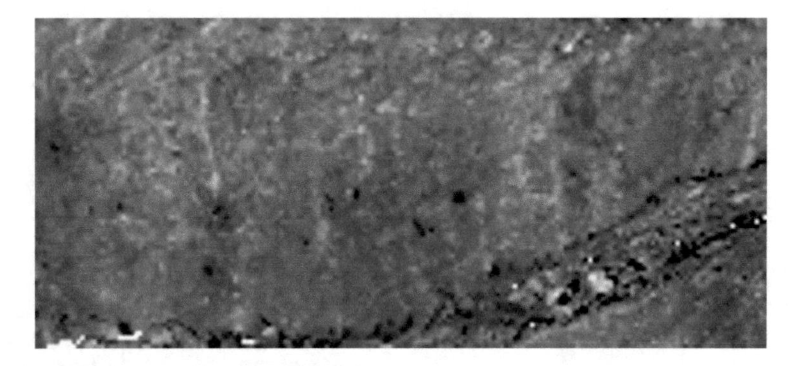

Fig. 1 Scanned rock massif

2 Methodology

The main idea of our research was to determine if we can speed up process of GRID creation using MapReduce architecture on Apache Hadoop software.

According to Krishnan et al. (2010) creating GRID from LIDAR data with Hadoop should be very fast. Krishnan et al. (2010) also conclude that number of nodes help to speed up generation of GRID. According to Jian et al. (2015) the number of nodes should speed up generation of GRID as well.

2.1 LAS Format

LAS is a format defined by The American Society for Photogrammetry and Remote Sensing. LAS format is commonly used to store LIDAR data. There are several types of records that could be stored in LAS file. In our case was used the type 0 which is the simplest one. Its structure is defined in the following table (Table 1, Source: The American Society for Photogrammetry & Remote Sensing 2013).

For our purposes were important items X, Y, Z and Classification. Item X, Y were used to determine where is the point in 2D space. Item Z was used for building output GRID. Item Classification was used to determine type of point to exclude some types of points (for example vegetation).

2.2 MapReduce

We have developed prototype software name LAStoGRID based on MapReduce architecture. MapReduce architecture is based on two tasks.

Table 1 LAS record type 0

Item	Format	Size
X	long	4 bytes
Y	long	4 bytes
Z	long	4 bytes
Intensity	unsigned short	2 byte
Return number	3 bits (bits 0–2)	3 bits
Number of returns (given pulse)	3 bits (bits 3–5)	3 bits
Scan direction flag	1 bit (bit 6)	1 bit
Edge of flight line	1 bit (bit 7)	1 bit
Classification	unsigned char	1 byte
Scan angle rank (−90 to +90)—left side	char	1 byte
User data	unsigned char	1 byte
Point source ID	unsigned short	2 bytes

The first task is Map. In this part of data processing are data read and mapped to Key-Value pairs (KVP). The Key could be any Object and Value could be any Object as well. All keys must be from the same Class and all Values must be from the same Class as well.

In our case we have decided to use simple Long (8 bytes) for Key items. Long was used to identify pixel in GRID.

We have used simple formula to create pixelid:

$$\begin{aligned}\text{pixelid} = {} & \text{round}(\text{abs}(\text{YMAX} - \text{Y})/\text{RES}))\\ & * 1000000 + (\text{round}(\text{abs}(\text{X} - \text{XMIN})/\text{RES}));\end{aligned} \tag{1}$$

where:

- *YMAX* is maximum Y coordinate in studied area.
- *XMIN* is minimum X coordinate in studied area.
- *Y* is a Y coordinate of processed record.
- *X* is a X coordinate of processed record.
- *RES* is a resolution of produced GRID.

Key part of KVP was based on Integer (4 bytes) and it represented Z value of processed record.

Reduce task is focused to reduce KVP to obtain aggregated data. In our case we have simply compute average of Z (values) for each pixel in GRID.

Apache Hadoop orders the data according to keys to be able simply process values for each key in the Reduce part.

Finally the user has available full GRID in a case when there is available at least one point for each pixel or the user has available points that represents several GRID pixels in a case when there are pixels where the points are not available.

In a case of not fully covered GRID data the interpolation must be done. In a case of tested data each pixel had at least one point for larger datasets.

We would like to demonstrate MapReduce on the commented source code from LAStoGRID software that we have developed for purposes of our research. The code is available in the attachment of this paper.

2.3 Testing

We have tested four datasets as an input:

- with 10 millions of records
- with 100 millions of records
- with 500 millions of records
- with 1000 millions of records

We have tested processing on three platforms:

- single computer with desktop software used by our colleagues from the institute of geology
- single computer with Apache Hadoop
- cluster of one master and 3 slave computers with Apache Hadoop

All computers used for different platforms were same in a meaning of hardware capabilities (same CPU, same disks, same memory). We have used VMware to create same testing conditions.

2.4 Study Area

Testing area was 360 × 180 m wide with GRID resolution 0.1 m. Produced GRID was 3600 × 1800 px.

3 Results

We have discovered that the used desktop software is not able to use all 4 CPU cores. The software was not able to finish GRID calculation for more than 10 millions of records (Tables 2, 3, 4; Fig. 2).

Table 2 Single computer with desktop software

Number of records (in millions)	Time for import (s)	Time for building GRID (s)
10	151	537
100	255	Finished with error and no result
500	876	Finished with error and no result
1000	1643	Finished with error and no result

Table 3 Single computer with Apache Hadoop

Number of records (in millions)	Time for building GRID (s)
10	47
100	296
500	1100
1000	2650

Table 4 Cluster of 1 master and 3 slaves with Apache Hadoop

Number of records (in millions)	Time for building GRID (s)
10	71
100	424
500	1962
1000	3849

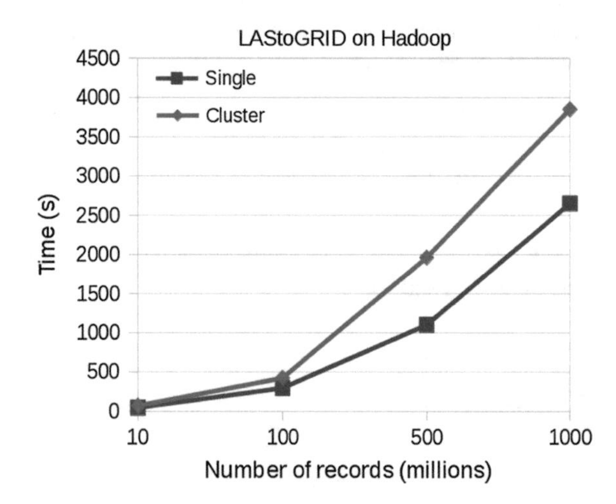

Fig. 2 LASToGRID on Hadoop

4 Conclusion

We have confirmed that large amount of data is not possible to process in one step with software used by geologists. The data must be divided to areas (tiles) that hold smaller amount of records.

We have confirmed that the LAS data could be processed with MapReduce architecture on Apache Hadoop and that the processing is quite fast in comparison to method used by geologists. You can compare 688–47 s in a case of 10 millions of records.

We must say that there could be additional time necessary to compute missing pixels values, but the count of these pixels in our case was small. In a case of 10 mil records it was less than 10 % of pixels. In a case of 100 mil. records it was less than 1 % of pixels.

We have discovered that when using cluster of computers then computation time is higher than in a case of a single node. So our results are different than results presented by Jian et al. or by Krishnan et al. This is discussed in the discussion section.

We can conclude that we can speed up creation of GRID with MapReduce architecture on Apache Hadoop. We can conclude that in a case of such simple computation is not efficient to use cluster of computers.

5 Discussion

We did not tested any other desktop software for building GRID that would be suitable for geologists. This may be done in future.

We did not tested computing time when the data are divided to areas that hold smaller amount of records. But we could expect that for example if we divide 100 millions of records to 10 areas that the processing will cost at least 10× time more for processing of 10 millions points. This is summarised in the following table (Table 5) and image (Fig. 3). But we could expect that the time spent would be higher.

We did not calculate time for post-processing of GRID created by LAStoGRID tool. This could be done in future.

We did not tested GRID creation based on similar technique like MapReduce, but without Apache Hadoop. This may be done in future.

When using cluster of computers then computation time is higher than in a case of single node. This is probably caused by simple computation that is used for both tasks (Map and Reduce). Distributing data between nodes consumes some time and this time is higher than time that is spent by computing on several computers. Jian et al. and Krishnan et al. used more complicated algorithm of GRID creating.

Table 5 LAStoGRID compared to desktop

Number of records (in millions)	LAStoGRID (s)	Desktop (expected smallest time) (s)
10	47	688
100	296	6880
500	1100	34,400
1000	2650	68,800

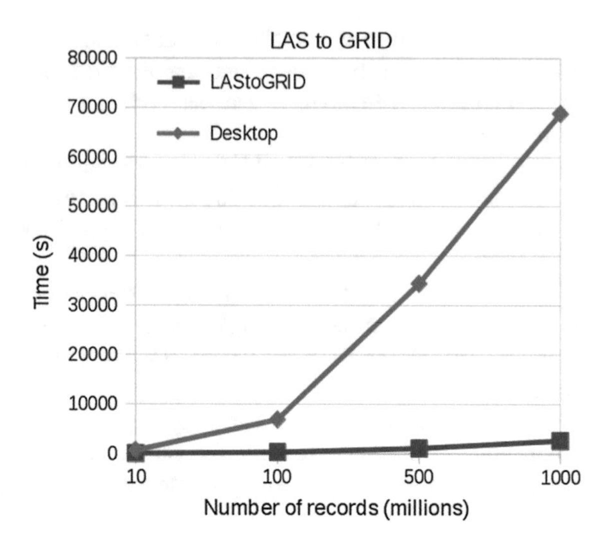

Fig. 3 LASToGRID

We would like to do more complicated computation on LIDAR data to use full advantage of MapReduce architecture in the future, for example to build TIN (Triangular Irregular Network).

Acknowledgments We would like to thank to all open source developers.

References

Cloudera (2015) http://www.cloudera.com/content/www/en-us/downloads.html

Jian et al (2015) A Hadoop-based algorithm of generating dem grid from point cloud data. In: The international archives of the photogrammetry, remote sensing and spatial information sciences, vol XL-7/W3, 2015 36th international symposium on remote sensing of environment, 11–15 May 2015, Berlin, Germany. http://www.int-arch-photogramm-remote-sens-spatial-inf-sci.net/XL-7-W3/1209/2015/isprsarchives-XL-7-W3-1209-2015.pdf

Krishnan et al (2010) Evaluation of MapReduce for Gridding LIDAR data. http://users.sdsc.edu/~sriram/publications/Krishnan_DEM_MR.pdf

The American Society for Photogrammetry and Remote Sensing (2013) LAS Specification. http://www.asprs.org/a/society/committees/standards/LAS_1_4_r13.pdf

Compression of 3D Geographical Objects at Various Level of Detail

Karel Janečka and Libor Váša

Abstract Compression of 3D objects has been recently discussed mainly in the domain of 3D computer graphics. However, more and more applications demonstrate that the third dimension plays an important role also in the domain of modelling and streaming of geographical objects. This is especially true for 3D city models and their distribution via internet. Despite the fact that compression of textual information related to geographical objects has a significant importance, in this paper we concentrate only on compression of geometry information and also on more complex geometries with irregular shapes. Considering the compression of 3D geographical objects, the 3D triangle meshes representation are used. 3D mesh compression is a way how to reduce the required cost of storage for triangle meshes without losing any details. The triangle is the basic geometric primitive for standard graphics rendering hardware. The compression algorithm aims at storing the input data into a binary file that is as small as possible. For encoding of the mesh connectivity, our compression implements the popular Edgebreaker algorithm. The character of geometry encoding is largely governed by the way connectivity is encoded. A popular choice of prediction for the Edgebreaker algorithm is the parallelogram predictor. It has been observed in Váša and Brunnett (IEEE Trans Vis Comput Graph 19(9):1467–1475, 2013) that such prediction can be further improved by taking a two-step approach, first transmitting the complete connectivity and only afterwards transmitting the geometry. We used this approach to compress geographical objects at various level of detail. It does not bring an improvement for all datasets, especially meshes with many parallelogram shape prediction stencils do not benefit from it. However for complex geographical

K. Janečka (✉)
Department of Mathematics, Faculty of Applied Sciences,
University of West Bohemia, Technická 8, 306 14 Pilsen, Czech Republic
e-mail: kjanecka@kma.zcu.cz

L. Váša
Department of Computer Science and Engineering, Faculty of Applied Sciences,
University of West Bohemia, Technická 8, 306 14 Pilsen, Czech Republic
e-mail: lvasa@kiv.zcu.cz

I. Ivan et al. (eds.), *The Rise of Big Spatial Data*, Lecture Notes
in Geoinformation and Cartography, DOI 10.1007/978-3-319-45123-7_26

objects (bridges in our case) the used algorithm works nicely and after the compression the amount of data is even lower than 4 % of the original file size.

Keywords 3D city models · Mesh compression · Level of detail

1 Introduction

Compression of 3D objects has been recently discussed mainly in the domain of 3D computer graphics. Some techniques for compression of such objects gives e.g. Peng et al. (2005). However, more and more applications demonstrate that the third dimension (3D) plays an important role also in the domain of modelling and streaming of geographical objects. This is especially true for 3D city models and their distribution via internet. Goetz and Zipf (2010) emphasize the need for advanced and efficient streaming possibilities for 3D city models in case of 3D indoor location based services. Schmidt (2009) argues that the speed and overall performance of mobile networks is still not sufficient, and that transferring huge amounts of data still often takes too much time.

For representing a 3D city model, CityGML (OGC 2012) is becoming popular. Since CityGML is a textual XML standard based scheme, textual compressions seem to be highly relevant. However, as CityGML stores spatial information specified for the 3D city model, compression in geometry and connectivity will also be useful. Therefore, compression algorithms for both semantics from CityGML as well as geometry information need to be improved to suit the 3D scenario. In textual compression, GZIP is often used as the compressor behind the scene (Siew and Rahman 2012). Sakr (2009) studied several textual based compression techniques which are specialized in XML-based compression and are of lossless type. Several studies have shown that a binary representation of XML tag also offers a better compression rate, e.g. (Bruce 2006; W3C 2014). Siew and Rahman (2012) tested various XML-based compression techniques on CityGML buildings and achieved 8.7 % of the original file size. Coors and Ewald (2005) compressed the CityGML files of the city of Stuttgart (3D buildings at LOD2 according to CityGML, no textures). They first converted the CityGML files to triangulated VRML Indexed Face Sets and then compressed the VRML files using the Delphi algorithm (Coors and Rossignac 2004) based on the Edgebreaker coder (Rossignac 1999; Rossignac et al. 2001). They reduced the amount of data to 7–8 % of the original file size and also stated, that for more complex meshes representing LOD3 buildings, even higher compression ratio could be achieved. In their study, a binary XML codec was used for compression of the non-geometric part of input CityGML files.

Despite the fact that compression of textual information related to geographical objects has a significant importance, in the rest of the paper we concentrate only on compression of geometry information and also on more complex geometries with irregular shapes (triangles).

Considering the compression of 3D geographical objects, the 3D triangle meshes representation are used. 3D mesh compression is a way how to reduce the required cost of storage for triangle meshes without losing any detail. The triangle is the basic geometric primitive for standard graphics rendering hardware and for simulation algorithms. The specification of a triangle mesh consists of combinatorial entities—vertices, edges and faces, and numerical quantities, such as vertex positions, normals etc. The connectivity describes the incidence between elements and is implied by the topology of the mesh.

In compression of 3D meshes we distinguish between single-rate and progressive compression techniques, depending on whether the model is decoded during or only after the transmission. In the case of single-rate lossless coding, the goal is to remove the redundancy present in the original description of the data (Alliez and Gotsman 2005). The compression algorithm (Váša and Brunnett 2013) that is used in this paper is also based on the Edgebreaker scheme. The mesh is traversed by conquering one triangle at the time, each new vertex position is predicted and the 3D residual vector is quantised and encoded. In contrast with other approaches based on Edgebreaker scheme (Gumhold 2000; Isenburg and Snoeyink 2000; King and Rossignac 1999; Szymczak et al. 2001), the algorithm uses more accurate way of estimating the shape of the prediction stencil. We applied the mesh compression algorithm on 3D geographical objects in various level of detail to demonstrate the general usability of the selected approach. Regarding the principle of the used algorithm, we expected good compression results for more complex geometries. In particular, tested datasets consisted of buildings at LOD1 (the city of Pilsen) and LOD2 (the city of Terezín). Furthermore, to test the compression algorithm on a more complex representation of geographical objects, we applied the compression also on bridges, which contain much more vertices (geometry) and triangles (connectivity) than buildings at LOD1 or LOD2 respectively.

2 Compression Framework

As stated in the Introduction section, our focus is the compression of geometry information. The surfaces of geographical objects are represented by 3D triangle meshes. From the topological point of view the surfaces of objects must be manifold. A mesh is a manifold if each edge is incident to only one or two faces and the faces incident to a vertex form a closed or an open fan. All non-manifolds geometries must be omitted from the testing datasets. Figure 1 demonstrates an object with non-manifold geometry.

Fig. 1 In reality two separate buildings (with highlighted footprints). In real data, these two touching buildings are stored as a one building, which leads to a non-manifold geometry

2.1 Testing Datasets

2.1.1 Buildings at LOD1

Despite the fact that LOD1 models are not particularly complex, they are widely used in many applications (van den Brink et al. 2013; Czerwinski et al. 2007; Ranjbar et al. 2012) or (Tomić et al. 2012). Another application of using LOD1 buildings could be potentially a 3D cadastre, e.g. 3D legal boundaries around the buildings. Currently, in the Czech Republic, the data of cadastre (still registered in 2D) are distributed as open data, so that anyone has an access to them. In particular, the data from the Registry of Territorial Identification, Addresses and Real Estate (RTIARE) (see http://www.cuzk.cz/Uvod/Produkty-a-sluzby/RUIAN/RUIAN.aspx) which is running since 2012 were used. Some inconsistencies can be identified in these data. More information on finding inconsistencies in RTIARE data give e.g. (Janečka and Hejdová 2014). Among others, RTIARE contains the 2D building footprints. There is no attribute with information on height of the building. In theory, the attribute number of floors could be used in some way. In our case we modelled the buildings uniformly with height of 10 metres. It must be said that for LOD1 geometries this has no influence on the final geometry in the sense of number of vertices and triangles forming the particular triangle meshes. The 2D building footprints for the city of Pilsen (the 4th biggest city in the Czech Republic) were used. The original RTIARE data were downloaded in an exchange format which is a XML based file format.

2.1.2 Buildings at LOD2

Similarly as in Coors and Ewald (2005), the buildings at LOD2 were also used. In particular we applied compression to the models of buildings which were created in the project GEPAM: Memorial Landscapes—Dresden and Terezín as places to remember the Shoah. Overall, 217 LOD2 buildings of the city of Terezín were used (Fig. 2). The models were created in SketchUp (Hájek et al. 2015).

2.1.3 Bridges

To extend the testing datasets to geographical objects with more complex geometries, we further decided to test the compression algorithm on two bridges. Also, no similar work is known to the authors. The models of bridges were taken again from the GEPAM project. The triangle surface meshes for such objects consist of a large amount of vertices and edges and also the triangles are more irregular than in the case of buildings LOD1 or LOD2. The compression algorithm we used in the paper was expected to work better for such complex geometries.

2.2 Transformation

The compression utility (that implements the used compression algorithm) works with input data in Object Files (OBJ) in ASCII format. In our case, OBJ files contain a list of vertex indices and triangle face elements. For triangulation of the geographical objects and their transformation into OBJ files the FME software was used.

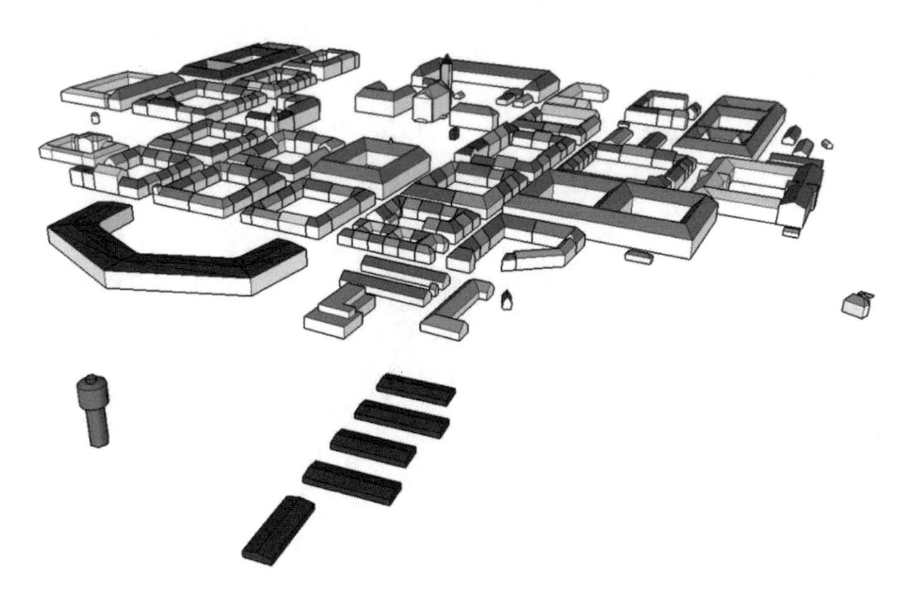

Fig. 2 City of Terezín at LOD2

2.2.1 Buildings at LOD1

The source RTIARE data were downloaded in the XML file and then imported into ESRI geodatabase. Next, only building footprints were exported into the ESRI shapefile data format. Subsequently, using FME, the building footprints were extruded into 3D, triangulated and transformed into OBJ files (see Fig. 3 to see an extent of the LOD1 data).

2.2.2 Buildings at LOD2 and Bridges

First, for the mesh compression purposes, we deleted from the models of LOD2 buildings and bridges information about the used materials and textures, so that only the pure geometry models remained. Second, for both LOD2 buildings and bridges, the FME software was used for a triangulation and transformation of input data stored in the SKP format into the OBJ format. Figure 4 demonstrates a building at LOD2 represented by triangle mesh and the corresponding OBJ file.

2.3 Compression

Having data without any topological inconsistencies, represented by 3D triangle meshes and stored in the OBJ format, we can run the compression. Used encoder utility implements the compression algorithm (Váša and Brunnett 2013) which is described in more detail in Chapter "Open Source First Person View 3D Point

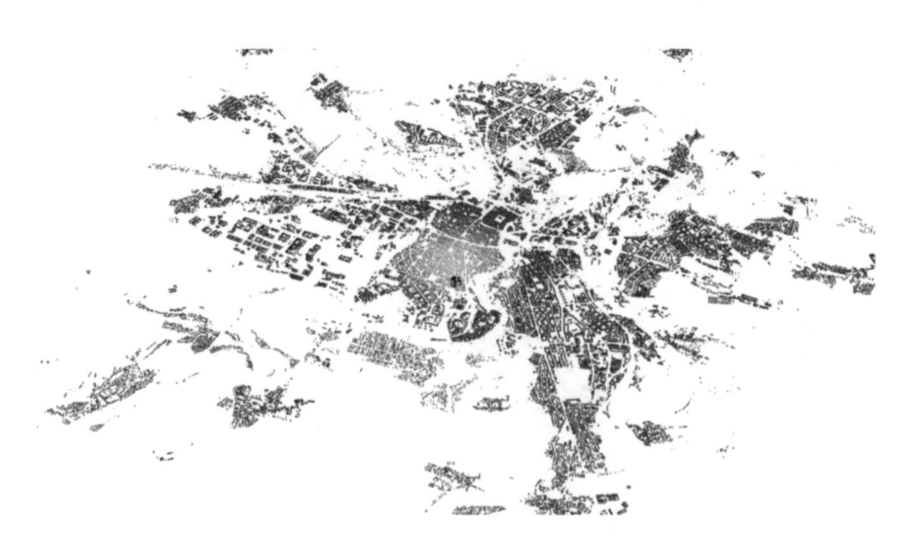

Fig. 3 LOD1 buildings for the city of Plzeň after extrusion of building footprints to 3D in FME

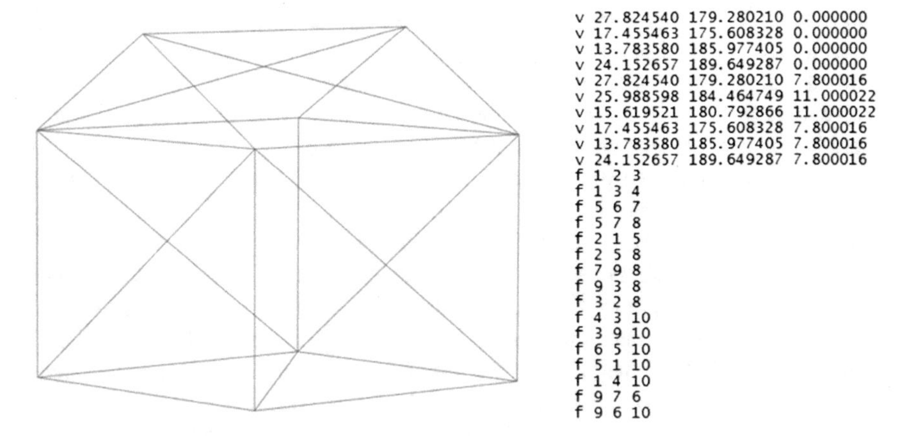

Fig. 4 The geographical object at LOD2 (a building with saddle-shaped roof) represented by a triangle mesh and corresponding OBJ file. It consists of 10 vertices and 16 faces

Cloud Visualizer for Large Data Sets". The compressed data are stored into a binary BIN file. The same utility can be used also for decompression.

3 Triangle Mesh Compression

The compression algorithm aims at storing the input data into a binary file that is as small as possible. In contrast to mesh simplification, compression always preserves the number of vertices and triangles, however, in most applications, a certain precision loss is allowed for vertex coordinates. This is also the case of geographical data, where precision higher than that of the data acquisition device is redundant. The compression algorithm is therefore lossy in geometrical precision, while lossless in connectivity.

The lossless nature of the connectivity encoding, however, does not guarantee complete identity of the input and output triangle lists, because it is not necessary. In particular, the exact numbering of vertices, ordering of triangles and ordering of vertices within triangles (apart from the clockwise/counterclockwise distinction) carries no spatial shape information, and can be therefore changed, if such change leads to a more efficient way of storing the geometric information.

3.1 Connectivity Encoding

For encoding of the mesh connectivity, our compression implements the popular Edgebreaker algorithm. It works in a progressive manner, starting by encoding a single triangle and then expanding the encoded portion of the mesh by one triangle in each step. First, a particular border edge (a gate) is selected using some convention, i.e. no information is stored, since the decoder will be able to use the same convention in order to select the same gate. Since we assume a manifold mesh, there is one triangle that should be attached to the gate.

At this point, the algorithm stores information that allows the decoder to attach the triangle correctly into the temporary connectivity. In particular, it stores a code that describes the tip vertex of the triangle. Either, it is a new vertex, not previously processed by either encoder or decoder (so-called C-case), or it lies somewhere on the border of the temporary connectivity. In the latter case, we further distinguish the cases L (on the border, immediately to the left), R (on the border, immediately to the right), E (both immediately to the left and to the right, i.e. it is a final triangle of a closed component) and S (somewhere else on the border).

The authors of Edgebreaker provide a smart way of determining the exact location of the vertex in the S case, without storing any additional information. The decoder is therefore able to reconstruct the connectivity only from a sequence of these symbols, one for each triangle. It can be shown that the cost of such connectivity encoding is never larger than 2 bits per triangle, i.e. 4 bits per vertex for closed manifold meshes of low genus.

3.2 Geometry Encoding

The character of geometry encoding is largely governed by the way connectivity is encoded. In particular, vertex positions are encoded when a new connected component of the input mesh is encountered (initial triangle), and for each vertex that gets attached when a C case is encountered by the Edgebreaker algorithm.

In order to encode the coordinates efficiently, one usually performs a common prediction at both the encoder and the decoder (i.e. using only the data already available at the decoder). This prediction is then subtracted from the actual value that is being encoded, and the resulting residual is quantized and stored into the output binary stream. The purpose of the prediction is to reduce the entropy of the resulting values, while their total number remains unchanged.

A popular choice of prediction for the Edgebreaker algorithm is the parallelogram predictor. It exploits the fact that in each C case, a complete neighbouring triangle is available at the decoder, and thus it can be used for the prediction. The prediction itself is then a simple linear combination of the vertices of the neighbouring triangle:

$$\bar{v}_o = \tilde{v}_l + \tilde{v}_r - \tilde{v}_b,$$

where $\tilde{v}_l$, $\tilde{v}_r$ and $\tilde{v}_b$ are the decoded coordinates of the vertices in a neighbouring triangle (see Fig. 5 for illustration) and $\bar{v}_o$ is the predicted position of the tip vertex.

It has been observed in Váša and Brunnett (2013) that such prediction can be further improved by taking a two-step approach, first transmitting the complete

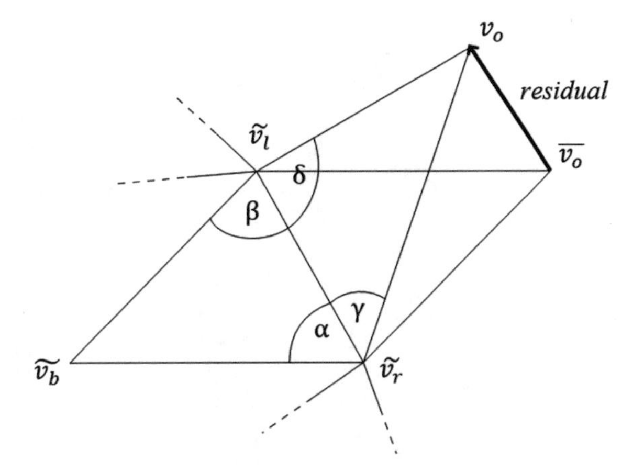

Fig. 5 Prediction stencil

connectivity and only afterwards transmitting the geometry. In such case, not only the vertex positions of the neighbouring triangle are available for the prediction, but also the vertex degrees (i.e. the total number of neighbouring vertices for each vertex). This information allows for a more precise prediction by using a more complex predictor:

$$\bar{v}_o = w_1\tilde{v}_l + w_2\tilde{v}_r + (1 - w_1 - w_2)\tilde{v}_b,$$

where the weights w_1 and w_2 are computed as

$$w_1 = \frac{\cot(\beta) + \cot(\delta)}{\cot(\delta) + \cot(\gamma)}, \quad w_2 = \frac{\cot(\alpha) + \cot(\gamma)}{\cot(\delta) + \cot(\gamma)}.$$

The angles α, β, γ and δ (see Fig. 5) can be estimated from the vertex degrees, assuming that they roughly sum up to 2π around each vertex. For more details on the derivation of the weighted predictor, see Váša and Brunnett (2013).

Finally, the connectivity codes are stored into the output binary file using an arithmetic coder which exploits their probability distribution. The prediction residuals are quantized (rounded) to a desired precision and encoded using a context adaptive arithmetic encoder. In the case of geographical data, the precision is dictated by the intended purpose, e.g. for buildings in a cadastre typically a precision of less than 1 cm is meaningless.

4 Experimental Results

4.1 Buildings at LOD1 and LOD2

The average size of the OBJ file for LOD1 building is approximately 1 kB (per building). Further, for the testing LOD1 data of the city of Pilsen, we have got 28

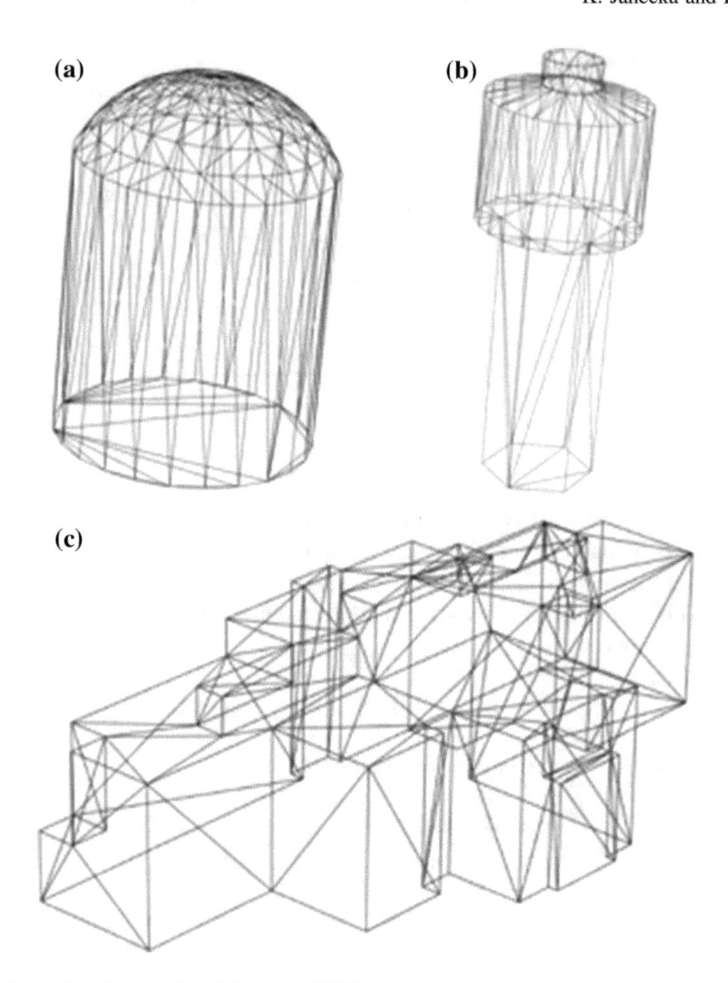

Fig. 6 Tringulated selected buildings at LOD2

triangles per building on average. If a building footprint is a square, then we need 8 vertices and 12 triangles to represent this LOD1 object as a 3D triangle mesh. However, data more complex footprints can be found in the testing (see Fig. 6) and therefore the average number of triangles per building is higher.

The average size of the OBJ file for LOD2 building is approximately 0.7 kB (per building). Additionally, for the testing LOD2 data of the Terezin city, we have got 23 triangles per building on average. Table 1 summarizes the obtained results for buildings at LOD1 and LOD2.

We also applied the compression algorithm separately on several selected buildings (see Fig. 6) with more vertices and faces from the LOD2 testing dataset (see Table 2 for results).

Table 1 Mesh compression results for buildings at LOD1 and LOD2

	LOD 1—City of Pilsen	LOD 2—City of Terezín
Buildings	38,177	217
Triangles	1,065,660	5095
OBJ files size	37,846 kB	158 kB
Compressed files size	6483 kB (17 %)	31 kB (19 %)

Table 2 Mesh compression results for selected objects at LOD2

	Object (a)	Object (b)	Object (c)
Vertices	169	108	95
Triangles	334	198	166
OBJ file size	10,625 bytes	6555 bytes	5527 bytes
Compressed file size	427 bytes (4 %)	472 bytes (7.2 %)	468 bytes (8.5 %)

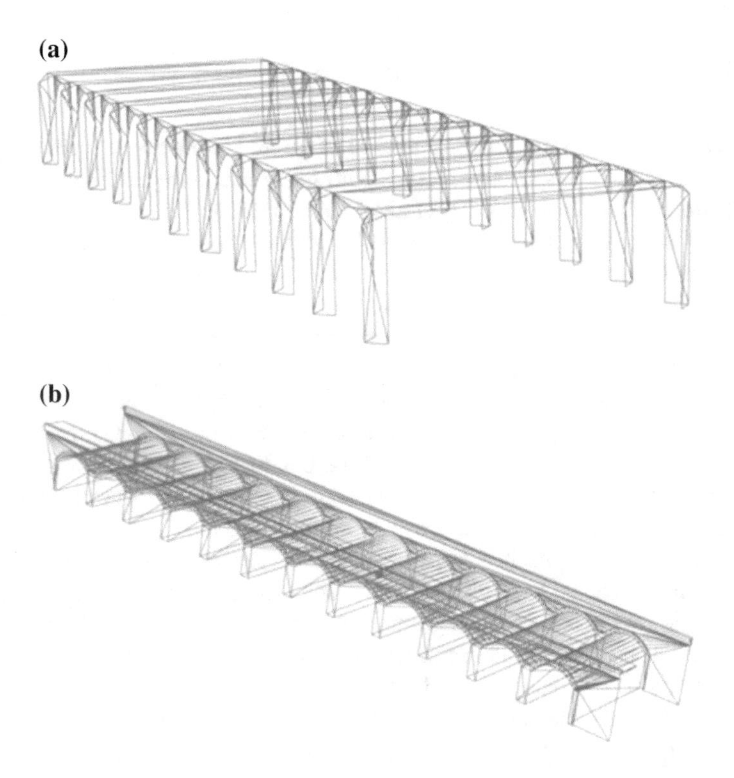

Fig. 7 Triangulated bridges

Table 3 Mesh compression results for bridges

	Bridge (a)	Bridge (b)
Vertices	393	1047
Triangles	453	2027
OBJ files size	19,942 bytes	65,311 bytes
Compressed files size	453 bytes (4.7 %)	2477 bytes (3.8 %)

4.1.1 Bridges

Finally, the compression algorithm was applied on two bridges which have more complex geometries than the tested buildings. Figure 7 shows these bridges represented by triangle meshes. Table 3 provides the statistics on compression. It can be seen that the size of the compressed file for bridge (b) is less than 4 % of the original file size.

5 Discussion and Conclusions

The used algorithm works for various level of detail, but it does not bring improvement for all datasets, especially meshes with many parallelogram shape prediction stencils do not benefit from it (Váša and Brunnett 2013). The experimental results show, that in the case when buildings at LOD1 have complex footprints, the compression results can be better than in the case of buildings at LOD2 if these LOD2 buildings have relatively simple geometry, e.g. a gabled roof (OGC 2012). We expected that if the proposed compression algorithm is applied to the LOD2 dataset, then we will receive better results than in the case of the LOD1 dataset, due to the more complex geometries of the LOD2 buildings. However, the compression ratio for LOD2 is worse than for LOD1. This is caused by the fact that the particular testing LOD2 dataset contains more than half of the buildings with a very simply shaped gabled roof and relatively simple footprints. To represent such geometries, we usually need only 10 vertices and 16 triangles per building (see Fig. 4).

Comparing the results in particular for LOD2 achieved here against the ones published in Coors and Ewald (2005) our results are not better. This can be caused by two main reasons. First, the algorithm we applied is suitable for more complex geometries. The more complex the geometry is, the better compression ratio we achieve. Second, as already stated, in the testing dataset (LOD2 buildings of city of Terezín), more than half of buildings had very simple geometry.

However, as supposed at the beginning of the study, for complex geographical objects (bridges) the used algorithm works nicely and after the compression the amount of data is even lower than 4 % of the original file size.

As the speed of geodata collection is still increasing the need for the effective geodata compression will be essential. The techniques like laser scanning (generating massive point clouds) or building information modeling (BIM) are producing big geodata which in case of web applications have to be delivered to the final user (e.g. 3D cadastre and visualization of registered buildings) within a reasonable time.

The compression method used in the paper works independently of LOD. If there is a strong demand to achieve a good compression ratio for a particular LOD, then in the future the algorithm can be modified to better fit to this LOD. As further work, we also plan to test the algorithm on digital terrain models.

Acknowledgments This publication was supported by the Project LO1506 of the Czech Ministry of Education, Youth and Sports.

References

Alliez P, Gotsman C (2005) Recent advances in compression of 3-D meshes. In: Advances in multiresolution in geometric modeling, Springer Berlin Heidelberg, pp 3–26

Bruce G (2006) OGC Binary Extensible Markup Language (BXML) encoding specification. OGC 03-002r9. OpenGIS Best Practices Document.http://portal.opengeospatial.org/files/?artifact_id=13636. Accessed 19 Nov 2015

Coors V, Ewald K (2005) Compressed 3D urban models for internet-based e-planning. In: Proceedings of the 1st international workshop on next generation 3D city models, Bonn 2005, EuroSDR Publication #49

Coors V, Rossignac J (2004) Delphi: geometry-based connectivity prediction in triangle mesh compression. Vis Comput 20:507–520

Czerwinski A, Sandmann S, Elke SM, Plümer L (2007) Sustainable SDI for EU noise mapping in NRW—best practice for INSPIRE. Int J Spat Data Infrastruct Res 2(1):1–18

Goetz M, Zipf A (2010) Open issues in bringing 3D to location based services (Lbs)—a review focusing on 3D data streaming and 3D indoor navigation. In: International archives of the photogrammetry, remote sensing and spatial information sciences, vol XXXVIII-4/W15

Gumhold S (2000) New bounds on the encoding of planar triangulations. Technical Report WSI-2000-1, Univ. of Tűbingen

Hájek P, Jedlička K, Kepka M, Fiala R, Vichrová M, Janečka K, Čada V (2015) 3D cartography as a platform for remindering important historical events: the example of the Terezín memorial. In: Brus J, Vondrakova A, Vozenilek V (eds) Modern trends in cartography. Lecture notes in geoinformation and cartography. Springer, Berlin, pp 425–437. ISBN 978-3-319-07925-7

Isenburg M, Snoeyink J (2000) Spirale Reversi: reverse decoding of the Edgebreaker encoding. In: Proceedings of 12th Canadian conference on computational geometry, pp 247–256

Janečka K, Hejdová J (2014) Validation of data of the basic register of territory identification, addresses and real estates. In: Proceedings of the 5th international conference on cartography and GIS, e-Proceedings. Bulgarian Cartographic Association, Riviera, Bulgaria, ISSN 1314-0604

King D, Rossignac J (1999) Guaranteed 3.67 V bit encoding of planar triangle graphs. In: 11th Canadian conference on computational geometry, pp 146–149

OGC (2012) OGC City Geography Markup Language (CityGML) En-coding Standard, open geospatial consortium, version 2.0.0, Publication Date: 2012-04-04. http://www.opengeospatial.org/standards/citygml. Accessed on 17 Nov 2015

Peng J, Kim ChS, Kuo CCJ (2005) Technologies for 3D mesh compression: a survey. J Vis Commun Image Represent 16(6):688–733

Ranjbar HR, Gharagozlou AR, Nejad ARV (2012) 3D analysis and investigation of traffic noise impact from Hemmat highway located in Tehran on buildings and surrounding areas. J Geogr Inf Syst 4(4):322–334

Rossignac J (1999) Edgebreaker: connectivity compression for triangle meshes. IEEE Trans Vis Comput Graph 5(1):47–61

Rossignac J et al (2001) 3D compression made simple: Edgebreaker on a corner-table. In: Invited lecture at the shape modelling international conference, Genoa, Italy, May 2001

Sakr S (2009) XML compression techniques: a survey and comparison. J Comput Syst Sci 75 (5):303–322

Schmidt K (2009) Aktuelle Entwicklungen des Mobilfunkmarkts. GRIN, Norderstedt

Siew ChB, Rahman AA (2012) Compression techniques for 3D SDI. In: FIG working week. Rome, Italy, 6–10 May

Szymczak A, King D, Rossignac J (2001) An Edgebreaker-based efficient compression scheme for regular meshes. Comput Geom 20(1–2):53–68

Tomić H, Roić M, Mastelić Ivić S (2012) Use of 3D cadastral data for real estate mass valuation in the urban areas. In: van Oosterom P (ed) Proceedings of the 3rd international workshop on 3D cadastres: developments and practices. International Federation of Surveyors, Shenzhen, pp 73–86

Van den Brink L, Stoter J, Zlatanova S (2013) Establishing a national standard for 3D topographic data compliant to CityGML. Int J Geogr Inf Sci 27(1):92–113

Váša L, Brunnett G (2013) Exploiting connectivity to improve the tangential part of geometry prediction. IEEE Trans Vis Comput Graph 19(9):1467–1475

W3C (2014) Efficient XML Interchange (EXI) Format 1.0. W3C Recommendation 11 February 2014. http://www.w3.org/TR/exi/. Accessed on 17 Nov 2015

Applicability of Support Vector Machines in Landslide Susceptibility Mapping

Lukáš Karell, Milan Muňko and Renata Ďuračiová

Abstract Landslides in Slovakia are followed by great economic loss and threat to human life. Therefore, implementation of landslides susceptibility models is essential in urban planning. The main purpose of this study is to investigate the possible applicability of Support Vector Machines (SVMs) in landslides susceptibility prediction. We have built a classification problem with two classes, landslides and stable areas, and applied SVMs algorithms in the districts Bytča, Kysucké Nové Mesto and Žilina. A spatial database of landslides areas and geologically stable areas from the State Geological Institute of Dionýz Štúr were used to fit SVMs models. Four environmental input parameters, land use, lithology, aspect and slope were used to train support vector machines models. During the training phase, the primal objective was to find optimal sets of kernel parameters by grid search. The linear, polynomial and radial basis function kernels were computed. Together 534 models were trained and tested with LIBLINEAR and LIBSVM libraries. Models were evaluated by Accuracy parameter. Then the Receiver Operating Characteristic (ROC) and landslides susceptibility maps were produced for the best model for every kernel. The best predictive performance was gained by radial basis function kernel. This kernel has also the best generalization ability. The results showed that SVMs employed in the presented study gave promising results with more than 0.90 (the area under the ROC curve (AUC) prediction performance.

Keywords Landslide susceptibility · Machine learning · Support vector machines · Validation · Receiver operating characteristic

---

L. Karell (✉) · M. Muňko · R. Ďuračiová
Department of Theoretical Geodesy, Faculty of Civil Engineering,
Slovak University of Technology, Radlinského 11, 810 05 Bratislava, Slovakia
e-mail: lukas.karell@stuba.sk

M. Muňko
e-mail: milan.munko@stuba.sk

R. Ďuračiová
e-mail: renata.duraciova@stuba.sk

I. Ivan et al. (eds.), *The Rise of Big Spatial Data*, Lecture Notes
in Geoinformation and Cartography, DOI 10.1007/978-3-319-45123-7_27

1 Introduction

Floods, soil erosion and landslides are the most dangerous geohazards in Slovakia and their activity is followed by great economic loss and threat to human life. Based on the research of slope deformation occurrence 5.25 % of the total land area of the Slovak Republic is affected by slope failures (Šimeková et al. 2014). According to Ministry of Environment of the Slovak Republic 45 million Euro will be invested into landslides prevention, identification, monitoring, registration and sanation in the time period between the years 2014 and 2020. Therefore, implementation of geological hazards models is essential in urban planning process for effective prevention and protection in high risk areas. In recent time, a machine learning became very popular phenomenon. Cheaper hardware allows wider ranges of companies and institutions to store and examine large datasets. This environment is very stimulating and new or upgraded machine learning techniques are developed continuously. This study is focused on the Support Vector Machines (SVMs) method and its usability in prediction modelling of landslides susceptibility in geographical information systems (GIS), which was never used before by professionals in Slovakia.

The objective of this study is to test linear, polynomial and radial basis function kernels and their inner parameters to find optimal model to create landslides susceptibility maps. For this purpose, we have built a classification problem with two classes "landslides" and "stable areas".

Further step is training the SVMs to compute probability that unobserved data belong to landslides class. First, these models were validated by their accuracy. Subsequently, for each kernel, the model with the best result was trained to predict probability output and validated by the Receiver Operating Characteristic (ROC) curve. Finally, landslides susceptibility maps were created.

There are number of different approaches to the numerical measurement of landslide susceptibility evaluation in the current literature, including direct and indirect heuristic approaches and deterministic, probabilistic and statistical approaches (Pradhan et al. 2010). Statistical analysis models for landslide susceptibility zonnation were developed in Italy, mainly by Carrara (1983) who later modified his methodology to the GIS environment (Carrara et al. 1990) using mainly the bivariate and mutlivariate statistical analysis. More recently, new techniques have been used for landslide susceptibility mapping. New landslide susceptibility assessment methods such as artificial neural networks (Lee 2004), neuro-fuzzy (Vahidnia 2010) or (Hyun-Joo and Pradhan 2011), SVMs (Yao et al. 2008) and decision tree methods (Nefeslioglu et al. 2010) have been tried and their performance have been assessed. In Slovakia, landslide susceptibility assessment has experienced a significant step forward especially due to Pauditš (2005) and Bednárik (2011) achievements in their work (Bednarik et al. 2014) using the multivariate conditional analysis method and bivariate statistical analysis with weighting.

2 Study Area, Data and Materials

The study area is located in north-west of Slovakia overlying the districts Bytča, Kysucké Nové Mesto, Žilina, and covers a surface area of 300 km^2 (20 km × 15 km). It lies between the S-JTSK coordinates of the y-axis 458,653.27–438,653.27 m and coordinates of the x-axis 1,162,115.32–1,177,115.32 m (S-JTSK is a national coordinate system in Slovakia, the corresponding coordinates of the study area in the WGS 84 are: latitude N49°18′59″ to N49°10′03″ and longitude E18°31′02″ to E18° 48′28″). The mean annual rainfall in the area varies from 650 to 700 mm. The prominent rainy season is during the months of June and July. The geomorphology of the area is characterized by a rugged topography with the hills ranges varying from 292 to 817 m a.s.l. The study area is largely formed by the Javorníky and Súľov Montains and the Žilina basin. The land use in the area is dominated by forest land and urban land. The high occurrence of landslides was mainly within Javorníky mountains mapped in 1981 and 1982 (Baliak and Stríček 2012).

2.1 *Landslide Conditioning Factors*

Identification and mapping of a suitable set of instability factors having a relationship with slope failures require a priori knowledge of main causes of landslides and it is a domain of geologists. These instability factors include surface and bedrock lithology and structure, seismicity, slope steepness and morphology, stream evolution, groundwater conditions, climate, vegetation cover, land use and human activity (Pradhan 2013). The acquirement and availability of these thematic data is often a difficult task. A grid based digital elevation model (DEM) was used to acquire geomorphometric parameters. The DEM was provided by Geodetic and Cartographic Institute Bratislava in the version "3.5". In this study, the calculated and extracted factors were converted to a spatial resolution of 10 × 10 m. This mapping unit was small enough to capture the spatial characteristics of landslide susceptibility and large enough to reduce computing complexity. Total four different input datasets are produced as the conditioning factors for occurrence of landslides. Slope and aspect were extracted using the DEM. The slope configuration plays an important role when considered in conjunction with a geological map. The geological map of the Slovak republic at scale 1:50,000 in the shapefile format was provided by State Geological Institute of Dionýz Štúr (Káčer et al. 2005). The last conditioning factor land use is extracted from Corine Land Cover 2006 with a spatial resolution 100 × 100 m.

In this study, geological and statistical evaluation of input conditional factors according to landslide susceptibility was not performed. The results from this evaluation are used to reclassify input parameters into the groups with similar properties causing slope failures. Reclassified input parameters are then used in further landslide susceptibility evaluation as in the case of bivariate and mutlivariate statistical analysis. According to the properties of the Support Vector Machines method it is not necessary to reclassify input parameters and they can be used directly in their untouched form. The only parameter that has to be reclassified is aspect because it contains continues values and special coded value for the flat. Values of aspect were reclassified into the categories N, NE, E, SE, S, SW, W, NW and Flat.

Input parameters were used in the following form. Geological map contains 90 distinct classes, land use contains 18 distinct classes and aspect was passed with 9 classes in the study area. The last used conditioning factor slope was used in its continuous form.

2.2 Landslide Inventory Map

The database includes vector and raster spatial datasets using the ArcGIS software package. As a groundwork of digital layers (landslides and stable areas) serve the data from the project of Atlas of Slope Stability Maps SR at 1:50,000 (Šimeková et al. 2006), which was completed in 2006 by State Geological Institute of Dionýz Štúr. Slope deformations were identified from archival materials or mapped in the field. Data were provided in shapefile format. Landslides were extracted from zos_zosuvy_sr.shp with value of attribute STUPAKT = A (activity level = active). Stable areas were extracted from zos_8002_SR.shp. Subsequently, the landslide vector map was transformed into a grid database with a cell size of 10 × 10 m. For the study area, 72 landslides (in the range 13,464–241,044 m^2) were extracted using the inventory map. The landslide inventory map was especially helpful in understanding the different conditioning factors that control slope movements. Distribution of landslides and stable areas is shown in the Fig. 1.

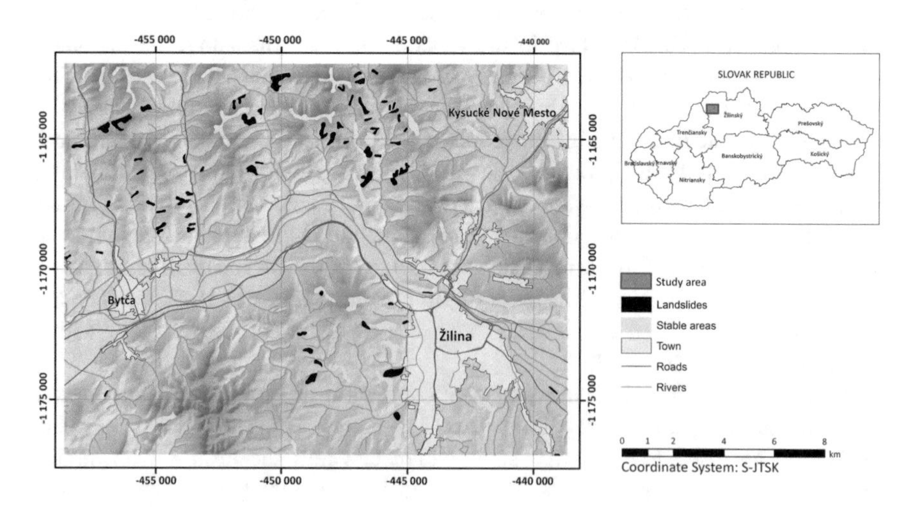

Fig. 1 Location map of the study area showing landslides and stable areas from Atlas of Slope Stability Maps SR at 1:50,000 (Šimeková et al. 2006) used in training and validation

3 Learning with Support Vector Machines

Machine learning algorithms are widely used in cases where the mapping function between input parameter(s) and output parameter(s) is complex, or even unknown. Definition of machine learning can be as following: "A computer is said to learn from experience E with respect to some test T and some performance measure P, if its performance on T, as measured by P, improves with experience E" (Mitchell 1997).

SVMs were introduced in 1995 by prof. Vladimir Naumovich Vapnik (Vapnik 1995). Since then, SVMs and their variants and extensions (also called kernel-based methods) have become one of the preeminent machine learning paradigms. Nowadays, SVMs are routinely used in wide range of areas e.g. handwriting recognition or bioinformatics.

3.1 *Two-Class Support Vector Machines*

In Two-Class SVMs the m-dimensional input x is mapped into l-dimensional (l · m) feature space z. Then in the feature space z the optimal separating hyperplane is found by solving the quadratic optimization problem. The optimal separating hyperplane is a boundary that separates classes with maximal generalization ability as shown in the Fig. 2.

Circles represent class 1 and squares are class −1. As shown in the picture (Fig. 2), there is infinite number of separating hyperplanes which separates classes

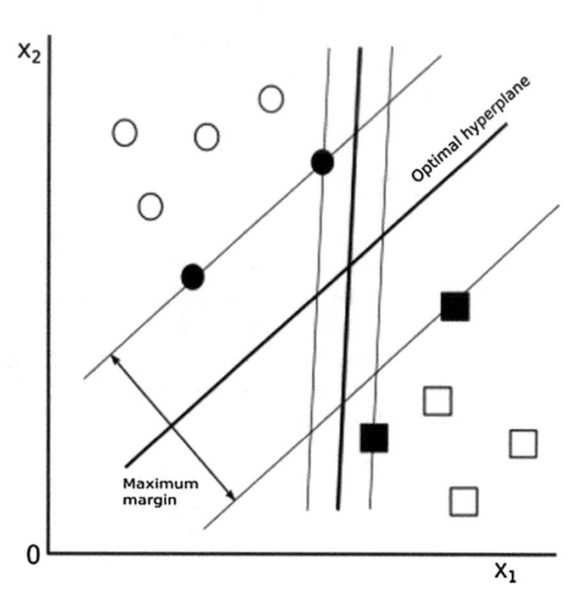

Fig. 2 Optimal hyperplane separating two classes with maximum margin

1 and −1. The separating hyperplane with maximal margin is called the optimal separating hyperplane. In this case, the separating hyperplane is given by equation:

$$\mathbf{w}^{\mathrm{T}}\mathbf{x} + b = 0 \tag{1}$$

where $\mathbf{w}$ is m-dimensional vector, b is bias term and $\mathbf{x}$ is m-dimensional input.

SVMs will then predict class 1 every time when $\mathbf{w}^{\mathrm{T}}\mathbf{x} + b > 0$, and class 2 when $\mathbf{w}^{\mathrm{T}}\mathbf{x} + b < 0$. Equation (1) can be converted to form:

$$y_i\left(\mathbf{w}^{\mathrm{T}}\mathbf{x} + b\right)\begin{cases} \geq 1 & \text{for every } y_i = 1 \\ \leq -1 & \text{for every } y_i = -1 \end{cases} \tag{2}$$

where y is sample label.

From Eq. (2) it is clear that we obtain exactly the same separating hyperplane, even if all the data that satisfy the inequalities are deleted. The separating hyperplane is defined by the data that satisfy the equalities (2) and this data are called support vectors. In the Fig. 2 the support vectors are marked with filled circles and squares. Example above is linearly separable. Many times the problem can not be separated linearly and we have to admit training errors. The influence of training errors is regulated with error cost parameter C. Maximizing the margin is a problem of constrained optimization, wich can be solved by Lagrange method. Final decision function is given by:

$$f(\mathbf{x}_{new}) = \operatorname{sign}\left(\sum_{i=1}^{\#SV} \alpha_i y_i \langle \mathbf{x}_i^{SV}, \mathbf{x}_{new} \rangle + b\right) \tag{3}$$

where $f(x_{new})$ is decision function, $\#SV$ is number of support vectors, x^{SV} is m-dimensional support vector input and x_{new} is m-dimensional input of unobserved data and α is Lagrange multiplier.

3.2 Kernel Trick

Very few problems are linearly separable in the input space, and therefore SVMs does not have high generalization ability. To overcome linearly inseparable problems we map original input space into high-dimensional dot-product feature space. To map m-dimensional input feature space into l-dimensional feature space we use nonlinear vector function $g(x) = (g_1(x), \ldots, g_n(x))^{\mathrm{T}}$ (Abe 2010). Using the Hilbert-Schmidt theory, the final solution transforms to following:

$$f(\mathbf{x}_{new}) = \operatorname{sign}\left(\sum_{i=1}^{\#SV} \alpha_i y_i K\left(\mathbf{x}_i^{SV}, \mathbf{x}_{new}\right) + b\right) \tag{4}$$

where $K\left(\mathbf{x}_i^{SV}, \mathbf{x}_{new}\right)$ is the kernel function or kernel.

Kernels allow us to map input space into high-dimensional (even infinite-dimensional) feature space without explicit treatment of variables.

In this study we are using three types of kernels:

1. Linear kernel $K(x_1,x_2) = <x_1,x_2>$
2. Polynomial kernel $K(x_1,x_2) = (\gamma <x_1,x_2> + c_0)^d$
3. Radial basis function (RBF) $K(x_1,x_2) = exp(-\gamma\|x_1 - x_2\|^2)$

where γ is width of radial basis function, coefficent in a polynomial, d is degree of a polynomial, c_0 is additive constant in polynomial a C is influence of training errors. Detailed information about SVMs definition can be found in Abe (2010), Hamel (2009) or Cambell and Ying (2011).

4 Implementation of Support Vector Machines

In this study, the software LIBSVM v3.20 (Chang and Lin 2011) and LIBLINEAR v1.96 (Fan at el. 2008), which are widely used SVMs libraries, were used to SVMs analysis. Input environmental parameters were processed using ArcGIS. Then, MATLAB was used to transform environmental parameters into the form suitable for LIBSVM and LIBLINEAR analysis. Categorical data land use, lithology and aspect were dummy coded. Finally, all input parameters were normalized into the $\langle -1, 1 \rangle$ range. Therefore, four input environmental parameters, namely aspect (9 categories), geology map (90 categories), land use (18 categories) and slope (one continuous value), formed 118-dimensional input space. Landslides areas cover 35,191 pixels against 59,304 pixels covered by stable areas in the study area. These pixel sets and values of the parameters on their locations formed experimental data matrix of shape 94,495 $\times$ 118 with landslides or stable area pixels in the rows and values of the input parameters in the columns. This data was then divided into training set used to train SVMs models and validation set used to validate performance of the models with ratio 50–50 %. This size of validation sample allows to perform more complex ROC testing over more test data and is not causing decrease of the predictive performance.

The optimal set of kernel parameters was gained by grid search. Unfortunately, there is no way of how to determine the kernel parameters a priori and they must be determined experimentally. The parameters were tested in the following grid:

- parameter C: 1, 3, 10, 30, 100, 300
- parameter γ: 1/118 (0.01, 0.03, 0.1, 0.3, 1, 3, 10, 30, 100)
- parameter c_0: 0, 10, 30, 100
- parameter d: 2, 3

For linear kernel there is one special parameter in LIBLINEAR library which is type of the solver. LIBLINEAR has implemented 8 types of solver [further information in Fan at el. (2008)]. This parameter will be called S: 0, 1, 2, 3, 4, 5, 6, 7.

In this case study, together 534 models were trained and tested.

5 The Validation of the Landslide Susceptibility Maps

The outputs of the SVMs models after their spatialization, are generally presented in the form of maps expressed quantitatively. Final landslides susceptibility maps are usually divided into the five susceptibility levels. On the other hand, reclassification to discreet classes causes the information loss and so the final outputs were left in the continuous form in order to produce smoother ROC curves. Primary outputs from the SVMs algorithm are predicted class labels. The probability outputs are available by transformation of sample distance to separating hyperplane, but this process is computationally costly and cannot be performed for every model during the best inner parameters search phase. The first models predictive performance was measured by parameter "Accuracy" (ACC) (Metz 1978), which was gained over the validation set (47,247 landslide grid cells that is not used in the training phase). The results of the ten best models for each kernel are shown in Tables 1, 2 and 3. Then each kernel (linear, polynomial and radial basis function) was retrained with the optimal set of inner parameters to produce probabilistic outputs. Finally, these outputs were used to validate models. The spatial effectiveness of models were checked by ROC curve. The ROC curve is a useful method of representing the quality of deterministic and probabilistic detection and forecast system (Swets 1988). The area under the ROC curve (AUC) characterizes the quality of a forecast system by describing the system's ability to anticipate correctly the occurrence or non-occurrence of pre-defined "events". The ROC curve plots the false positive rate (FP rate) on the x-axis and the true positive rate (TP rate) on the y-axis. It shows the trade-off between the two rates (Negnevitsky 2002). If the AUC is close to 1, the

Table 1 The results of the ten best models of linear kernel

S	C	TP rate	FP rate	ACC	Training time (s)
4	1	0.8045	0.0856	0.8734	32.0
4	3	0.8031	0.0847	0.8734	95.4
4	30	0.8051	0.0864	0.8731	838.6
4	10	0.7960	0.0811	0.8730	314.0
3	1	0.8094	0.0893	0.8729	10.7
4	100	0.8123	0.0912	0.8728	1937.5
3	3	0.7917	0.0796	0.8724	23.2
3	10	0.7706	0.0746	0.8676	39.1
7	1	0.8092	0.1023	0.8647	8.0
0	1	0.8076	0.1012	0.8647	4.3

Table 2 The results of the ten best models of polynomial kernel

d	γ	c0	C	TP rate	FP rate	ACC	Training time (s)
3	1/118 × 10	0	300	0.8801	0.0475	0.9255	2446.5
3	1/118 × 30	10	10	0.8794	0.0471	0.9254	3118.7
3	1/118 × 30	0	10	0.8790	0.0471	0.9253	2512.2
3	1/118 × 10	0	100	0.8820	0.0492	0.9251	1620.7
3	1/118 × 10	10	30	0.8923	0.0554	0.9251	5189.6
3	1/118 × 30	10	1	0.8825	0.0496	0.9251	1040.5
3	1/118 × 10	10	10	0.8864	0.0519	0.9250	1410.5
3	1/118 × 30	10	30	0.8834	0.0502	0.9250	2666.8
3	1/118 × 10	0	30	0.8801	0.0484	0.9249	901.1
3	1/118 × 30	10	300	0.8802	0.0485	0.9249	3164.7

Table 3 The results of the ten best models of radial basis function kernel

γ	C	TP rate	FP rate	ACC	Training time (s)
1/118 × 30	300	0.8826	0.0490	0.9255	311.2
1/118 × 30	100	0.8817	0.0487	0.9253	225.0
1/118 × 10	300	0.8817	0.0494	0.9249	218.0
1/118 × 30	30	0.8810	0.0492	0.9247	180.4
1/118 × 10	100	0.8791	0.0483	0.9246	183.6
1/118 × 100	10	0.8796	0.0489	0.9244	197.8
1/118 × 3	300	0.8791	0.0494	0.9239	158.3
1/118 × 10	30	0.8777	0.0486	0.9239	162.8
1/118 × 30	10	0.8785	0.0496	0.9236	155.2
1/118 × 100	3	0.8772	0.0494	0.9232	179.0

result of the validation is excellent. The predictor which is simply guessing will score 0.5 (AUC), because probability of right guess from two classes is 50 %. The results of the ROC for the best models are illustrated in the Fig. 3.

According to ROC testing, the best predictive performance was gained by polynomial kernel followed by radial basis function kernel and linear kernel. Values of AUC for each kernel point to very good prediction ability over validation set. On the other hand, only numerical results of validation are not sufficient to measure predictive performance. As we can see in the maps below (Figs. 4, 5), polynomial and linear kernels failed to generalize predictive performance on areas where there was no training data and their characteristics differ from the areas covered by the training data significantly. The most problematic region for both linear and polynomial kernel is region of city Žilina. This is caused by distribution of training data in the study area which is not homogeneous. As we can see in the Fig. 1, the most of the stable areas pixels and active landslisdes pixels that form training and validation datasets are distributed in mountainous parts of the study area. The lack of training data from areas with significantly different characteristics (e.g. from region

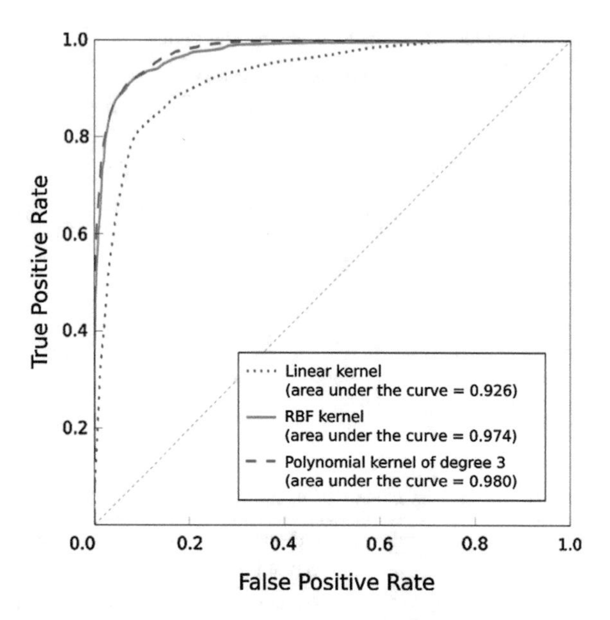

Fig. 3 ROC plots for the susceptibility maps showing false positive rate (x-axis) vs. true positive rate (y-axis) for the best model of each kernel

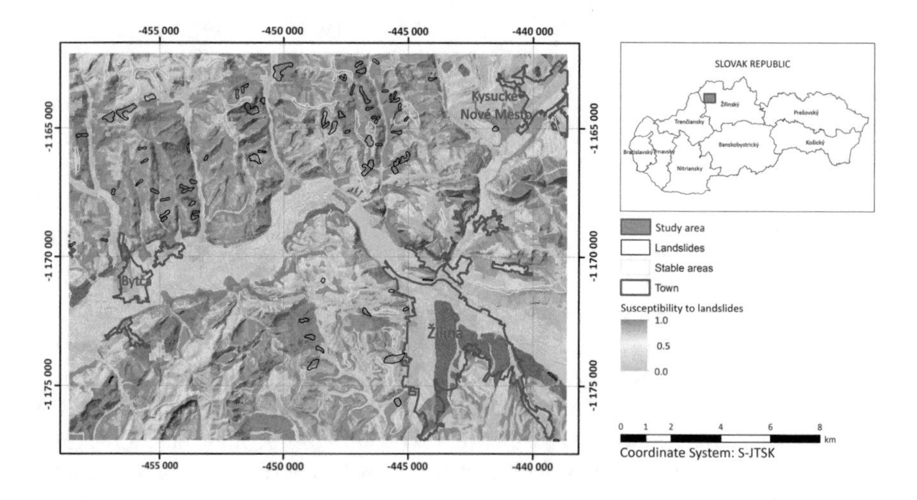

Fig. 4 The classification results of the study area produced by linear kernel

of city Žilina) causes poor predictive performance of linear and polynomial kernels. This shortage cannot be detected by means of statistical validation and outputs need to be also visually validated.

After numerical and visual validation of the classification results, the radial basis function kernel (Fig. 6) was chosen as the optimal kernel for landslides

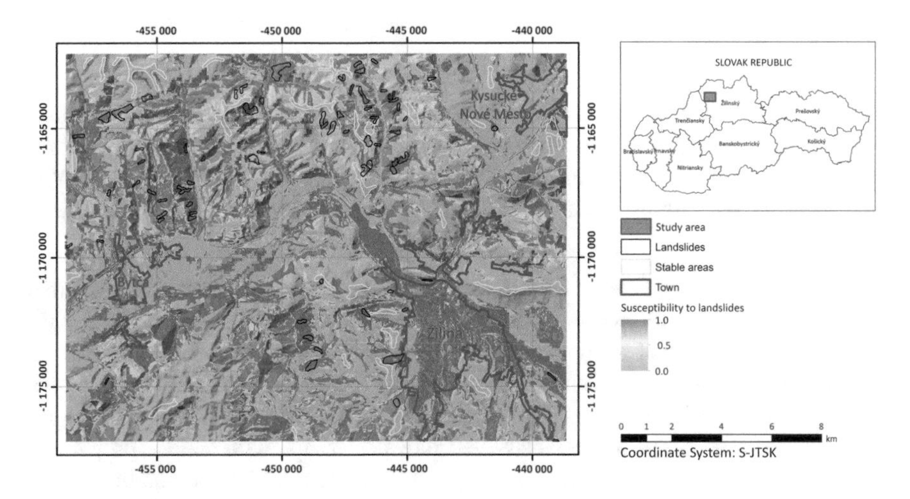

Fig. 5 The classification results of the study area produced by polynomial kernel

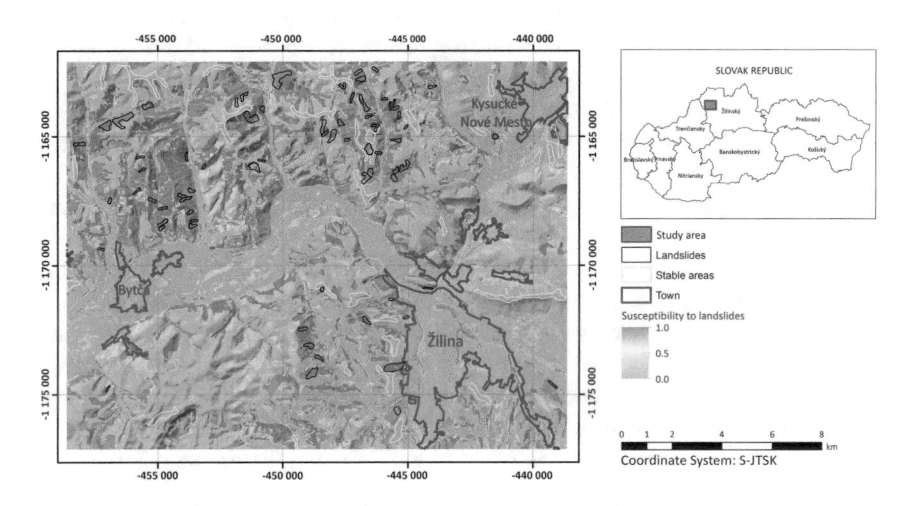

Fig. 6 The classification results of the study area produced by radial basis function kernel

susceptibility mapping, the model's AUC validation result is 0.974 what is only 0.006 less than the best result gained by polynomial kernel in statistical testing. Radial basis function kernel has also the best generalization ability from all kernels. Another advantage of radial basis function kernel is that the kernel has only two parameters to be set and the training time is also lower than training time of polynomial kernel with four parameters to set.

6 Conclusion

This paper presents an applicability case study of the SVMs method in landslides susceptibility mapping based on landslides recorded in districts Bytča, Kysucké Nové Mesto and Žilina. We have applied the SVMs algorithm using the LIBSVM and LIBLINEAR libraries and made a SVMs application work-flow that is as following:

1. *Data preparation*. One of the advantages of the SVMs method is the fact that there is no need to reclassify the input environmental parameters. The SVMs method is capable of using the continuous parameters such as slope steepness in the original form. The SVMs algorithm is designed for classification problems based on hundreds of input parameters. Therefore, environmental input parameters need not to be reclassified into the groups with similar properties, which prevents information loss and can be passed into the computation in the original form.
2. *Inner parameters grid search*. Support Vector Machines has various inner parameters for every kernel that need to be set correctly in order to gain the best predictive performance. Unfortunately, there is no way of choosing the parameters values a priory and values of parameters have to be search by training and testing with model's accuracy.
3. *Support Vector Machines retraining* with the best set of parameters from grid search with probability outputs enabled.
4. *Statistical validation* over validation dataset not used in grid search and final model training.
5. *Visual validation*. Support Vector Machines algorithm is able to fit to area of interest closely and is prone to training data distribution. If the area of interest has parts with significantly different characteristics (e.g. mountainous forest and urbanized land), it is very important to ensure that the training and validation datasets are evenly distributed over all parts of study area. If this need is not met, the predictive performance will be optimized only to characteristics bounded to training data and predictions in other parts of region may not be reliable.

We have tested three types of kernels and their inner parameters. The results were first validated numerically with ROC curves over the validation set, then we produced the landslides susceptibility maps and validated them visually. From the results, we can draw following conclusions:

(a) The results showed that the SVMs method employed in the present study gave promising results with more than 0.90 (AUC) prediction performance.
(b) The best predictive performance 0.980 (AUC) was gained by polynomial kernel, but this kernel failed to generalize predictive performance over all study area. This type of kernel is prone to training data distribution. Another disadvantage of this kernel is more kernel parameters to be set, which means more combinations of kernel parameters to be tested in order to gain the best predictive performance.

(c) After the visual validation of the landslides susceptibility maps, the radial basis function kernel was chosen as the optimal kernel for landslides susceptibility mapping. This kernel gained 0.974 (AUC) prediction performance. Advantages of this kernel are:

  - two parameters to be set, γ and error cost parameter C, which results to less combinations of kernel parameters to be set than in case of polynomial kernel,
  - very good generalization ability,
  - low training time.

(d) Linear kernel scored the lowest prediction performance 0.926 (AUC). The advantages of this kernel is very low training time. The disadvantage is poor generalization ability caused by nature of linear kernel which is the simplest that does not map input space into feature space. This kernel is very effective when the problem is linearly separable or there is very large amount of input parameters which form high-dimensional input space.

(e) Among disadvantages of the SVMs method belong computational cost and the biggest disadvantages is hard interpretation of kernels and their parameters and incapability of input parameters weights determination.

As we presented in this case study, the SVMs algorithm with radial basis function kernel gives promising results in landslides susceptibility mapping. Next we suggest further testing of the SVMs method by the professionals from geology and detailed results testing from the lithology, slope, land use and other conditioning factors points of view by the domain experts.

Acknowledgments This work was supported by the Grants Nos. 1/0954/15 and 1/0682/16 of the Grant Agency of Slovak Republic VEGA.

References

Abe S (2010) Support vector machines for pattern classification, 2nd edn. Springer, Kobe University, Kobe

Baliak F, Stríček I (2012) 50 rokov od katastrofálneho zosuvu v Handlovej/50 years since the catastrophic landslide in Handlova (in Slovak only) Mineralia Slovaca, 44:119–130

Bednarik M (2011) Štatistické metódy pri hodnotení zosuvného hazardu a rizika. Habilitation Thesis. Bratislava: Comenius University, Faculty of Natural Sciences

Bednarik M, Pauditš P, Ondrášik R (2014) Rôzne spôsoby hodnotenia úspešnosti máp zosuvného hazardu: bivariačný verzus multivariačný štatistický model/Various techniques for evaluating landslide hazard maps reliability: Bivariate vs. multivariate statistical model (in Slovak only) Acta Geologica Slovaca 6(1):71–84

Cambell C, Ying Y (2011) Learning with Support Vector Machines. Morgan & Claypool Publishers, San Rafael

Carrara A (1983) Multivariate models for landslide hazard evaluation. Math Geol 15(3):403–427

Carrara A, Cardinalli M, Detti R, Guzzetti F, Pasqui V, Reichenbach P (1990) Geographical information systems and multivariate models in landslide hazard evaluation. In: Cancell A (ed) ALPS 90 (Alpine Landslide Practical Seminar) Proceedings of the 6th international conference and field workshop on landslide. Universita degli Studi di Milano, Milano, pp 17–28

Chang C, Lin C (2011) A library for support vector machines. ACM Trans Intell Syst Technol. http://www.csie.ntu.edu.tw/~cjlin/libsvm

Fan RE, Chang KW, Hsieh CJ, Wang XR, Lin CJ (2008) A library for large linear classification. J Mach Learn Res 9:1871–1874. http://www.csie.ntu.edu.tw/~cjlin/liblinear

Hamel L (2009) Knowledge discovery with Support Vector Machines. Wiley, London

Hyun-Joo O, Pradhan B (2011) Application of a neuro-fuzzy model to landslide-susceptibility mapping for shallow landslides in a tropical hilly area. Comput Geosci 37:1264–1276

Káčer Š, Antalík M, Lexa J, Zvara I, Fritzman R, Vlachovič J, Bystrická G, Brodianska M, Potfaj M, Madarás J, Nagy A, Maglay J, Ivanička J, Gross P, Rakús M, Vozárová A, Buček S, Boorová D, Šimon L, Mello J, Polák M, Bezák V, Hók J, Teťák F, Konečný V, Kučera M, Žec B, Elečko M, Hraško Ľ, Kováčik M, Pristaš J (2005) Digitálna geologická mapa Slovenskej republiky v M 1:50,000 a 1:500,000/Digital geological map of the Slovak Republic at 1:50,000 and 1: 500,000. (in Slovak only) MŽP SR, ŠGÚDŠ, Bratislava

Lee S, Ryu J, Won J, Park H (2004) Determination and application of the weights for landslide susceptibility mapping using an artificial neural network. Eng Geol 71:289–302

Metz CE (1978) Basic principles of ROC analysis. Semin Nucl Med 8(4):283–298

Mitchell T (1997) Machine learning. McGraw-Hill Science, New York

Nefeslioglu HA, Sezer E, Gokceoglu C, Bozkir AS, Duman TY (2010) Assessment of landslide susceptibility by decision trees in the metropolitan area of Istanbul, Turkey. Math Problems Eng. doi:10.1155/2010/901095

Negnevitsky M (2002) Artificial intelligence: a guide to intelligent systems. Addison-Wesley/Pearson Education, Harlow

Pauditš P (2005) Landslide susceptibility assessment using statistical methods within GIS environment. PhD Thesis. Comenius University, Faculty of Natural Sciences 2005, Bratislava

Pradhan B (2013) A comparative study on the predictive ability of the decision tree, support vector machine and neuro-fuzzy models in landslide susceptibility mapping using GIS. Comput Geosci 51:350–365

Pradhan B, Lee S, Buchroithner M (2010) A GIS-based back-propagation neural network model and its cross-application and validation for landslide susceptibility analyses. Comput Environ Urban Syst 34:216–235

Šimeková J, Martinčeková T (eds) Baliak PAF, Caudt L, Gejdoš T, Grencíková A, Grman MD, Jadron DH, Kopecký M, Kotrcová E, Liščák P, Malgot J, Masný M, Mokrá M, Petro L, Polašcinová E, Rusnák M, Sluka V, Solciansky R, Wanieková D, Záthurecký A, Žabková E (2006) Atlas máp stability svahov Slovenskej republiky 1:50,000/Atlas of Slope Stability Maps of the Slovak Republic at scale 1:50,000). Technical report, Vyd. MŽP SR Bratislava/Ingeo-IGHP s.r.o., (in Slovak only), Žilina

Šimeková J, Liščák P, Jánová V, Martinčeková T (2014) Atlas of slope stability maps of the Slovak Republic at Scale 1:50,000—its results and use in practice. Slovak Geological Magazine, vol 14, SGÚDŠ, Bratislava, pp 19–31

Swets JA (1988) Measuring the accuracy of diagnostic systems. Science 240:1285–1293

Vahidnia MH, Alesheikh A, Alimohammadi A, Hosseinali F (2010) A GIS-based neuro-fuzzy procedure for integrating knowledge and data in landslide susceptibility mapping. Comput Geosci 36:1101–1114

Vapnik V (1995) The nature of statistical learning theory. Springer, New York

You X, Tham LG, Dai FC (2008) Landslide susceptibility mapping based on Support Vector Machine: a case study on natural slopes of Hong Kong, China. Geomorphology 101:572–582

Integration of Heterogeneous Data in the Support of the Forest Protection: Structural Concept

Jana Faixová Chalachanová, Renata Ďuračiová, Juraj Papčo, Rastislav Jakuš and Miroslav Blaženec

Abstract The basic precondition for effective management and protection of a forest is a concept built on modern methods of collection, processing, analysis and publication of spatial data about forest coverage, as well as its health status. This paper provides a structural concept for integration of heterogeneous data in the support of the forest protection with implementation of the latest methods of remote sensing data collection. The main principles of the structural concept of Forest protection management system (FPMS) result from the assessment of the current and new available data sources (represented by the *Diagram of data sources*), data analysis and development of innovative mathematical techniques of image processing (represented by the *Diagram of methods and tools*). Mind maps of the proposed diagrams were created in the free mind mapping application FreeMind. The structural concept is represented by an analytical model of the co-operation of data sources, tools and applications. The dynamic structure was proposed using Unified Modelling Language (UML). The *Diagram of the use cases* is represented by a Use Case Diagram in UML. The *Diagram of the processes*, which describes the main processes realized within forest protection management, is represented by an Activity Diagram in UML. The UML diagrams were created in open-source

J. Faixová Chalachanová (✉) · R. Ďuračiová · J. Papčo
Department of Theoretical Geodesy, Faculty of Civil Engineering,
Slovak University of Technology in Bratislava, Radlinského 11,
810 05 Bratislava, Slovak Republic
e-mail: jana.chalachanova@stuba.sk

R. Ďuračiová
e-mail: renata.duraciova@stuba.sk

J. Papčo
e-mail: juraj.papco@stuba.sk

R. Jakuš · M. Blaženec
Institute of Forest Ecology, Slovak Academy of Science, Ľudovíta Štúra 2,
960 53, Zvolen, Slovak Republic
e-mail: jakus@savzv.sk

M. Blaženec
e-mail: blazenec@savzv.sk

I. Ivan et al. (eds.), *The Rise of Big Spatial Data*, Lecture Notes
in Geoinformation and Cartography, DOI 10.1007/978-3-319-45123-7_28

software StarUML. The resulted structural concept of FPMS is the basis of a predictive model improvement and a web application development for the forest protection from the bark beetle.

Keywords Data modelling · Satellite data · UAV technology · Forest protection · Predictive model

1 Introduction

Considering the increasing impacts of adverse climatic conditions and to increasing calamity situations, forest protection from the bark beetle (*Ips typographus*) is a particularly topical issue. It is subject to the ruling of the Ministry of Agriculture and Rural Development of the Slovak Republic No. 453/2006 on forest management and protection. The measures to protect forest management and forest protection are defined in § 43 by measures to protect the forest against damage caused by insects and other live organisms. It is more specifically covered in the norm (STN 48 27 11). In the context of our universe of discourse, the concept of forest protection is reduced to protection in relation to the risks arising from bark beetle threats in the territory of the Slovak republic.

The goal of this paper is to design a suitable structure to support decisions in the management and protection of forests from bark beetle attacks. It is based on the assessment of the current and new available data sources and selection of relevant methods of processing and filtering image data obtained by remote sensing data capture technologies. In this paper, we present a structural concept of all potential data sources in combination with modelling the relevant methods and tools for their processing which can improve the modelling and prediction of bark beetle attacks and thereby it supports decision making processes in forest management. The meaning of this concept is especially in determination of system boundary, i.e. definition of system functionality and roles and tasks for various types of users which is essential to saving costs (time, finance and human resources).

In order to optimize data for spatial analyses of forest management and protection and to reduce the costs of data capturing and updating, it is necessary to test new satellite and spatial data capture technologies. It means particularly aerial survey images and hyper-spectral images with high resolution, radar images, thermal images and data from laser scanning (terrestrial and airborne), as well as unmanned aerial vehicle (UAV) systems (Comolina and Molina 2014; Whalin 2012). These new kinds of spatial data have the potential to improve forest management, but it is necessary to follow the main principles of data integration considering data usage, quality and spatial resolution. Current methods for forest health diagnosing do not enable quality and fast enough evaluation of the present state of larger areas. This causes enormous difficulties in implementing protective measures against bark beetles in spruce stands. Active terrestrial identification of bark beetle trees is time consuming. Some features of the attack are not obvious so errors occur often (Jakuš et al. 2013). Infrared imaging can bring a partial solution to this issue. But there are still disadvantages which limit the practical using of an infrared camera, such as the lack of accurately determining the health of spruce trees, or insufficient, respectively delayed, differentiation between healthy trees and trees attacked by bark beetles (Arnberg and Wastenson 1973; Jones and Schofield 2008). The use of standard aerial infrared imaging is often limited by weather conditions. Due to costs and weather limitations, aerial imaging is taken usually just once a year. The output maps are mainly useful for mapping bark beetle attacks that happened in the same year. It is similar in the case of satellite imaging. Yearly it is usually taken just one proper image of the territory. Another partial solution to this problem is usage of a small UAV, equipped with a simple infrared camera (Whalin 2012). This technology has been experimentally used at the Swedish University of Agricultural. In this case, images of research areas have been taken 3 times per season to monitoring bark beetle attacks on spruce stands (SmartPlanes 2006).

The design of a suitable method of data integration for the purposes of creating a predictive model needs to be based on the main principles of the *Ips typographus* attack mechanism. Part of this mechanism is connected with the effects of solar radiation and surface temperatures on individual trees and forest stands (Jakuš et al. 2003b; Grodzki et al. 2006; Akkuzu et al. 2009). This is also in agreement with Schopf and Köhler (1995). Drought predisposes spruce trees to bark beetle attacks (Arnberg and Wastenson 1973; Rouault et al. 2006). Thermal sensing is a proxy of insolation and water stress effects on trees. According to Jones, Schofield (2008), currently it is primarily used method to study plant water relations.

The results presented in Jakuš et al. (2003a) also showed the importance of parameters related to spruce crown transformation and host resistance (Jakuš et al. 2011). The results correspond with the results from the Bohemian Forest mentioned in Moravec et al. (2002). In Malenovský et al. (2008) it is suggested the use of remote sensing for mapping crown transformation. One type of marker of spruce host resistance are phenolic compounds (Brignolas et al. 1998), which may affect host colonization (Schlyter 2001) and are active as antifeedant semi chemicals (Faccoli and Schlyter 2007). According to Soukupová et al. (2001), it is possible to assess the content of phenolic compounds in spruce needles by using remote sensing techniques. Niemann et al. (2015) have shown possibilities of the hyper spectral data use for early detection of bark beetle attacks.

To predict the development of bark beetle populations several database or geographical information systems (GIS) under laid models were developed in the USA. The management group of the National Park Bavarian Forest has developed a simulator of damage to spruce stands by *Ips typographus* under various ways of care for the protected areas (Fahse and Heurich 2003). In the High Tatras, a model simulating the development of *Ips typographus* based on measurements of air temperature and on a digital model of the terrain was tested (Netherer 2003). Kissiyar et al. (2005) have developed a GIS for the evaluation of forest stand predisposition to bark beetle attacks. The system is based on the known causal links between bark beetle outbreaks and environmental parameters. A module on the prognosis of a bark beetle stand infestation is also a part of the system. The basis for the model is the analysis of the vegetation changes (image differentiation), digital terrain model and stand characteristics. The basis for the study was the time series of Landsat images with the resolution of 30 m (Jakuš et al. 2003a). A system of partial models has been used, where each partial model produced an output, which is used in the main model later. One-year forecast of *Ips typographus* invasion was based on modelling the two processes related to outbreak dissemination of bark beetles: the initiation of bark beetle spots and spot spreading (Jakuš et al. 2003a). The current risk of invasion was estimated using a model of the bark beetles spreading in forest stands (Netherer and Pennerstorfer 2001; Baier et al. 2007). Time series of satellite images were analysed in order to obtain information on the distribution of wind thrown trees and the trees attacked by bark beetles. The meteorological data were used to predict the development of populations as well as the development of water stress. This information was then integrated into the system on the base of GIS. Predictions of future forest stand attacks by *Ips typographus* were subsequently made, but methodology and automation of the prediction processes are not yet sufficiently developed.

2 Materials and Methods

Spatial analysis realized in GIS environment is a key factor in forest protection decision making process. Therefore, focus should be on the quality of interpreted inputs into spatial analysis leading to modelling of the mechanism of the bark beetles spreading in a forest and to predict its future development.

2.1 Data Sources Used in the Domain of the Forest Protection

Data used in the forest management arise from many heterogeneous data sources, e.g. *in situ data* (such as data from geodetic surveying, hydrological data, meteorological data and forestry data about tree parameters, bark beetle outbreaks, etc.), *reference spatial data* (data about digital elevation model (DEM), spatial localization of the objects, cadastral administration, etc.) and *ex situ* data (i.e., spatial data collected by the latest remote sensing technologies and processing methods). Since the first two datasets are relatively well known (more detailed description is available at http://www.tanabbo.org/tiki-index.php?page=TANABBO%20DB%20Inputs), the main principles, specifications and usability of not so well known *ex situ data* are mentioned below.

Mountain forests are specific for the data collection because of the variation in terrain morphology and vegetation cover, which causes certain issues when ex situ data are collecting and processing (e.g., terrestrial scanning, UAV flight route planning or georeferencing). On the other hand, the usage of UAV technology has great potential in terms of high spatial resolution (decimetre to centimetre) and temporal resolution (observations can be realized if necessary the daily). Then in case of good weather, it is possible to collect the data within a few hours. Digital hyper-spectral high resolution records enable derivation of information about the number of trees, density, storage, area, tree species composition, health condition and forest structure much more precisely than with analogue images with lower resolution. Furthermore, they can be used to monitor the effectiveness of spraying of agricultural crops and forests and to detect and monitor fire danger, etc. Albrechtová and Rock (2003) were dealing with the detection of health conditions of the Norway spruce in the Ore Mountains in the Czech Republic in order to find changes that indicate a threat.

Data from aerial laser scanning (ALS) and terrestrial laser scanning (TLS) are used in forestry mainly for the determination of parameters of trees and a forest (tree height, maximum height, mean height, median, canopy base height, etc.) and for more accurately determining the position and parameters of tree canopies (Andersen et al. 2005). The TLS uses a fundamentally different scanning geometry than ALS, and it provides better resolution (of the order of centimetre accuracy). TLS allows deriving the observed parameters more precisely. Besides these parameters, it also allows obtaining detailed information about the shape of the trunk, limbs and branches and measures for direct derivation of dendrometric characteristics. The potential of ALS and TLS for forestry applications is presented by Van Luewen and Nieuwenhuis (2010), or Smreček and Tuček (2011).

At present, the following satellite data sources and data services are used in the forest management:

- Landsat-8/LDCM (Landsat Data Continuity Mission), which has the spatial resolution and the sensitivity to pick up some of the smaller features than older Landsat missions were able to. Ground sample distance (GSD) for Landsat-8/LDCM ranges from 15 m (for panchromatic data) to 30 m (for multispectral data) (https://directory.eoportal.org/web/eoportal/satellite-missions/l/landsat-8-ldcm).
- GFW (Global Forest Watch) service providing global forest monitoring network with an interactive map of tree cover changes (http://www.globalforestwatch.org/).
- STALES (Forest cover change detection by satellite scenes) service for visualization of actual and historical satellite compositions of the Slovak territory to monitor forest stands state (http://www.nlcsk.org/stales/).

New available relevant satellite data sources are as follows:

- Sentinel-1—(The SAR Imaging Constellation for Land and Ocean Services) include Sentinel-1A as well as the latest Sentinel-1B. The default mode over land is 250 km swath (Interferometric Wide Swath mode) with ground resolution of 5 × 20 m (https://directory.eoportal.org/web/eoportal/satellite-missions/c-missions/copernicus-sentinel-1).

- Sentinel-2—(The Optical Imaging Mission for Land Services) reaches the spatial resolution of 10 m in 290 km swath with 13 spectral channels. Detailed specifications of satellite data are mentioned in (https://directory.eoportal.org/web/eoportal/satellite-missions/c-missions/copernicus-sentinel-2).

The importance of the radar images (Sentinel-1) is that global view from satellites, which it provides, is complementary with local view of study area obtained from UAV technology. Several studies, e.g. (Ortiz et al. 2013), declare the suitability of a combination of radar and optical techniques (Sentinel-2, UAV technology etc.) due to their complementarities and sensitivity to another aspect of the bark beetle impacts. An important contribution of our concept is first application of the dual radar polarimetry technology in the environment of the Slovak Republic and evaluation of a potential of the radar data sources. Our study presents remote sensing method, based on the principle of polarization observed from radar images (e.g., Radarsat-2, Sentinel-1A), which enables information to be gathered on the health of forests and the amount of forest biomass etc. The basic products of the dual radar technique used in the study area of Bratislava on 1.7.2015 are presented in Fig. 1. Radar image with two bands (VV: vertical-vertical radar pulses, VH: vertical horizontal radar pulses) and two types of measurement (amplitude or intensity and phase) was combined in several variations to show the potential of dual radar polarimetry for various purposes. In Fig. 1 the stated combination are

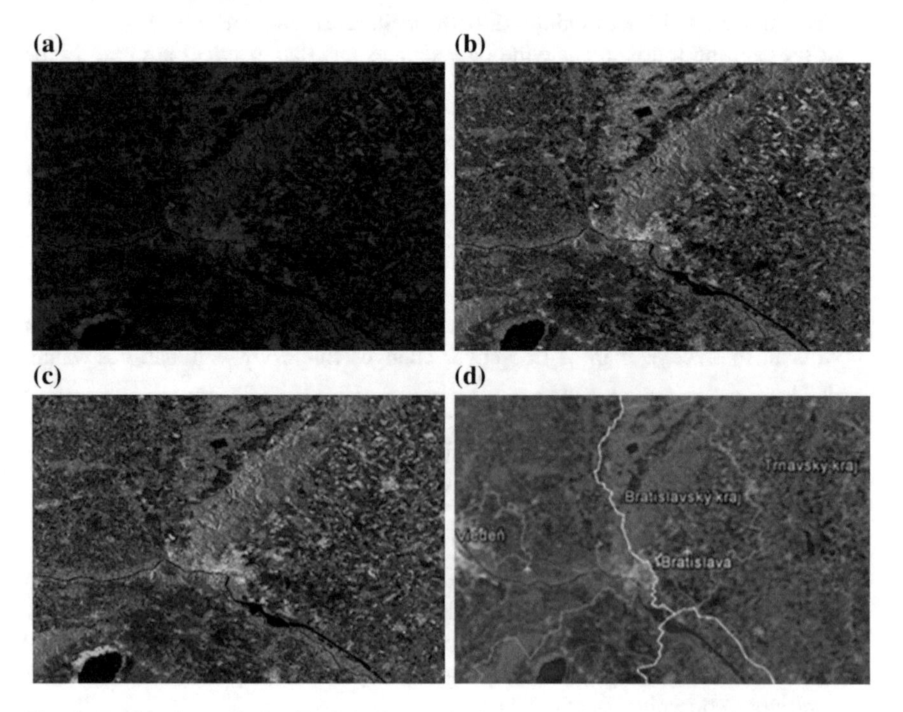

Fig. 1 Combination and visualization of the radar images

shown: (a) VH-VV/VH-VV, (b) VV-VH-VV/VH, (c) VH-VV-VV-VH and the complex visualization at (d) coverage preview of the Sentinel-1A image over the Bratislava area.

Detailed part of the study area is processed in R-G-B (Fig. 2). According to Lee and Pottier (2009), in the next pictures we can see the basic parameters:

a) *The alpha parameter*—it indicates the nature of the scattering (single or double bounce reflection or volume scattering over anisotropic media). Red indicates low values. Over the water it indicates a single bounce reflection, characteristic for surfaces. Blue indicates double bounce reflection, which can be found over built-up areas, due to the ground-wall interaction. Green indicates areas where alpha reaches an intermediate values, depicting volumetric media over vegetated areas.
b) *The entropy parameter*—it provides information on the scattering degree of randomness. Blue indicates scattering over the water which is characterized by a low degree of randomness. Red indicates the mixing of different scattering mechanisms over built up areas which results in intermediate entropy values. Green indicates the random scattering process on forested zones.
c) *The anisotropy parameter*—it provides information on the relative importance of secondary mechanisms. This parameter cannot be interpreted separately from

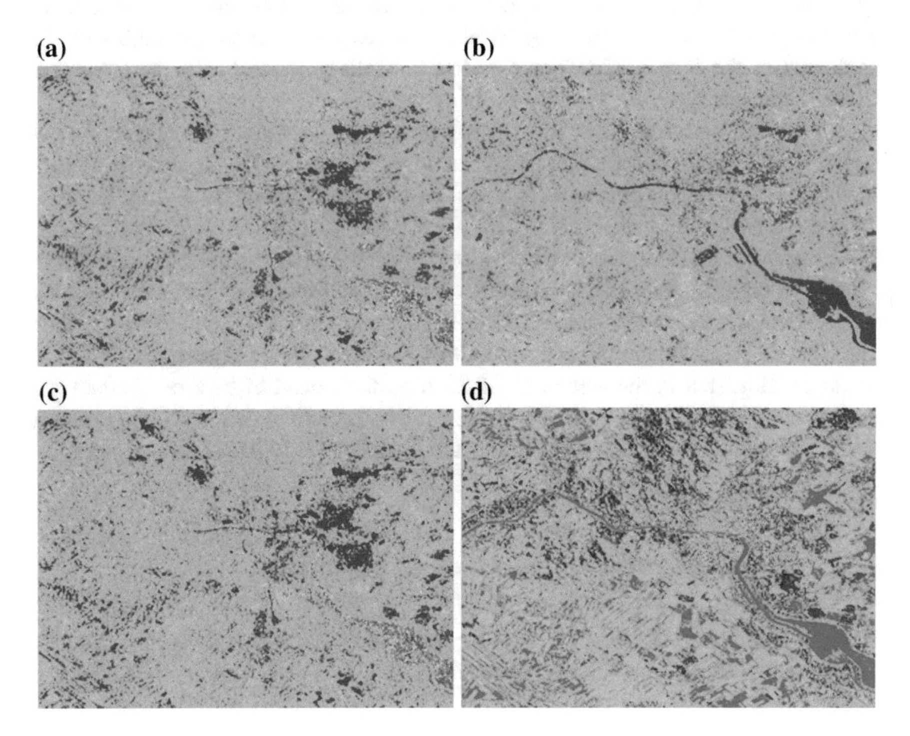

Fig. 2 Decomposition of the radar images

the entropy. It is then interesting to observe the different combinations between the entropy and the anisotropy.

d) *The H-alpha wishart classification parameter*—it provides important information about a smooth surface (e.g., water) which is considered as a deterministic surface. In urban areas, double bounce scattering dominates. Over vegetated areas, the scattering mechanism is more random and is associated to anisotropic scatterers.

Presented outcomes from dual radar polarimetry images processing (Figs. 1, 2) need to be discussed moreover in the forestry, agricultural and environmental community. However, due to update period, availability, spatial resolution and information value of radar images, it is clearly the technology with a high potential for the forest management.

2.2 Principles of Predictive Modelling in the Forest Protection

The basic precondition for effective management and protection of a forest is the Forest protection management system (FPMS) based on spatial structures obtained by remote sensing data capture technologies and others integrated heterogeneous data sources. The development of FPMS is based on the main principles of the prediction of the bark beetle impact on forest stands processed over the integrated relevant data sources.

Our background field work for the structural concept is involved in our original research and development of the spatio-temporal model for the prediction of bark beetle threats to forest stands. It is a part of the international innovative decision support systems (DSS) "TANABBO 2" (www.tanabbo.org). The study area includes locations in the Western Tatras where insect infestation began after extensive storm damage in May 2014. The online meteorological sensors associated with the bark beetle development model PHENIPS will be installed there. Similarly to system (http://www.borkenkaefer.ch), this will allow modelling the current state of the bark beetle development in the area and will forecast the spread of bark beetle damage. Furthermore, based on the modelling of forest transpiration deficit it will be possible to determine the time and locations of stands susceptible to bark beetle attacks.

2.3 Visual Modelling of Information Systems

Mind mapping is one of the best ways to visualize thoughts, but it can be used also for visual modelling of available information. In this case, a mind map is a tool to organize a structural concept of relevant data sources. The mind map is considered to be a special type of spider diagram. It is usually created around a single concept in the centre of a blank page. Major ideas are connected directly to this central concept, and other ideas branch out from those. A simple mind map can be drawn by hand, but for modelling of more complex data structures is better to use mind mapping software (e.g., open source software FreeMind (Freemind.Sourceforge.net), XMind (XMind.net) or Mind42 (Mind42.com)).

The Unified Modeling Language (UML) is a general-purpose modelling language with a semantic specification, a graphical notation, an interchange format, and a repository query interface." (ISO/IEC 19505 1). The UML specification is most-used specification of the Object Management Group (OMG), which is used for modelling of system structure, behaviour, and architecture, but also business processes and data structure (http://www.uml.org). The version 2.4.1 has been formally published by ISO as the 2012 edition standard ISO/IEC 19505-1 and 19505-2.

It is appropriate to support UML modelling by Computer-aided software engineering (CASE), which is the useful tool to ensure tasks in the system modelling, developing and managing. In this paper, open-source software StarUML was used for modelling of a use case diagram (UCD) and an activity diagram (AD).

3 Results

Based on the assessment of the current and new available data sources, methods of data processing and users requirements on forest protection we proposed a structural concept of FPMS which is represented by the model of several diagrams to optimize the forest protection measures:

- the Diagram of data sources (the mind map, Figs. 3, 4, 5, 6, 7),
- the Diagram of methods and tools (the mind map, Figs. 8, 9),
- the Diagram of use cases (UCD, Fig. 10),
- the Diagram of processes (AD, Fig. 11).

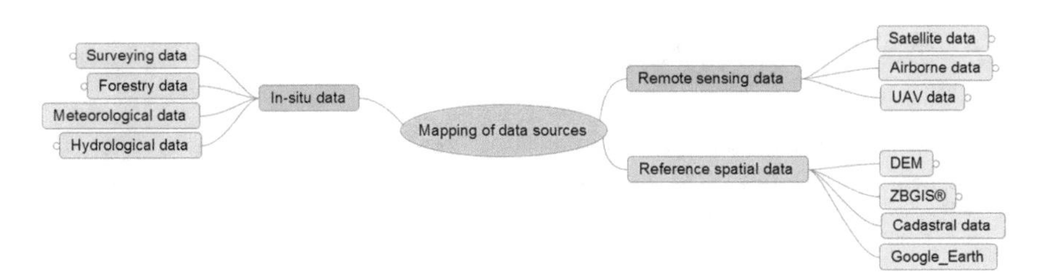

Fig. 3 Main structure of the diagram of data sources

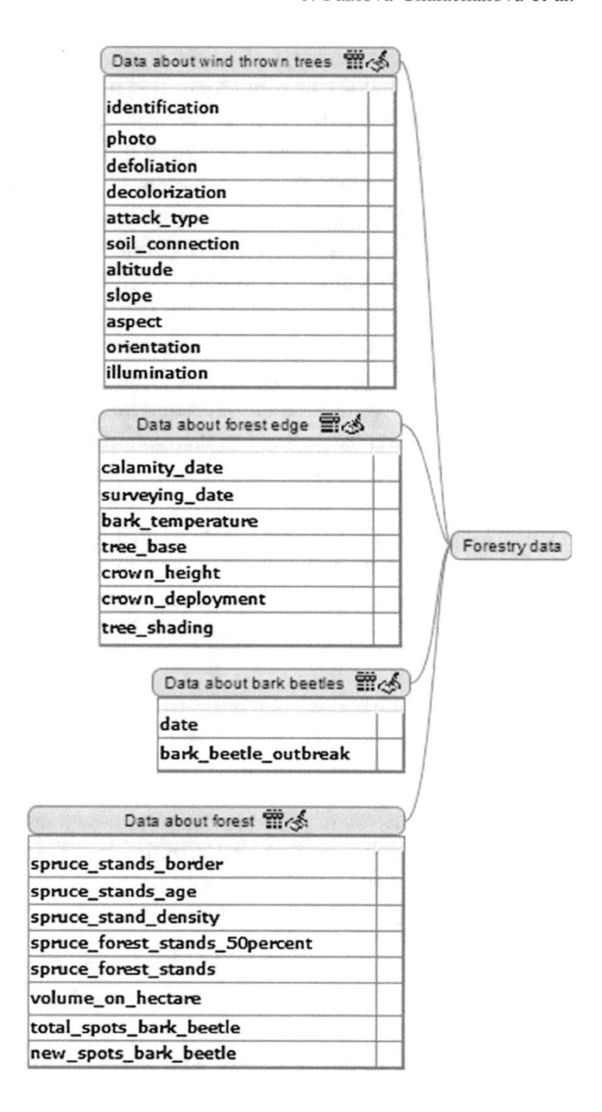

Fig. 4 The diagram of data sources—structure of forest data sources

3.1 *The Diagram of Data Sources*

Based on the differences and advantages of mentioned heterogeneous data, it is appropriate to combine and integrate them from the same territory, time, and conditions of scanning. The connection of heterogeneous data will permit new and more accurate parameters to be inputted into the predictive model of the bark beetles spreading in forest stands. The main structure of all currently used data sources is represented by the Diagram of data sources in the mind map (Fig. 3).

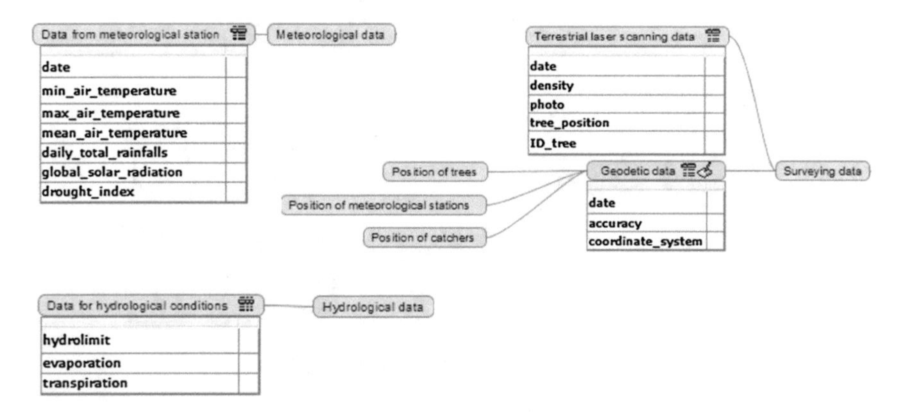

Fig. 5 The Diagram of data sources—structure of meteorological, hydrological and surveying data sources

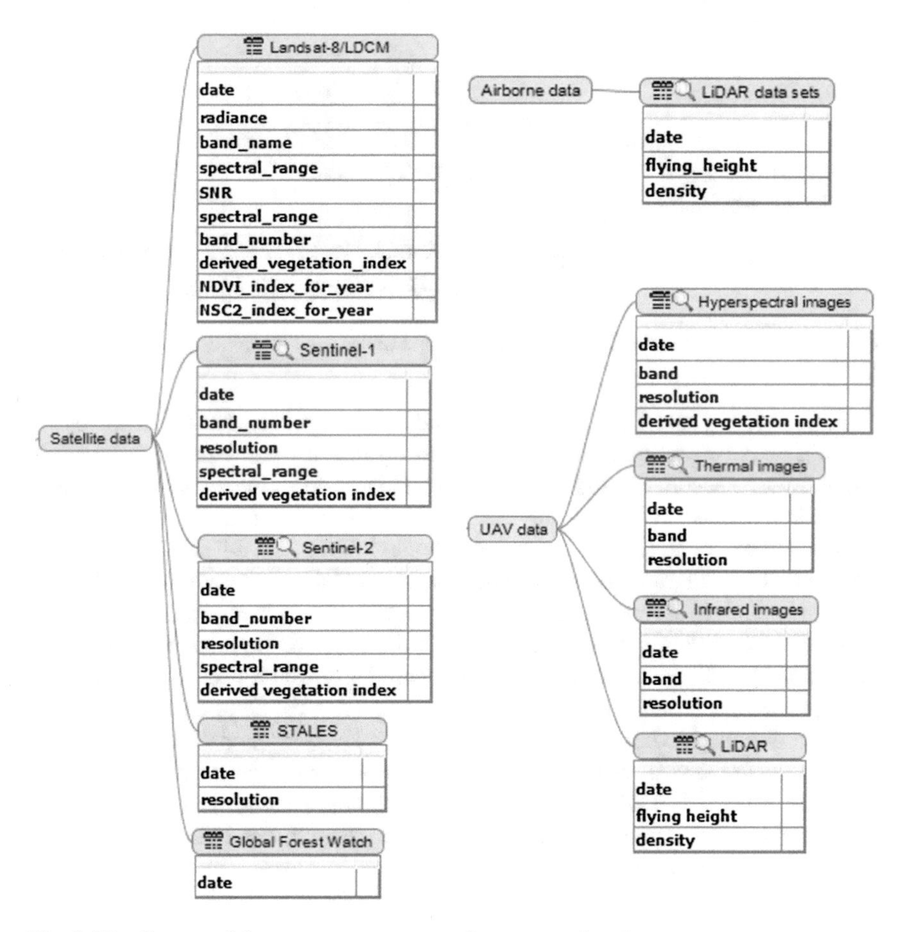

Fig. 6 The diagram of data sources—structure of remote sensing data sources

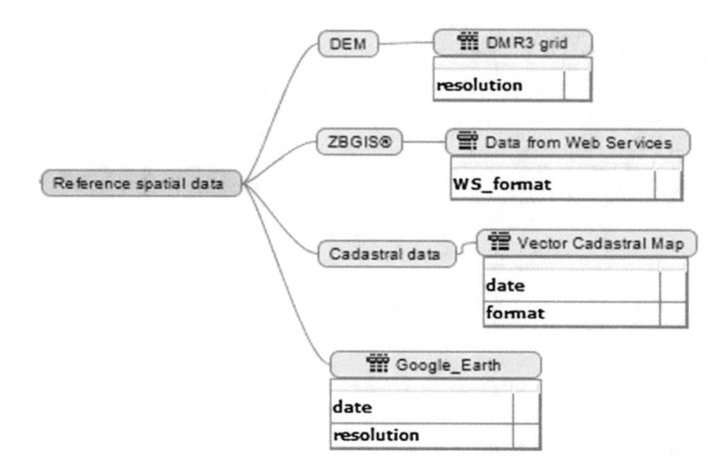

Fig. 7 The diagram of data sources—structure of reference spatial data sources

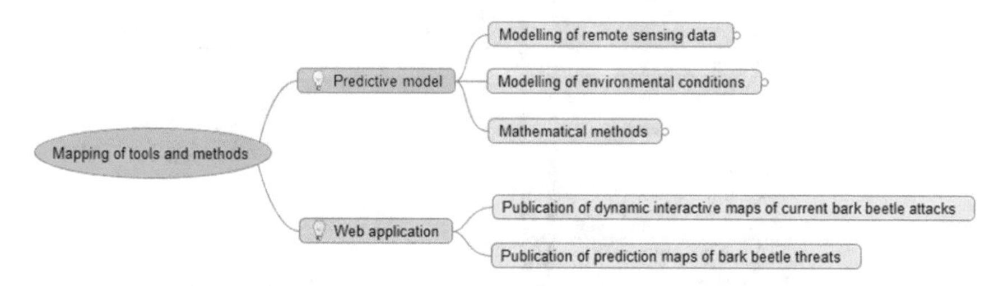

Fig. 8 Main structure of the diagram of methods and tools

Figures 4, 5, 6 and 7 present the proposed diagram of integrated data sources in detail. Extensive part of in situ data centred on forestry properties is itemized in Fig. 4. Many specific forestry attributes are resulting from the actual requirements on predictive model and are closely described by Jakuš et al. (2003b), Grodzki et al. (2006), Akkuzu et al. (2009), Jones and Schofield (2008) etc. Meaning of the items used in the Diagram of data sources is explained in www.tanabbo.org. Meteorological, hydrological and surveying data sources are concretized in Fig. 5. Remote sensing data sources are summarized in Fig. 6. New available types of satellite data and data collected by UAV technology, which need to be discussed and assessed, are marked by a magnifier icon (Fig. 6). Reference spatial data sources, giving the information about digital elevation model (DEM), spatial localization (ZB*GIS*®), cadastral administration and real view of the objects, are shown in Fig. 7. The advantage of designed structure based on mentioned data sources is that it is possible to choose a required resolution and also to set detailed information on individual trees.

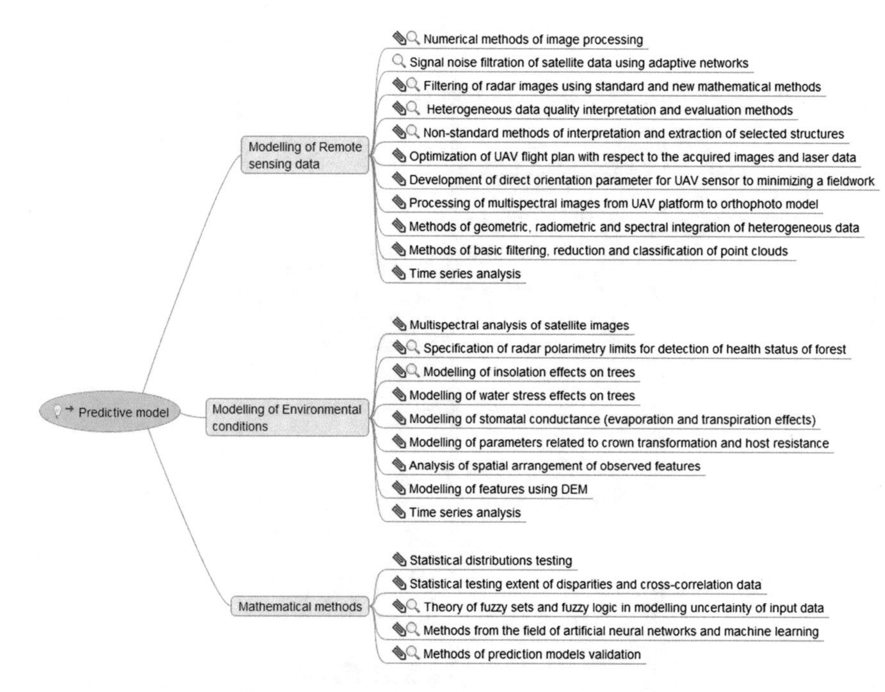

Fig. 9 Extended methods in the predictive model

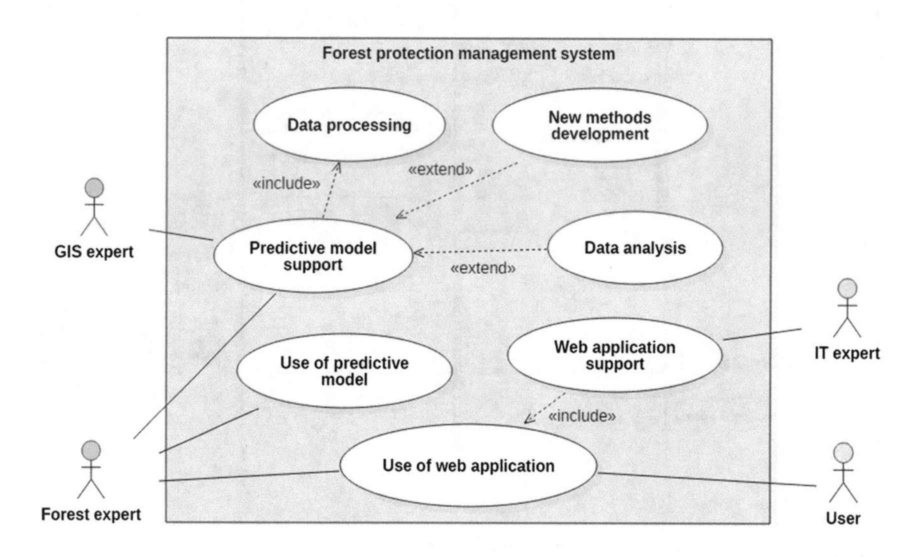

Fig. 10 The forest protection management system—the diagram of use cases

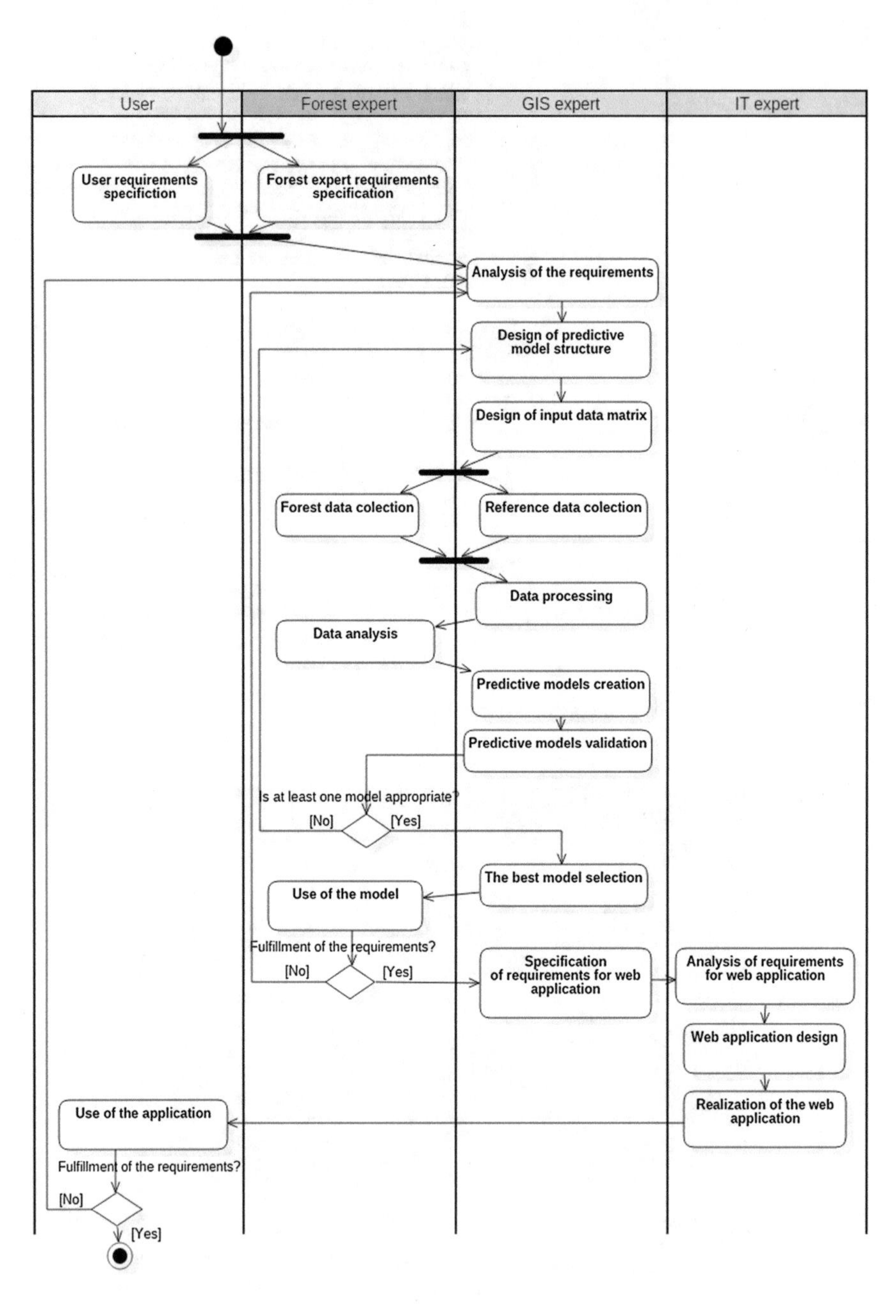

Fig. 11 The forest protection management system—the diagram of processes

3.2 The Diagram of Methods and Tools

The Diagram of methods and tools is divided into two main tools proposed for FPMS: *Predictive model* and *Web application* (Fig. 8).

The predictive model is formed by standard methods of data processing, filtering and analysing (Fig. 9). Standard methods are extended by innovative mathematical techniques (in Fig. 9 it is marked by a magnifier icon). The various methods can be divided into three groups: methods of remote sensing data modelling, methods of environmental conditions modelling and mathematical methods. Several authors refer to standard methods of processing and filtering image, e.g. Meigs et al. (2011) deal with Landsat time series approach to study the impact of the bark beetles, or Meddens et al. (2013) deal with the evaluation of methods to detect bark beetle using Landsat imagery. The assessment of hyperspectral remote sensing detection of bark beetle-infested areas is mentioned in Niemann et al. (2015). The main principles of the proposed innovative methods are discussed in several publications, e.g. Krivá et al. (2015) deal with method for diffusion equations with application to images filtering. Lieskovský et al. (2013) deal with uncertainty modelling by fuzzy sets and predictive models validation. Modelling of insolation effects on trees is mentioned in Jakuš et al. (2013). Forecasting bark beetle outbreaks based on hyperspectral remote sensing techniques is discussed in Lausch et al. (2013). Ortiz et al. (2013) deal with early detection of bark beetle attack using data from synthetic aperture radar (SAR), etc.

3.3 The Diagram of Use Cases

The concept of FPMS from user's point of view is modelled by UCD in UML (Fig. 10). UCD in UML is a simple representation of an interaction of different user's types within the system (FPMS). The rectangle bounding use cases represents system boundary, i.e. it specifies only relevant functions in FPMS. Each additional function unnecessarily increases the costs of implementation and maintenance of the system without increasing the effectiveness of the resulting measures protecting from bark beetles threats.

3.4 The Diagram of Processes

Each of the ongoing processes in the FPMS with their relations and continuities is modelled as an AD in UML (Fig. 11). AD is an UML behaviour diagram which describes dynamic aspect of the system and shows the workflow from a start point (represented by a fat black dot) to the finish point (the black dot in a white circle). It specifies a sequence of behaviour from one activity to another (indicated by arrows). Activities of each actor are represented by the so-called swim lanes (user, forest expert, GIS expert and IT specialist). In contrast to UCD (Fig. 10), AD (Fig. 11) describes the procedural logic of the processes and it is used to identify dataflow and process modelling in the FPMS to optimize its implementation and utilization.

4 Conclusion and Discussion

The significant technical development of satellite data and UAV technology increases the amount of image data acquired from these systems. Automation of their interpretation is therefore an essential prerequisite for their effective processing. In this regard, creation of a methodology for various data sources integration (first of all satellite images, radar images, laser scanning, multispectral, infrared and thermal images) is a key factor in the forest protection management.

Advantages of integrating heterogeneous satellite data obtained from satellite and UAV technology in complex web based application to support forest protection are as follows:

a) *Time and space optimization of UAV technology*—the modelling of bark beetle development will allow focusing UAV on stands attacked by swarming bark beetles. Modelling of transpiration deficit will prioritize scanning of weakened stands.
b) *Optimization of the sanitary logging timing*—on-line maps of bark beetle attacks in combination with the PHENIPS model can be the basis for the specification of limit biotechnological terms for attacked tree treatment.
c) *Increasing speed of identifying infested trees*—it can lead to a higher value of processed wood because late processing could cause wood deterioration. It also leads to a reduction of the negative effects of insect infestation on spruce stands. This allows carrying out forestry measures, including salvage cutting and timber sales at a time suitable to the user, allowing him to increase economic opportunity and improve yields and more permanent incomes.

The next step will be the creation of the interactive dynamic maps of current bark beetle attacks and the predictive maps of bark beetle threats. This can be the basis for a web application providing a real time predictive system of the threat of bark beetles to forest coverage. The basic prerequisite of effective predictive modelling is the subsequent publication of data and predictive maps in the form of a web based application, compatible with globally recognized standards for spatial data. The aim of the web based application will be to support fast and relevant decision-making in forest management and protection. This is especially related to the source data about the impact of the changes of the climatic parameters on the current state of the predictive model (hydrological parameters, solar radiation, health conditions of the forest vegetation, etc.). These requirements can be satisfied by distributed systems that use web services. It provides the exchange, sharing, and publishing of dynamically changing spatial data, which illustrates current bark beetle infestations, as well as modelling its future development. A significant trend nowadays is also the opening and linking of datasets (Open Linked Data), from which the general public can also benefit. It is not limited to working with the application, but can work directly with the data. The emphasis is thus put on the publication of using tools supporting European (INSPIRE) and international (W3C, OGC) standard specifications, as well as generally accepted open source technologies (OpenLayers, GeoServer, GeoWebCache). A high priority is also put on the modularity of the solution. The combination of these factors allows fast and easy development of custom extensions, independent from the authors of the application. With the latest web technologies (ECMAScript 2015, HTML5, CSS3, and HTTP 2.0), it is possible to create an application that will work without additional installation of accessories in different types of devices, whether desktop or mobile. Such applications can quickly visualize complex data with conversion to information for employees, activists, researchers, or the general public.

The proposed structural concept includes data filtering from various sources with emphasis on the usage of latest technologies designed to increase the quality and timeliness of inputs. It is the basis of the predictive model being developed for the purposes of forest protection. This concept optimizes the preventive or remedial environmental measures in forest management, as well as the automation of satellite data processing and the spatiotemporal optimization of the collection methods based on UAV technology.

Acknowledgments This work was supported by the Grants Nos. 1/0682/16 and 1/0954/15 of the Grant Agency of the Slovak Republic—VEGA and Grant No. APVV-0297-12 of The Slovak Research and Development Agency—APVV.

References

Akkuzu E, Sariyildiz T, Kucuk M, Duman A (2009) *Ips typographus* (L.) and *Thanasimus formicarius* (L.) populations influenced by aspect and slope position in Artvin-Hatila valley national park, Turkey. Afr J Biotechnol 8:877–882

Albrechtová J, Rock B (2003) Remote sensing of the Krusne hory Mts. Forests. From microscope to macroscope and back again. Vesmír 82, 323(6):322–325

Andersen H-E, Mcgaughey RJ, Reutebuch SE (2005) Estimating forest canopy fuel parameters using LIDAR data. Remote Sens Environ 94(4):441–449

Arnberg W, Wastenson L (1973) Use of aerial photographs for early detection of bark beetle infestations of spruce. Ambio 2:77–83

Baier P, Pennerstorfer J, Schopf A (2007) PHENIPS—a comprehensive phenology model of *Ips typographus* (L.) (Col., Scolytinae) as a tool for hazard rating of bark beetle infestation. For Ecol Manag 249:171–186

Barka I, Bucha T (2010) Satellite-based regional system for observation of forest response to global environmental changes. In: Horák J, Halounová L, Hlásny T, Kusendová D, Voženílek V (eds) Advances in geoinformation technologies 2010. Technical University of Ostrava, Ostrava, pp 1–14

Brignolas F, Lieutier F, Sauvard D, Christiansen E, Berryman A (1998) Phenolic predictors for Norway spruce resistance to the bark beetle *Ips typographus* (Coleoptera: Scolytidae) and an associated fungus, *Ceratocystis polonica*. Can J For Res 28:720–728

Comolina I, Molina P (2014) Unmanned aerial systems for photogrammetry and remote sensing: a review. ISPRS J Photogram Remote Sens 92:79–97

Faccoli M, Schlyter F (2007) Conifer phenolic resistance markers are bark beetle antifeedant semiochemicals. Agric For Entomol 9:237–245

Fahse L, Heurich M (2003) Simulation and analysis of outbreaks of bark beetle infestations and their management at the stand level. Ecol Model 222(11):1833–1846

Grodzki W, Jakuš R, Lajzová E, Sitková Z, Maczka T, Škvarenina J (2006) Effects of intensive versus no management strategies during an outbreak of the bark beetle *Ips typographus* (L.) (Col.: Curculionidae, Scolytinae) in the Tatra Mts. In Poland Slovakia Ann For Sci 63:55–61

ISO/IEC 19505-1:2012(E) Information technology—Object Management Group Unified Modeling Language (OMG UML), Infrastructure

Jakuš R, Grodzki W, Ježik M, Jachym M (2003a) Definition of spatial patterns of bark beetle *Ips typographus* (L.) outbreak spreading in Tatra Mountains (Central Europe), using GIS. In: Mc Manus M, Liebhold A (eds) Ecology, Survey and management of forest insects, proceedings of the conference GTR NE-311. USDA Forest Service, Newtown Square, pp 25–32

Jakuš R, Schlyter F, Zhang Q-H, Blaženec M, Vavercák R, Grodzki W, Brutovsky D, Lajzová E, Bengtsson M, Blum Z, Turcáni M, Gregoiré J-C (2003b) Overview of development of anti-attractant based technology for spruce protection against *Ips typographus*: from past failures to future success. J Pest Sci 76:89–99

Jakuš R, Zajíčkova L, Cudlín P, Blaženec M, Turčani M, Ježík M, Lieutier F, Schlyter F (2011) Landscape-scale *Ips typographus* attack dynamics: from monitoring plots to GIS-based disturbance models. Forest 4:256–261

Jakuš R, Cudlin P, Blaženec M, Brenkus T (2013) Evaluation of Norway spruce crown and trunk in relation to spruce bark beetle (*Ips typographus*). In: Vliv abiotických a biotických stresorů na vlastnosti rostlin 2013, Praha

Jones HG, Schofield P (2008) Thermal and other remote sensing of plant stress. Gen Appl Plant Physiol Special Issue 34:19–32

Kissiyar O, Blaženec M, Jakuš R, Willekens A, Ježík M, Baláž P, Van Valckenborg J, Celer S, Fleischer P (2005) TANABBO model: a remote sensing based early warning system for forest decline and bark beetle outbreaks in Tatra Mts. Overview. In: Grodzki W (ed) GIS and databases in the forest protection in Central Europe. Forest Reseach Institute, Krakow, pp 15–34

Krivá Z, Papčo J, Vanko, J (2015) Quad-tree based finite volume method for diffusion equations with application to SAR imaged filtering. In: Acta Universitatis Palackianae Olomucensis. Facultas Rerum Naturalium. Mathematica, vol. 54 (2015), issue 2, pp. 41–61. ISSN: 0231-9721

Lausch A, Heurich M, Gordalla D, Dobner H-J, Gwillym-Margianto S, Salbach C (2013) Forecasting potential bark beetle outbreaks based on spruce forest vitality using hyperspectral remote-sensing techniques at different scales. For Ecol Manag 308:76–89

Lee J-S, Pottier E (2009) Polarimetric radar imaging: from basics to applications. CRC Press, Boca Raton, 438 p. ISBN 9781420054972

Lieskovský T, Ďuračiová R, Karell L (2013) Selected mathematical principles of archaeological predictive models creation and validation in the GISenvironment. Interdisciplinaria Archaeologica—Natural Sciences in Archaeology, vol. IV, issue 2/2013, s. 33–46. ISSN 2336-1220 (online), ISSN 1804-848X (print)

Malenovský Z, Martin E, Homolova L, Gastellu-Etchegorry J, Zurita-Milla R, Schaepman M, Pokorný R, Clevers J, Cudlin P (2008) Influence of woody elements of a Norway spruce canopy on nadir reflectance simulated by the DART model at very high spatial resolution. Remote Sens Environ 112:1–18

Meddens AJH, Hicke JA, Vierling LA, Hudak AT (2013) Evaluating methods to detect bark beetle-caused tree mortality using single-date and multi-date Landsat imagery. Remote Sens Environ 132:49–58

Meigs GW, Kennedy RE, Cohen WB (2011) A Landsat time series approach to characterize bark beetle and defoliator impacts on tree mortality and surface fuels in conifer forests. Remote Sens Environ 115:3707–3718

Moravec I, Cudlín P, Polák T, Havlíček F (2002) Spruce bark beetle (Ips typographus L.) infestation and Norway spruce status: is there a causal relationship? Silva Gabreta 8:255–264

Netherer S, Pennerstorfer J (2001) Parameters relevant for modelling the potential development of *Ips typographus* (Coleoptera: Scolytidae). Integr Pest Manag Rev 6:177–184

Netherer S (2003) Modelling of bark beetle development and of site- and stand-related predisposition to *Ips typographus* (L.) (Coleoptera; Scolytidae). University of Natural Resources and Life Science Vienna. Thesis

Niemann KO, Quinn G, Stephen R, Visintini F, Parton D (2015) Hyperspectral remote sensing of mountain pine beetle with an emphasis on previsual assessment. Can J Remote Sens 41:191–202

No. 453/2006 (2006) Decree of Ministry of agriculture and rural development on forest management and forest protection, Bratislava, Slovakia

Ortiz SM, Breidenbach J, Kändler G (2013) Early detection of bark beetle green attack using TerraSAR-X and RapidEye Data. Remote Sens 5:1912–1931

Rouault G, Candau J, Lieutier F, Nageleisen L, Martin J, Warzée N (2006) Effects of drought and heat on forest insect populations in relation to the 2003 drought in Western Europe. Ann For Sci 63:613–624

Schlyter F (2001) Integrated risk assessment and new pest management technology in ecosystems affected by forest decline and bark beetle outbreaks "TATRY". Final report of EU INCO Copernicus project IC15-CT98-0151 2001 [cited 17.12.2015]. http://www-vv.slu.se/fs/tatry/fin_rapp/FR_Outl.htm

Schopf R, Köhler U (1995) Untersuchungen zur Populationsdynamik der Fichtenborkenkäfer im Nationalpark Bayerischer Wald. In: Biberlriether H (ed) Nationalpark Bayerischer Wald-25 Jahre auf dem Weg zum Naturwald. Passavia Druckerei GmbH, Passau, pp 88–110

SmartPlanes (2006) ASA wind throw and insect attack [cited 17.12.2015]. http://www.smartplanes.se/2006/09/asa-wind-throw-and-insect-attack/

Smreček R, Tuček J (2011) Posúdenie presnosti merania hrúbky stromu pomocou pozemného laserového skenovania. In: Suchomel J, Tuček J, Gejdoš M, Jurica J (eds) Progresívne postupy spracovania náhodných ťažieb. TU Zvolen, Zvolen, pp 109–112

Soukupová J, Rock B, Albrechtová J (2001) Comparative study of two spruce species in a polluted mountainous region. New Phytol 150:133–145

STN 48 27 11 (2012) Forest protection. Forest protection against main bark beetles on spruce. SUTN, Bratislava, Slovakia

Van Leeuwen M, Nieuwenhuis M (2010) Retrieval of forest structural parameters using LiDAR remote sensing. Eur J For Res 129(4):749–770

Whalin B (2012) Unmanned aircraft systems' remote sensing technology used against bark beetles in national forests [cited 17.12.2015]. http://www.suasnews.com/2012/02/11985/unmanned-aircraft-systems%E2%80%99-remote-sensing-technology-used-against-bark-beetles-in-national-forests/

Author Index

I. Ivan et al. (eds.), *The Rise of Big Spatial Data*, Lecture Notes in Geoinformation and Cartography, DOI 10.1007/978-3-319-45123-7

Printed in the United States
By Bookmasters